UNITEXT

La Matematica per il 3+2

Volume 151

The **UNITEXT - La Matematica per il 3+2** series is designed for undergraduate and graduate academic courses, and also includes books addressed to PhD students in mathematics, presented at a sufficiently general and advanced level so that the student or scholar interested in a more specific theme would get the necessary background to explore it.

Originally released in Italian, the series now publishes textbooks in English addressed to students in mathematics worldwide.

Some of the most successful books in the series have evolved through several editions, adapting to the evolution of teaching curricula.

Submissions must include at least 3 sample chapters, a table of contents, and a preface outlining the aims and scope of the book, how the book fits in with the current literature, and which courses the book is suitable for.

For any further information, please contact the Editor at Springer: francesca.bonadei@springer.com

THE SERIES IS INDEXED IN SCOPUS

***

UNITEXT is glad to announce a new series of free webinars and interviews handled by the Board members, who rotate in order to interview top experts in their field.

Access this link to subscribe to the events:

https://cassyni.com/events/TPQ2UgkCbJvvz5QbkcWXo3

Manuel Benz • Thomas Kappeler

Linear Algebra for the Sciences

 Springer

Manuel Benz
Department of Mathematics
Literargymnasium Raemibuehl
Zurich, Switzerland

Thomas Kappeler *(deceased)*
Institute of Mathematics
University of Zurich
Zurich, Switzerland

ISSN 2038-5714   ISSN 2532-3318   (electronic)
UNITEXT
ISSN 2038-5722   ISSN 2038-5757   (electronic)
La Matematica per il 3+2
ISBN 978-3-031-27219-6   ISBN 978-3-031-27220-2   (eBook)
https://doi.org/10.1007/978-3-031-27220-2

Cover illustration: © Sozh / Stock.adobe.com Linear algebra is a broad and complex topic, but in its essence, it can be reduced to fundamental and easily understandable concepts. Such a concept, namely systems of linear equations, inspired the cover: the image has a simple quality but simultaneously shows richness in its details. From left to right, it is well-balanced, like a system of linear equations, beautifully realized by the artist. Many fields, including neural networks and artificial intelligence, use linear algebra at their core. Like images sketched by contemporary artificial intelligence, this cover is not only slightly chaotic but also shows beauty; in this case, the beauty of linear algebra.

This Springer imprint is published by the registered company Springer Nature Switzerland AG
The registered company address is: Gewerbestrasse 11, 6330 Cham, Switzerland

Preface

This book is based on notes of a course, designed for first semester students in the sciences, which had been given by the second author at the University of Zurich several times. The goal of the course, with a format of three lectures classes and one exercise class per week, is twofold: to have students learn a basic working knowledge of linear algebra as well as acquire some computational skills, and to familiarize them with mathematical language with the aim of making mathematical literature more accessible. Therefore, emphasis is given on precise definitions, helpful notations, and careful statements of the results discussed. No proofs of these results are provided, but typically, they are illustrated with numerous examples, and for the sake of better understanding, we quite frequently give some supporting arguments for why they are valid.

Together with the course *Analysis for the Sciences* for second semester students, for which a book is in preparation, the course *Linear Algebra for the Sciences* constitutes a basic introduction to mathematics. One of the main novelties of this introduction consists in the order in which the courses are taught. Since students have acquired a basic knowledge in linear algebra in the first semester, they are already familiar with functions of several variables at the beginning of the second semester. Differential and integral calculus is then developed at each stage for functions of one and of several variables. Furthermore, numerous illustrations of concepts from linear algebra in the differential and integral calculus are pointed out and examples and problems are discussed which involve notions from linear algebra. Nevertheless, by and large, the two books can be studied independently from each other.

The table of contents of *Linear Algebra for the Sciences* describes the topics covered in the book. The book has six chapters and each of them three sections. Each chapter and each section begin with a short summary. Below we only give a brief overview.

Chapter 1 treats finite systems of linear equations with real coefficients and explains in detail the algorithm of Gaussian elimination for solving such systems. This algorithm is used in all subsequent chapters.

Chapter 2 introduces the notion of matrices, discusses how they operate, and explains how they are connected to systems of linear equations. In addition, we define the notion of linear (in)dependence of vectors in $\mathbb{R}^k$ and the one of a basis in $\mathbb{R}^k$ and show how systems of linear equations can be used for computations. Finally, we define the notion of the determinant of a square matrix and discuss its most important properties.

In Chapter 3, we introduce complex numbers and discuss their basic properties, state the fundamental theorem of algebra and extend the notions and results, which were introduced in the previous chapters on the basis of real numbers.

Chapters 4 and 5 constitute the core of the course. In Chap. 4, the notions of vector spaces (over the real as well as over the complex numbers) and the notions of a basis and of the dimension of such spaces are introduced. Furthermore, we define the notion of linear maps between such spaces and discuss the matrix representation of linear maps between finite dimensional vector spaces with respect to bases. In particular, we discuss in detail the change of bases of finite dimensional vector spaces and introduce special notation to describe it. The last section of Chap. 4 introduces inner product spaces and discusses linear maps between such spaces, which leave the inner products invariant.

In Chap. 5, we discuss the basics of the spectral theory of linear maps on a finite dimensional vector space over the complex numbers. In a separate section, we examine the spectral theory of a linear map on a finite dimensional vector space over the real numbers. In the last section of Chap. 5, we introduce quadratic forms and, as an application, discuss conic sections.

Chapter 6 is an application of results of linear algebra, treated in the earlier chapters. We study systems of linear ordinary differential equations with constant coefficients, with the main focus on systems of first order. One of the aims of this chapter is to illustrate the use of linear algebra in other fields of mathematics such as analysis and to showcase the fundamental idea of linearity in the sciences. In this final chapter, some basic knowledge of analysis is assumed.

In each section, numerous examples are discussed and problems solved with the aim of illustrating the introduced concepts and methods. Each section ends with a set of five problems, except for Sects. 1.1 and 6.1, which both are of an introductory nature. These problems are of the same type as the ones solved in the course of the corresponding section with the purpose that students acquire some computational skills and get more familiar with the concepts introduced. They are intentionally kept simple. Solutions of these problems are given in chapter "Solutions" at the end of the book.

We would like to thank Riccardo Montalto, who was involved at an early stage, but, due to other commitments, decided not to participate further in the project. Over the years, the notes were used by colleagues of the Department of Mathematics of the University of Zurich. We would like to thank them and all the students for their feedback. Finally, we thank Camillo De Lellis for encouraging us to turn these notes into a book.

Zurich, Switzerland   Manuel Benz
                                                                        Thomas Kappeler

Addendum

It is with sadness that I must inform the reader that Thomas Kappeler suddenly passed away during the final stages of this book.

On a personal level, I lost a friend and mentor. On a professional level, the mathematical world has lost a great mind. His approach to mathematics was precise and careful and yet his presentation of it stayed very accessible. His way of teaching mathematics was thoughtful, smart and sophisticated. Additionally, I feel that his work shows his enthusiasm and joy for the subject, but also him as a person with great respect for human intelligence.

I hope this book can serve as a testament to all these good qualities of Thomas and remind everyone of him as a mathematician, but also as a person enjoying mathematics.

Zurich, Switzerland   Manuel Benz

Contents

Chapter 1
Systems of Linear Equations

1.1 Introduction

One of the most important problems in linear algebra is to solve systems of linear equations, also referred to as *linear systems*. For illustration, let us consider a simple example of a system with two linear equations and two unknowns x, y,

$$(S) \quad \begin{cases} x + 2y = 5 & (1.1) \\ 2x + 3y = 8. & (1.2) \end{cases}$$

We say that the pair (u, v) of real numbers u and v is a *solution of* (S) if

$$u + 2v = 5 \quad \text{and} \quad 2u + 3v = 8.$$

The basic questions are the following ones:

($Q1$) Do linear systems such as (S) have a solution (*existence*)?
($Q2$) Do linear system such as (S) have at most one solution (*uniqueness*)?
($Q3$) Do there exist (efficient) methods to find all solutions of (S)?

It is convenient to introduce the set L of all solutions of (S),

$$L := \left\{ (u, v) \in \mathbb{R}^2 \mid u + 2v = 5; 2u + 3v = 8 \right\}$$

Then the questions ($Q1$) and ($Q2$) can be rephrased as follows:

($Q1'$) Is the set of solutions of a linear system such as (S) a nonempty set?
($Q2'$) Does the set of solutions of a linear system such as (S) have at most one element?

© The Author(s), under exclusive license to Springer Nature Switzerland AG 2023
M. Benz, T. Kappeler, *Linear Algebra for the Sciences*, La Matematica
per il 3+2 151, https://doi.org/10.1007/978-3-031-27220-2_1

Let us now discuss a method for finding the set L of solutions of (S). It is referred to as the *method of substitution* and consists in the case of (S) of the following four steps:

Step 1. Solve (1.1) for x,

$$x = 5 - 2y. \tag{1.3}$$

Step 2. Substitute the expression (1.3) for x into equation (1.2),

$$2(5 - 2y) + 3y = 8 \qquad \text{or} \qquad 10 - y = 8. \tag{1.4}$$

Step 3. Determine y by (1.4),

$$y = 2. \tag{1.5}$$

Step 4. Determine x by using (1.3) and (1.5),

$$x = 5 - 2y = 5 - 2 \cdot 2 = 1.$$

Hence the set L of solutions of (S) is given by

$$L = \{(1, 2)\}.$$

It consists of *one element*, meaning that $(1, 2)$ is the unique solution of (S).

However it is not always the case that a linear system of two equations with two unknowns has a unique solution. To see this consider

$$\begin{cases} x + y = 4 & (1.6) \\ 2x + 2y = 5. & (1.7) \end{cases}$$

By the same method, one obtains

$$x = 4 - y \tag{1.8}$$

and hence by substitution

$$2(4 - y) + 2y = 5,$$

or

$$8 - 2y + 2y = 5, \qquad \text{i.e.,} \qquad 8 = 5.$$

But $8 \neq 5$. What does this mean? It means that the set of solutions L of (1.6)–(1.7) is empty, $L = \emptyset$, i.e., there are *no solutions of* (1.6)–(1.7).

Finally let us consider the following linear system

$$\begin{cases} x + 3y = 2 & (1.9) \\ 2x + 6y = 4. & (1.10) \end{cases}$$

Again by the method of substitution, one has $x = 2 - 3y$ and hence

$$2(2 - 3y) + 6y = 4 \qquad \text{or} \qquad 4 = 4.$$

As a consequence, the unknown y can take any value and the set L of solutions of (1.9)–(1.10) is given by

$$L = \{(2 - 3v, v) \mid v \in \mathbb{R}\}.$$

The set L can thus be viewed as the graph of the function $\mathbb{R} \to \mathbb{R}, v \mapsto 2 - 3v$. In particular, there are infinitely many solutions.

Summarizing our considerations so far, we have seen that the set of solutions can be empty, consist of one element, or of infinitely many elements. It turns out that this is true in general: The set of solutions of any given linear system fits into one of these three possibilities.

In practical applications, systems of linear equations can be very large. One there needs to develop theoretical concepts and appropriate notation to investigate such systems and to find efficient algorithms to solve them numerically. In this chapter, we present such an algorithm, referred to as *Gaussian elimination*.

1.2 Systems with Two Equations and Two Unknowns

In preparation of treating general linear systems, the goal of this section is to study first the case of two equations with two unknowns in full generality, to discuss a method for finding the set of solutions and to point out connections with geometry.

Consider a system of two equations and two unknowns x, y,

$$\begin{cases} p(x, y) = 0 \\ q(x, y) = 0 \end{cases}$$

where p and q are real valued functions on $\mathbb{R}^2 = \mathbb{R} \times \mathbb{R}$. Such a system is said to be *linear* (or, more precisely, $\mathbb{R}$-linear) if p and q are polynomials in x and y of degree one,

$$p(x, y) = ax + by - e, \qquad q(x, y) = cx + dy - f$$

where a, b, c, d, e, f are real numbers. In such a case we customarily write

$$\begin{cases} ax + by = e & \text{(1.11)} \\ cx + dy = f. & \text{(1.12)} \end{cases}$$

The system (1.11)–(1.12) being a general linear system means that the coefficients a, b, e of $p(x, y)$ and c, d, f of $q(x, y)$ can be arbitrary real numbers. In applications, they can be assigned specific values, but since we want to consider a general linear system, they are denoted with the letters a, b, e and respectively c, d, f.

We begin by pointing out some connections with geometry. Denote by L_1 the set of solutions of $ax + by = e$,

$$L_1 := \{(x, y) \in \mathbb{R}^2 \mid ax + by = e\}.$$

To describe L_1, we have to distinguish between different cases. Note that in the case $a = 0$, the equation $ax + by = e$ cannot be solved for x and in case $b = 0$, the equation cannot be solved for y. The following four cases arise:

Case 1. $a = 0, b = 0, e = 0$. Then $L_1 = \mathbb{R}^2$.

Case 2. $a = 0, b = 0, e \neq 0$. Then $L_1 = \emptyset$.

Case 3. $b \neq 0$. In this case, the equation $ax + by = e$ can be solved for y and we get

$$y = \frac{e}{b} - \frac{a}{b}x \qquad \text{and} \qquad L_1 = \left\{ \left(x, \frac{e}{b} - \frac{a}{b}x\right) \,\middle|\, x \in \mathbb{R} \right\}.$$

The set L_1 can be viewed as the graph of the function

$$\mathbb{R} \to \mathbb{R}, \qquad x \mapsto \frac{e}{b} - \frac{a}{b}x.$$

Its graph is sketched in Fig. 1.1.

Case 4. $b = 0, a \neq 0$. In this case, the equation $ax + by = e$ reads $ax = e$ and can be solved for x,

$$x = \frac{e}{a} \qquad \text{and} \qquad L_1 = \left\{ \left(\frac{e}{a}, y\right) \,\middle|\, y \in \mathbb{R} \right\}.$$

A graphical representation of L_1 can be seen in Fig. 1.2. Similarly, we denote by L_2 the set of solutions of the equation $cx + dy = f$,

$$L_2 := \{(x, y) \in \mathbb{R}^2 \mid cx + dy = f\}.$$

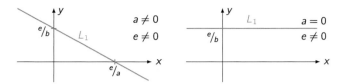

Fig. 1.1 For $a \neq 0$, $e \neq 0$ (left) and for $a = 0$, $e \neq 0$ (right) the set L_1 in Case 3 can be schematically pictured as shown in the two figures above

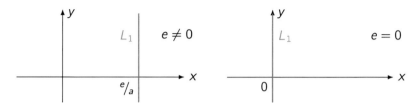

Fig. 1.2 For $e \neq 0$ (left) and for $e = 0$ (right) the set L_1 in Case 4 can be schematically pictured as shown in the two figures above

Since an element (x, y) in the set L of solutions of the system (1.11)–(1.12) is a solution of (1.11) *and* a solution of (1.12), the set L is given by the *intersection* of L_1 with L_2,

$$L = L_1 \cap L_2.$$

Combining the four cases described above for L_1 and the corresponding ones for L_2, one can determine the set of solutions in each of these cases. We leave it to the reader to do that and consider instead only the case where L_1 and L_2 are both straight lines. Then the following three possibilities for the intersection $L_1 \cap L_2$ can occur,

$$L_1 \cap L_2 = \{\text{point in } \mathbb{R}^2\}, \qquad L_1 \cap L_2 = \emptyset, \qquad L_1 = L_2.$$

Note that in the case where $L_1 \cap L_2 = \emptyset$, the two straight lines L_1 and L_2 are parallel. The three possibilities are illustrated in Fig. 1.3.

In the remaining part of this section we describe an algorithm how to determine the set L of solutions of the linear system (1.11)–(1.12), yielding explicit formulas for the solutions. Recall that (1.11)–(1.12) is the system (S) given by

$$\begin{cases} ax + by = e & \text{(1.11)} \\ cx + dy = f. & \text{(1.12)} \end{cases}$$

We restrict ourselves to the case where

$$a \neq 0. \tag{1.13}$$

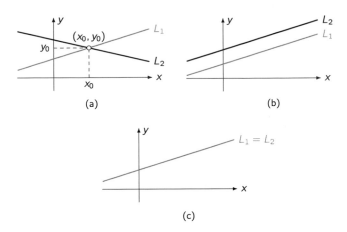

Fig. 1.3 The three possible cases for L_1 and L_2 in a graphical representation. Please note the solution set for each case. (**a**) $L_1 \cap L_2 = \{(x_0, y_0)\}$. (**b**) $L_1 \cap L_2 = \emptyset$. (**c**) $L_1 \cap L_2 = L_1$

Before we describe the algorithm, we need to introduce the notion of equivalent systems. Assume that we are given a second system (S') of two linear equations,

$$\begin{cases} a'x + b'y = e' \\ c'x + d'y = f' \end{cases}$$

with real coefficients a', b', e', and c', d', f'. Denote by L' its set of solutions. We say that (S) and (S') are *equivalent*, if $L = L'$. We are now ready to describe the algorithm to determine L of (S).

Step 1. Eliminate x from Eq. (1.12). To achieve this, we replace (1.12) by (1.12) $- \frac{c}{a}(1.11)$. It means that the left hand side of (1.12) is replaced by

$$cx + dy - \frac{c}{a}(ax + by) \quad \text{(use that } a \neq 0)$$

whereas the right hand side of (1.12) is replaced by

$$f - \frac{c}{a}e.$$

The new system of equations then reads as follows,

$$\begin{cases} ax + by = e & (1.11) \\ \left(d - \dfrac{c}{a}b\right)y = f - \dfrac{c}{a}e. & (1.14) \end{cases}$$

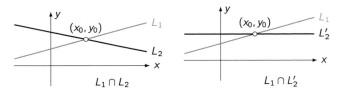

Fig. 1.4 The solution of the new system (1.11) and (1.14) still corresponds to the same point, but the elimination of x in (1.14) leads to the second line changed to horizontal

It is straightforward to see that under the assumption (1.13), the set of solutions of (1.11)–(1.12) coincides with the set of solutions of (1.11)–(1.14), i.e. that the two systems are equivalent.

Graphically, the elimination of x in Eq. (1.12) can be interpreted as follows. The unique solution of the original Eqs. (1.11) and (1.12) is given as the intersection point between two lines as shown in Fig. 1.4.

Step 2. Solve (1.14) for y and then (1.11) for x. We distinguish three cases.

Case 1. $d - \frac{c}{a} b \neq 0$. Then (1.14) has the unique solution

$$y = \frac{f - \frac{c}{a} e}{d - \frac{c}{a} b} = \frac{af - ce}{ad - bc}$$

and when substituted into (1.11) one obtains

$$ax = e - b\left(\frac{af - ce}{ad - bc}\right) \qquad \text{or} \qquad x = \frac{de - bf}{ad - bc}.$$

Hence the set L of solutions of (S) consists of one element

$$L = \left\{ \left(\frac{de - bf}{ad - bc}, \frac{af - ce}{ad - bc} \right) \right\} \tag{1.15}$$

Case 2. $d - \frac{c}{a} b = 0$, $f - \frac{c}{a} e \neq 0$. Then Eq. (1.14) has no solutions and hence $L = \emptyset$.

Case 3. $d - \frac{c}{a} b = 0$, $f - \frac{c}{a} e = 0$. Then any $y \in \mathbb{R}$ is a solution of (1.14) and the solutions of (1.11) are given by $x = \frac{e}{a} - \frac{b}{a} y$. Hence L is given by

$$L = \left\{ \left(\frac{e}{a} - \frac{b}{a} y, y \right) \,\middle|\, y \in \mathbb{R} \right\}.$$

Motivated by formula (1.15) for the solutions of (1.11), (1.14) in the case where $d - \frac{c}{a} b \neq 0$ we make the following definitions:

Definition 1.2.1

(i) An array A of real numbers of the form

$$A := \begin{pmatrix} a & b \\ c & d \end{pmatrix}, \quad a, b, c, d \in \mathbb{R},$$

is said to be a *real* 2×2 *matrix* (plural: *matrices*).

(ii) The determinant $\det(A)$ of a 2×2 matrix A is the real number, defined by

$$\det(A) := ad - bc.$$

The determinant can be used to characterize the solvability of the linear system (S), given by the two Eqs. (1.11) and (1.12), and to obtain formulas for its solutions. We say that the 2×2 matrix A, formed by the coefficients $a, b, c,$ and d in Eqs. (1.11) and (1.12),

$$A := \begin{pmatrix} a & b \\ c & d \end{pmatrix}$$

is the *coefficient matrix* of (S). We state without proof the following.

Theorem 1.2.2

(i) *The system of linear Eqs.* (1.11) *and* (1.12) *has a unique solution if and only if the determinant of its coefficient matrix does not vanish,*

$$\det \begin{pmatrix} a & b \\ c & d \end{pmatrix} \neq 0.$$

(ii) *If* $\det \begin{pmatrix} a & b \\ c & d \end{pmatrix} \neq 0$, *then the unique solution* (x, y) *of* (1.11), (1.12) *is given by the following formula (Cramer's rule)*

$$x = \frac{\det \begin{pmatrix} e & b \\ f & d \end{pmatrix}}{\det \begin{pmatrix} a & b \\ c & d \end{pmatrix}}, \quad y = \frac{\det \begin{pmatrix} a & e \\ c & f \end{pmatrix}}{\det \begin{pmatrix} a & b \\ c & d \end{pmatrix}}.$$

Examples

(i) Analyze the following system $(S1)$ of linear equations

$$\begin{cases} 2x + 4y = 3 \\ x + 2y = 5 \end{cases}$$

and find its set of solutions.

Solution The coefficient matrix A of $(S1)$ reads

$$A = \begin{pmatrix} 2 & 4 \\ 1 & 2 \end{pmatrix}.$$

Then $\det(A) = 0$. Hence Theorem 1.2.2 says that the system $(S1)$ does not have a unique solution. It is straightforward to see that actually $(S1)$ has no solutions.

(ii) Analyze the following system $(S2)$ of linear equations

$$\begin{cases} 2y + 4x = 3 \\ x + 2y = 5 \end{cases}$$

and find its set of solutions.

Solution First we rewrite the equation $2y + 4x = 3$ as $4x + 2y = 3$. The coefficient matrix A of $(S2)$ thus reads

$$A = \begin{pmatrix} 4 & 2 \\ 1 & 2 \end{pmatrix}.$$

Then $\det(A) = 8 - 2 = 6 \neq 0$. Hence according to Theorem 1.2.2, the system of equations has a unique solution given by

$$x = \frac{\det \begin{pmatrix} 3 & 2 \\ 5 & 2 \end{pmatrix}}{6} = \frac{6 - 10}{6} = -\frac{2}{3}, \qquad y = \frac{\det \begin{pmatrix} 4 & 3 \\ 1 & 5 \end{pmatrix}}{6} = \frac{20 - 3}{6} = \frac{17}{6}.$$

Problems

1. Determine the sets of solutions of the linear systems

   (i) $\begin{cases} x + \pi y = 1 \\ 2x + 6y = 4, \end{cases}$
   (ii) $\begin{cases} x + 2y = e \\ 2x + 3y = f. \end{cases}$

2. Consider the system (S) of two linear equations with two unknowns.

   $$\begin{cases} 2x + y = 4 & \qquad (S1) \\ x - 4y = 2 & \qquad (S2) \end{cases}$$

   (i) Determine the sets L_1 and L_2 of solutions of $(S1)$ and $(S2)$, respectively and represent them geometrically as straight lines in $\mathbb{R}^2$.

(ii) Determine the intersection $L_1 \cap L_2$ of L_1 and L_2 from their geometric representation in $\mathbb{R}^2$.

3. Compute the determinants of the following matrices and decide for which values of the parameter $a \in \mathbb{R}$, they vanish.

(i) $A = \begin{pmatrix} 1-a & 4 \\ 1 & 1-a \end{pmatrix}$ 　　　 (ii) $B = \begin{pmatrix} 1-a^2 & a+a^2 \\ 1-a & a \end{pmatrix}$

4. Solve the system of linear equations

$$\begin{cases} 3x - y = 1 \\ 5x + 3y = 2 \end{cases}$$

by Cramer's rule.

5. Decide whether the following assertions are true or false and justify your answers.

(i) For any given values of the coefficients $a, b, c, d \in \mathbb{R}$, the linear system

$$\begin{cases} ax + by = 0 \\ cx + dy = 0 \end{cases}$$

has at least one solution.

(ii) There exist real numbers a, b, c, d so that the linear system of (a) has infinitely many solutions.

(iii) The system of equations

$$\begin{cases} x_1 + x_2 = 0 \\ x_1^2 + x_2^2 = 1 \end{cases}$$

is a linear system.

1.3   Gaussian Elimination

Gaussian elimination is the name of an algorithm which determines the set of solutions of a general system of $m \geq 1$ linear equations and $n \geq 1$ unknowns. The n unknowns are customarily denoted by $x_1, \ldots, x_n$.

A system (S) of m equations and n unknowns is a system of equations of the form

$$f_1(x_1, \ldots, x_n) = 0, \qquad f_2(x_1, \ldots, x_n) = 0, \qquad \ldots, \qquad f_m(x_1, \ldots, x_n) = 0$$

where for any $1 \leq i \leq m$, f_i is a real valued function of n real variables

$$f_i : \mathbb{R}^n \to \mathbb{R}, (x_1, x_2, \ldots, x_n) \mapsto f_i(x_1, x_2, \ldots, x_n).$$

Definition 1.3.1 The system (S) is said to be *linear* if for any $1 \leq i \leq m$, the function f_i is a polynomial in $x_1, \ldots, x_n$ of degree 1 with real coefficients.

Since the number of equations and the one of unknowns can be arbitrarily large, one has to find an appropriate way how to denote the coefficients of the polynomials f_i. The following notation turns out to be very practical,

$$f_i(x_1, x_2, \ldots, x_n) = a_{i1}x_1 + a_{i2}x_2 + \cdots + a_{in}x_n - b_i, \quad 1 \leq i \leq m$$

where $a_{i1}, a_{i2}, \ldots, a_{in}$, and b_i are real numbers. The subscript i stands for the ith equation whereas we use the subscript j to list the unknowns x_j with $1 \leq j \leq n$. The system (S), in the case of being linear, is then customarily written as

$$\begin{cases} a_{11}x_1 + a_{12}x_2 + \cdots + a_{1n}x_n = b_1 \\ a_{21}x_1 + a_{22}x_2 + \cdots + a_{2n}x_n = b_2 \\ \quad \vdots \\ a_{m1}x_1 + a_{m2}x_2 + \cdots + a_{mn}x_n = b_m \end{cases}$$

A compact way of writing the above system of equations is achieved using the symbol $\sum$ (upper case Greek letter sigma) for the sum,

$$\sum_{j=1}^{n} a_{ij}x_j = b_i, \quad 1 \leq i \leq m. \tag{1.16}$$

Of course we could also use different letters than i and j to list the equations and the unknowns. We say that $(u_1, \ldots, u_n)$ in $\mathbb{R}^n$ is a solution of (1.16) if

$$\sum_{j=1}^{n} a_{ij}u_j = b_i, \quad 1 \leq i \leq m.$$

In the sequel, we often will not distinguish between $(x_1, \ldots, x_n)$ and $(u_1, \ldots, u_n)$. We denote by L the set of solutions of (1.16),

$$L := \Big\{(x_1, \ldots, x_n) \in \mathbb{R}^n \mid \sum_{j=1}^{n} a_{ij}x_j = b_i, 1 \leq i \leq m\Big\}.$$

Note that

$$L = \bigcap_{i=1}^{m} L_i, \quad L_i := \left\{ (x_1, \ldots, x_n) \in \mathbb{R}^n \mid \sum_{j=1}^{n} a_{ij} x_j = b_i \right\}, \quad 1 \le i \le m.$$

Before we describe Gaussian elimination, we need to introduce the notion of equivalent linear systems in full generality, which we already encountered in Sect. 1.2 in the case of systems of two equations and two unknowns.

Definition 1.3.2 We say that the system (S) of m linear equations and n unknowns $x_1, \ldots, x_n$,

$$\sum_{j=1}^{n} a_{ij} x_j = b_i, \quad 1 \le i \le m,$$

is *equivalent* to the system (S') with p equations and unknowns $x_1, \ldots, x_n$,

$$\sum_{j=1}^{n} c_{kj} x_j = d_k, \quad 1 \le k \le p,$$

if their sets of solutions coincide. If this is the case we write $(S) \equiv (S')$.

Example Consider the system (S) of two equations with three unknowns,

$$\begin{cases} 4x_1 + 3x_2 + 2x_3 = 1 \\ x_1 + x_2 + x_3 = 4 \end{cases}$$

and the system (S') of three equations, also with three unknowns,

$$\begin{cases} 4x_1 + 3x_2 + 2x_3 = 1 \\ x_1 + x_2 + x_3 = 4 \\ 2x_1 + 2x_2 + 2x_3 = 8 \end{cases}$$

The systems (S) and (S') are equivalent, since the first two equations in the latter system coincide with (S) and the third equation $2x_1 + 2x_2 + 2x_3 = 8$ of (S') is obtained from the second equation $x_1 + x_2 + x_3 = 4$ by multiplying left and right hand side by the factor 2.

The idea of Gaussian elimination is to replace a given system of linear equations in a systematic way by an equivalent one which is easy to solve. In Sect. 1.2 we presented this algorithm in the case of two equations ($m = 2$) and two unknowns ($n = 2$) under the additional assumption (1.13).

Gaussian elimination uses the following basic operations, referred to as *row operations* $(R1)$, $(R2)$, and $(R3)$, which leave the set of solutions of a given system of linear equations invariant:

$(R1)$ Exchange of two equations (rows).

    Example

$$\begin{cases} 5x_2 + 15x_3 = 10 \\ 4x_1 + 3x_2 + x_3 = 1 \end{cases} \quad \rightsquigarrow \quad \begin{cases} 4x_1 + 3x_2 + x_3 = 1 \\ 5x_2 + 15x_3 = 10 \end{cases}$$

    It means that the equations get listed in a different order.

$(R2)$ Multiplication of an equation (row) by a real number $\alpha \neq 0$.

    Example

$$\begin{cases} 4x_1 + 3x_2 + x_3 = 1 \\ 5x_2 + 15x_3 = 10 \end{cases} \quad \rightsquigarrow \quad \begin{cases} 4x_1 + 3x_2 + x_3 = 1 \\ x_2 + 3x_3 = 2 \end{cases}$$

    We have multiplied left and right hand side of the second equation by $1/5$.

$(R3)$ An equation (row) gets replaced by the equation obtained by adding to it the multiple of another equation. More formally, this can be expressed as follows: the kth equation $\sum_{j=1}^{n} a_{kj}x_j = b_k$ is replaced by the equation

$$\sum_{j=1}^{n} a_{kj}x_j + \alpha \sum_{j=1}^{n} a_{\ell j}x_j = b_k + \alpha b_\ell$$

    where $1 \leq \ell \leq m$ with $\ell \neq k$. In a more compact form, the new kth equation reads

$$(a_{k1} + \alpha a_{\ell 1})x_1 + \cdots + (a_{kn} + \alpha a_{\ell n})x_n = b_k + \alpha b_\ell.$$

    Example

$$\begin{cases} x_1 + x_2 = 5 \\ 4x_1 + 2x_2 = 3 \end{cases} \quad \overset{(R3)}{\rightsquigarrow} \quad \begin{cases} x_1 + x_2 = 5 \\ -2x_2 = -17 \end{cases}$$

It is not difficult to verify that these basic row operations lead to equivalent linear systems. We state without proof the following.

Theorem 1.3.3 *The basic row operations $(R1)$, $(R2)$, and $(R3)$ lead to equivalent systems of linear equations.*

The Gaussian algorithm consists in using these basic row operations in a systematic way to transform an arbitrary linear system (S) into an equivalent one, which is easy to solve, namely one which is in *row echelon form*.

To make this more precise, we first need to introduce some more notation. Let (S) be a system of linear equations of the form

$$\begin{cases} a_{11}x_1 + a_{12}x_2 + \cdots + a_{1n}x_n = b_1 \\ a_{21}x_1 + a_{22}x_2 + \cdots + a_{2n}x_n = b_2 \\ \qquad\qquad\qquad\vdots \\ a_{m1}x_1 + a_{m2}x_2 + \cdots + a_{mn}x_n = b_m \end{cases}$$

where a_{ij} ($1 \le i \le m$, $1 \le j \le n$) and b_i ($1 \le i \le m$) are real numbers. To (S) we associate an array A of real numbers with m rows and n columns,

$$A = \begin{pmatrix} a_{11} & \cdots & a_{1n} \\ \vdots & & \vdots \\ a_{m1} & \cdots & a_{mn} \end{pmatrix}.$$

Such an array of real number is called an $m \times n$ matrix with coefficients a_{ij}, $1 \le i \le m$, $1 \le j \le n$ and written in a compact form as

$$A = (a_{ij})_{\substack{1 \le i \le m \\ 1 \le j \le n}}$$

The matrix A is referred to as the *coefficient matrix* of the system (S). The *augmented coefficient matrix of* (S) is the following array of real numbers

$$\left[\begin{array}{ccc|c} a_{11} & \cdots & a_{1n} & b_1 \\ \vdots & & \vdots & \vdots \\ a_{m1} & \cdots & a_{mn} & b_m \end{array} \right].$$

In compact form it is written as $[A \parallel b]$.

Definition 1.3.4 $[A \parallel b]$ is said to be in *row echelon form* if it is of the form where $0 \leq n_1 \leq n$, $0 \leq m_1 \leq m$, and where the symbol $\star$ stands for a non-zero coefficient of A.

Remark In the case $n_1 = 0$ and $m_1 = 0$, the above augmented coefficient matrix reads

$$\left[\begin{array}{cccc|c} \star & & & & b_1 \\ 0 & \cdots & 0 & \star & \\ \vdots & & & \vdots & \vdots \\ 0 & \cdots & & 0 & \star & b_m \end{array}\right].$$

Let us express in words the features of an augmented coefficient matrix in echelon form:

(i) the coefficients of A below the echelon vanish; in particular, any zero row of A has to be at the bottom of A;

(ii) at each echelon, the corresponding coefficient of A is nonzero; but otherwise, there are no further conditions on the coefficients above the echelon;

(iii) there are no conditions on the coefficients $b_1, \ldots, b_m$.

Examples

(i) Examples of augmented coefficient matrices in row echelon form

$$\left[\begin{array}{cc||c} 0 & 1 & 1 \\ 0 & 0 & 0 \end{array}\right], \quad \left[\begin{array}{cc||c} 1 & 0 & 0 \\ 0 & 0 & 1 \end{array}\right], \quad \left[\begin{array}{ccc||c} 0 & 0 & 1 & 1 \\ 0 & 0 & 0 & 1 \end{array}\right],$$

$$\left[\begin{array}{ccc||c} 4 & 0 & 0 & 1 \\ 0 & 3 & 0 & 5 \end{array}\right], \quad \left[\begin{array}{ccc||c} 4 & 5 & 2 & 1 \\ 0 & 0 & 3 & 5 \end{array}\right].$$

(ii) Examples of augmented coefficient matrices *not* in echelon form

$$\left[\begin{array}{cc||c} 1 & 0 & 1 \\ 1 & 1 & 1 \end{array}\right], \quad \left[\begin{array}{cc||c} 0 & 1 & 1 \\ 1 & 1 & 1 \end{array}\right], \quad \left[\begin{array}{cc||c} 0 & 0 & 1 \\ 0 & 1 & 1 \end{array}\right].$$

If the augmented coefficient matrix of a linear system is in row echelon form, it can easily be solved. To illustrate this let us look at a few examples.

Examples

(i) The linear system

$$\begin{cases} 2x_1 + x_2 = 2 \\ \phantom{2x_1 + {}} 3x_2 = 6 \end{cases} \quad \rightsquigarrow \quad \left[\begin{array}{cc||c} 2 & 1 & 2 \\ 0 & 3 & 6 \end{array}\right]$$

is in row echelon form. Solve it by first determining x_2 by the second equation and then solving the first equation for x_1 by substituting the obtained value of x_2:

$$3x_2 = 6 \quad \leadsto \quad x_2 = 2;$$

$$2x_1 = 2 - x_2 \quad \leadsto \quad 2x_1 = 0 \quad \leadsto \quad x_1 = 0.$$

Hence the set of solutions is given by $L = \{(0, 2)\}$.

(ii) The linear system

$$\begin{cases} 2x_1 + x_2 + x_3 = 2 \\ 3x_3 = 6 \end{cases} \quad \leadsto \quad \left[\begin{array}{ccc|c} 2 & 1 & 1 & 2 \\ 0 & 0 & 3 & 6 \end{array}\right]$$

is in row echelon form. Solve the second equation for x_3 and then the first equation for x_1:

$$3x_3 = 6 \quad \leadsto \quad x_3 = 2;$$

$$x_2 \text{ is a free variable;}$$

$$2x_1 = 2 - x_2 - x_3 \quad \leadsto \quad x_1 = -\frac{1}{2} x_2.$$

Hence

$$L = \left\{ \left(-\frac{1}{2} x_2, x_2, 2\right) \;\middle|\; x_2 \in \mathbb{R} \right\},$$

which is a straight line in $\mathbb{R}^3$ trough $(0, 0, 2)$ in direction $(-1, 2, 0)$.

(iii) The augmented coefficient matrix

$$\left[\begin{array}{cc|c} 2 & 1 & 1 \\ 0 & 3 & 6 \\ 0 & 0 & 3 \end{array}\right]$$

is in row echelon form. The corresponding system of linear equations can be solved as follows: since

$$0 \cdot x_1 + 0 \cdot x_2 = 3$$

has no solutions, one concludes that $L = \emptyset$.

(iv) The augmented coefficient matrix

$$\left[\begin{array}{cc|c} 2 & 1 & 1 \\ 0 & 3 & 6 \\ 0 & 0 & 0 \end{array}\right]$$

is in row echelon form. The corresponding system of linear equations can be solved as follows:

$$0 \cdot x_1 + 0 \cdot x_2 = 0$$

is satisfied for any $x_1, x_2 \in \mathbb{R}$;

$$3x_2 = 6 \quad \rightsquigarrow \quad x_2 = 2;$$

$$2x_1 = 1 - x_2 \quad \rightsquigarrow \quad x_1 = -\frac{1}{2}.$$

Hence $L = \left\{ \left(-\frac{1}{2}, 2 \right) \right\}$.

(v) The augmented coefficient matrix

$$\left[\begin{array}{ccc|c} 0 & 1 & 2 & 1 \\ 0 & 0 & 1 & 6 \end{array}\right]$$

is in row echelon form and one computes

$$x_3 = 6, \qquad x_2 = 1 - 2x_3 = -11, \qquad x_1 \text{ free variable.}$$

Hence

$$L = \left\{ (x_1, -11, 6) \mid x_1 \in \mathbb{R} \right\},$$

which is a straight line in $\mathbb{R}^3$ through the point $(0, -11, 6)$ in direction $(1, 0, 0)$.

(vi) The augmented coefficient matrix

$$\left[\begin{array}{cccc|c} 1 & 2 & 0 & 1 & 0 \\ 0 & 0 & 0 & 3 & 6 \end{array}\right]$$

is in row echelon form and one computes

$$3x_4 = 6 \quad \rightsquigarrow \quad x_4 = 2;$$

x_3 and x_2 are free variables;

$$x_1 = -2x_2 - x_4 = -2x_2 - 2.$$

Hence

$$L = \{(-2x_2 - 2, x_2, x_3, 2) \mid x_2, x_3 \in \mathbb{R}\},$$

which is a plane in $\mathbb{R}^4$ containing the point $(-2, 0, 0, 2)$ and spanned by the vectors $(-2, 1, 0, 0)$ and $(0, 0, 1, 0)$.

(vii) The augmented coefficient matrix

$$\begin{bmatrix} 1 & 2 & 1 & 0 \\ 0 & 0 & 3 & 0 \end{bmatrix} \left\|\begin{matrix} 0 \\ 6 \end{matrix}\right]$$

is in row echelon form and one computes

$$x_4 \text{ is a free variable;}$$

$$3x_3 = 6 \quad \rightsquigarrow \quad x_3 = 2;$$

$$x_2 \text{ is a free variable;}$$

$$x_1 = -2x_2 - x_3 = -2x_2 - 2.$$

Hence

$$L = \{(-2x_2 - 2, x_2, 2, x_4) \mid x_2, x_4 \in \mathbb{R}\},$$

which is a plane in $\mathbb{R}^4$ containing the point $(-2, 0, 2, 0)$ and spanned by $(0, 0, 0, 1)$ and $(-2, 1, 0, 0)$.

As already mentioned, Gaussian elimination is an algorithm which transforms a given augmented coefficient matrix with the help of the three basic row operations $(R1)$–$(R3)$ into row echelon form. Rather than describing the algorithm in abstract terms, we illustrate how it functions with a few examples. It is convenient to introduce for the three basic row operations $(R1)$, $(R2)$, and $(R3)$ the following notations:

$(R1)$ $R_{i \leftrightarrow k}$: exchange rows i and k;
$(R2)$ $R_k \rightsquigarrow \alpha R_k$: replace kth row R_k by αR_k, $\alpha \neq 0$;
$(R3)$ $R_k \rightsquigarrow R_k + \alpha R_\ell$: replace kth row by adding to it αR_ℓ where $\ell \neq k$ and $\alpha \in \mathbb{R}$.

Examples

(i) The augmented coefficient matrix

$$\begin{bmatrix} 0 & 3 \\ 2 & 1 \end{bmatrix} \left\|\begin{matrix} 6 \\ 2 \end{matrix}\right]$$

is not in row echelon form. Apply $R_{1\leftrightarrow2}$ to get

$$\begin{bmatrix} 2 & 1 \;\bigg\| \; 2 \\ 0 & 3 \;\bigg\| \; 6 \end{bmatrix}.$$

(ii) The augmented coefficient matrix

$$\begin{bmatrix} 2 & 1 \;\bigg\| \; 1 \\ 4 & 3 \;\bigg\| \; 0 \end{bmatrix}$$

is not in row echelon form. The first row is ok; in the second row we have to eliminate 4; hence $R_2 \rightsquigarrow R_2 - 2R_1$ yielding

$$\begin{bmatrix} 2 & 1 \;\bigg\| \; 1 \\ 0 & 1 \;\bigg\| \; -2 \end{bmatrix}.$$

(iii) The augmented coefficient matrix

$$\begin{bmatrix} 1 & 1 & 1 \;\bigg\| \; 1 \\ 2 & 1 & 1 \;\bigg\| \; 0 \\ 4 & 1 & 2 \;\bigg\| \; 0 \end{bmatrix}$$

is not in row echelon form. The first row is ok; in the second and third row we have to eliminate 2 and 4, respectively. Hence $R_2 \rightsquigarrow R_2 - 2R_1$ and $R_3 \rightsquigarrow R_3 - 4R_1$ leads to

$$\begin{bmatrix} 1 & 1 & 1 \;\bigg\| \; 1 \\ 0 & -1 & -1 \;\bigg\| \; -2 \\ 0 & -3 & -2 \;\bigg\| \; -4 \end{bmatrix}.$$

Now R_1 and R_2 are ok, but we need to eliminate -3 from the last row. Hence $R_3 \rightsquigarrow R_3 - 3R_2$, yielding

$$\begin{bmatrix} 1 & 1 & 1 \;\bigg\| \; 1 \\ 0 & -1 & -1 \;\bigg\| \; -2 \\ 0 & 0 & 1 \;\bigg\| \; 2 \end{bmatrix}$$

which is in row echelon form.

(iv) The augmented coefficient matrix

$$\begin{bmatrix} 1 & 1 & 1 & 1 & 1 \;\bigg\| \; 1 \\ -1 & -1 & 0 & 0 & 1 \;\bigg\| \; -1 \\ -2 & -2 & 0 & 0 & 3 \;\bigg\| \; 1 \\ 0 & 0 & 1 & 1 & 3 \;\bigg\| \; -1 \\ 1 & 1 & 2 & 2 & 4 \;\bigg\| \; 1 \end{bmatrix}$$

is not in row echelon form. R_1 is ok, but we need to eliminate the first coefficients from the subsequent rows: $R_2 \rightsquigarrow R_2 + R_1$, $R_3 \rightsquigarrow R_3 + 2R_1$, $R_5 \rightsquigarrow R_5 - R_1$, yielding

$$
\left[
\begin{array}{ccccc|c}
1 & 1 & 1 & 1 & 1 & 1 \\
0 & 0 & 1 & 1 & 2 & 0 \\
0 & 0 & 2 & 2 & 5 & 3 \\
0 & 0 & 1 & 1 & 3 & -1 \\
0 & 0 & 1 & 1 & 3 & 0
\end{array}
\right].
$$

Rows R_1, R_2 are ok, but we need to eliminate the third coefficients in the rows R_3, R_4, R_5. $R_3 \rightsquigarrow R_3 - 2R_2$, $R_4 \rightsquigarrow R_4 - R_2$, $R_5 \rightsquigarrow R_5 - R_2$, yielding

$$
\left[
\begin{array}{ccccc|c}
1 & 1 & 1 & 1 & 1 & 1 \\
0 & 0 & 1 & 1 & 2 & 0 \\
0 & 0 & 0 & 0 & 1 & 3 \\
0 & 0 & 0 & 0 & 1 & -1 \\
0 & 0 & 0 & 0 & 1 & 0
\end{array}
\right].
$$

Now R_1, R_2, R_3 are ok, but we need to eliminate the last coefficients in R_4 and R_5, i.e., $R_4 \rightsquigarrow R_4 - R_3$, $R_5 \rightsquigarrow R_5 - R_3$, leading to

$$
\left[
\begin{array}{ccccc|c}
1 & 1 & 1 & 1 & 1 & 1 \\
0 & 0 & 1 & 1 & 2 & 0 \\
0 & 0 & 0 & 0 & 1 & 3 \\
0 & 0 & 0 & 0 & 0 & -4 \\
0 & 0 & 0 & 0 & 0 & -3
\end{array}
\right]
$$

which is in row echelon form.

(v) The augmented coefficient matrix

$$
\left[
\begin{array}{ccc|c}
1 & 1 & 1 & 0 \\
-1 & -1 & 0 & 0 \\
-2 & 1 & 0 & 1
\end{array}
\right]
$$

is not in row echelon form. R_1 is ok, but we need to eliminate the first coefficients in R_2, R_3, i.e., $R_2 \rightsquigarrow R_2 + R_1$, $R_3 \rightsquigarrow R_3 + 2R_1$, yielding

$$
\left[
\begin{array}{ccc|c}
1 & 1 & 1 & 0 \\
0 & 0 & 1 & 0 \\
0 & 3 & 2 & 1
\end{array}
\right].
$$

To bring the latter augmented coefficient matrix in row echelon form we need to exchange the second and the third row, $R_{2\leftrightarrow3}$, leading to

$$\left[\begin{array}{ccc|c} 1 & 1 & 1 & 0 \\ 0 & 3 & 2 & 1 \\ 0 & 0 & 1 & 0 \end{array}\right]$$

which is in row echelon form.

To simplify the computing of the set of solutions of a system of linear equations even more, one can go one step further and transform the augmented coefficient matrix of a given linear system into a reduced form. We begin by making some preliminary considerations. Consider the system (S)

$$\begin{cases} x_1 + 4x_2 = b_1 \\ 5x_1 + 2x_2 = b_2. \end{cases}$$

The corresponding augmented coefficient matrix is given by

$$[A \,\|\, b] = \left[\begin{array}{cc|c} 1 & 4 & b_1 \\ 5 & 2 & b_2 \end{array}\right].$$

Let us compare it with the system (S'), obtained by exchanging the two columns of A. Introducing as new unknowns y_1, y_2 this system reads

$$\begin{cases} 4y_1 + y_2 = b_1 \\ 2y_1 + 5y_2 = b_2 \end{cases}$$

and the corresponding augmented coefficient matrix is given by $[A' \,\|\, b]$ where

$$A' = \begin{pmatrix} 4 & 1 \\ 2 & 5 \end{pmatrix}.$$

Denote by L and L' the set of solutions of (S) respectively (S'). It is straightforward to see that the map

$$(x_1, x_2) \mapsto (y_1, y_2) := (x_2, x_1)$$

defines a bijection between L and L'. It means that any solution (x_1, x_2) of (S) leads to the solution $y_1 := x_2$, $y_2 := x_1$ of (S') and conversely, any solution (y_1, y_2) of (S') leads to a solution $x_1 := y_2$, $x_2 := y_1$ of (S). Said in words, by renumerating the unknowns x_1, x_2, we can read off the set of solutions of (S') from the one of (S). This procedure can be used to bring an augmented coefficient matrix $[A \,\|\, b]$ in row echelon form into an even simpler form. Let us explain the procedure with the following example.

Example Consider the augmented coefficient matrix $[A \parallel b]$, given by

$$\left[\begin{array}{ccccc|c} 1 & 4 & 1 & 2 & 1 & b_1 \\ 0 & 0 & 5 & 1 & 2 & b_2 \\ 0 & 0 & 0 & 0 & 6 & b_3 \\ 0 & 0 & 0 & 0 & 0 & b_4 \end{array}\right]$$

which is in row echelon form. Note that x_2 and x_4 are free variables. In a first step, we move the column C_2 to the far right of A,

$$\begin{pmatrix} C_1 & C_2 & C_3 & C_4 & C_5 \end{pmatrix} \quad \rightsquigarrow \quad \begin{pmatrix} C_1 & C_3 & C_4 & C_5 & C_2 \end{pmatrix},$$

and then move the column C_4 to the far right of A,

$$\begin{pmatrix} C_1 & C_3 & C_4 & C_5 & C_2 \end{pmatrix} \quad \rightsquigarrow \quad \begin{pmatrix} C_1 & C_3 & C_5 & C_2 & C_4 \end{pmatrix}.$$

The corresponding augmented coefficient matrix $[A' \parallel b]$ then reads

$$\left[\begin{array}{ccccc|c} 1 & 1 & 1 & 4 & 2 & b_1 \\ 0 & 5 & 2 & 0 & 1 & b_2 \\ 0 & 0 & 6 & 0 & 0 & b_3 \\ 0 & 0 & 0 & 0 & 0 & b_4 \end{array}\right].$$

Note that the latter echelon form has echelons with height and length equal to one and that the permuted variables, $y_1, \ldots, y_5$ are given by $y_1 = x_1$, $y_2 = x_3$, $y_3 = x_5$, $y_4 = x_2$, and $y_5 = x_4$. Furthermore, the coefficients $b_1, \ldots, b_5$ remain unchanged.

In a second step we use the row operation $(R2)$ to transform $[A' \parallel b]$ into an augmented coefficient matrix $[A'' \parallel b'']$ where the coefficients a''_{11}, a''_{22}, and a''_{33} are all 1. Note that $a'_{11} = 1$ and hence we can leave R_1 as is, whereas R_2 and R_3 are changed as follows

$$R_2 \rightsquigarrow \frac{1}{5} R_2, \qquad R_3 \rightsquigarrow \frac{1}{6} R_3.$$

We thus obtain

$$\left[\begin{array}{ccccc|c} 1 & 1 & 1 & 4 & 2 & b_1 \\ 0 & 1 & 2/5 & 0 & 1/5 & b_2/5 \\ 0 & 0 & 1 & 0 & 0 & b_3/6 \\ 0 & 0 & 0 & 0 & 0 & b_4 \end{array}\right].$$

In a third step we use the row operation $(R3)$ to transform $[A'' \,\|\, b'']$ to $[\widehat{A} \,\|\, \widehat{b}]$ where $\widehat{a}_{12} = 0$, $\widehat{a}_{13} = 0$, and $\widehat{a}_{23} = 0$. First we remove a''_{12} by the operation $R_1 \rightsquigarrow R_1 - R_2$ to obtain

$$
\left[
\begin{array}{ccccc|c}
1 & 0 & 3/5 & 4 & 9/5 & b_1 - b_2/5 \\
0 & 1 & 2/5 & 0 & 1/5 & b_2/5 \\
0 & 0 & 1 & 0 & 0 & b_3/6 \\
0 & 0 & 0 & 0 & 0 & b_4
\end{array}
\right].
$$

Then we apply the row operations

$$
R_1 \rightsquigarrow R_1 - \frac{3}{5} R_3 \qquad R_2 \rightsquigarrow R_2 - \frac{2}{5} R_2
$$

to obtain the *reduced echelon form* $[\widehat{A} \,\|\, \widehat{b}]$, given by

$$
\left[
\begin{array}{ccccc|c}
1 & 0 & 0 & 4 & 9/5 & b_1 - b_2/5 - 3/5 \cdot b_3/6 \\
0 & 1 & 0 & 0 & 1/5 & b_2/5 - 2/5 \cdot b_3/6 \\
0 & 0 & 1 & 0 & 0 & b_3/6 \\
0 & 0 & 0 & 0 & 0 & b_4
\end{array}
\right].
$$

The corresponding system of linear equations is given by

$$
\begin{cases}
y_1 + 4y_4 + \dfrac{9}{5} y_5 = b_1 - \dfrac{b_2}{5} - \dfrac{b_3}{10} \\[2mm]
y_2 + \dfrac{1}{5} y_5 = \dfrac{b_2}{5} - \dfrac{b_3}{15} \\[2mm]
y_3 = \dfrac{b_3}{6} \\[2mm]
0 = \displaystyle\sum_{j=1}^{5} 0 \cdot y_j = \widehat{b}_4 = b_4,
\end{cases}
$$

whose set of solutions can be easily described. We will discuss this for a general system in what follows.

Let us now consider the general case. Assume that A has m rows and n columns and is in row echelon form. By permuting the columns of A in the way explained in the above example, $[A \parallel b]$ can be transformed into the augmented coefficient matrix $[A' \parallel b]$ of the form

$$
\left[
\begin{array}{ccccccc}
a'_{11} & a'_{12} & a'_{13} & \cdots & a'_{1k} & a'_{1(k+1)} & \cdots & a'_{1n} \\
0 & a'_{22} & a'_{23} & \cdots & a'_{2k} & a'_{2(k+1)} & \cdots & a'_{2n} \\
0 & 0 & a'_{33} & \cdots & a'_{3k} & a'_{3(k+1)} & \cdots & a'_{3n} \\
\vdots & & \vdots & & \vdots & \vdots & & \vdots \\
0 & \cdots & & 0 & a'_{kk} & a'_{k(k+1)} & \cdots & a'_{kn} \\
0 & \cdots & & & & & & 0 \\
\vdots & & & & & & & \vdots \\
0 & \cdots & & & & & & 0
\end{array}
\right.
\left\|
\begin{array}{c}
b_1 \\ b_2 \\ b_3 \\ \vdots \\ b_k \\ b_{k+1} \\ \vdots \\ b_m
\end{array}
\right]
$$

where k is an integer with $0 \le k \le \min(m, n)$ and $a'_{11} \neq 0, a'_{22} \neq 0, \ldots, a'_{kk} \neq 0$. If $k = 0$, then A' is the matrix whose entries are all zero. Note that now all echelons have height *and* length equal to 'one' and that k can be interpreted as the *height* of the echelon.

Using the row operations $(R2)$ and $(R3)$, $[A' \parallel b]$ can be simplified. First we apply $(R2)$ to the rows R_i, $1 \le i \le k$,

$$
R_i \rightsquigarrow \frac{1}{a'_{ii}} R_i
$$

yielding

$$
\left[
\begin{array}{ccccccc}
1 & a''_{12} & a''_{13} & \cdots & a''_{1k} & a''_{1(k+1)} & \cdots & a''_{1n} \\
0 & 1 & a''_{23} & \cdots & a''_{2k} & a''_{2(k+1)} & \cdots & a''_{2n} \\
0 & 0 & 1 & & a''_{3k} & a''_{3(k+1)} & \cdots & a''_{3n} \\
\vdots & & & & \vdots & \vdots & & \vdots \\
0 & \cdots & & 0 & 1 & a''_{k(k+1)} & \cdots & a''_{kn} \\
0 & \cdots & & & & & & 0 \\
\vdots & & & & & & & \vdots \\
0 & \cdots & & & & & & 0
\end{array}
\right.
\left\|
\begin{array}{c}
b''_1 \\ b''_2 \\ b''_3 \\ \vdots \\ b''_k \\ b''_{k+1} \\ \vdots \\ b''_m
\end{array}
\right]
$$

and then we apply $(R3)$ to remove first the coefficient a''_{12} in the second column of A, then the coefficients a''_{13}, a''_{23} in the third column of A, $\ldots$, and finally the coefficients $a_{1k}'', a_{2k}'', \ldots a_{(k-1)k}''$ in the kth column of A. In this way one obtains an augmented coefficient matrix $[\widehat{A} \,\|\, \widehat{b}]$ of the form

$$
\left[\begin{array}{ccccccccc|c}
1 & 0 & 0 & \cdots & & 0 & \widehat{a}_{1(k+1)} & \cdots & \widehat{a}_{1n} & \widehat{b}_1 \\
0 & 1 & 0 & \cdots & & 0 & \widehat{a}_{2(k+1)} & \cdots & \widehat{a}_{2n} & \widehat{b}_2 \\
0 & 0 & 1 & & & 0 & \widehat{a}_{3(k+1)} & \cdots & \widehat{a}_{3n} & \widehat{b}_3 \\
\vdots & & & \vdots & & & & \vdots & & \vdots \\
0 & & \cdots & & 0 & 1 & \widehat{a}_{k(k+1)} & \cdots & \widehat{a}_{kn} & \widehat{b}_k \\
0 & & \cdots & & & & & & 0 & \widehat{b}_{k+1} \\
\vdots & & & & & & & & \vdots & \vdots \\
0 & & \cdots & & & & & & 0 & \widehat{b}_m
\end{array}\right]
\qquad (1.17)
$$

referred to as being in *reduced echelon form*. The system of linear equations corresponding to this latter augmented coefficient matrix $[\widehat{A} \,\|\, \widehat{b}]$ is then the following one:

$$
\left\{
\begin{aligned}
y_1 + \sum_{j=k+1}^{n} \widehat{a}_{1j} y_j &= \widehat{b}_1 \\
&\ \ \vdots \\
y_k + \sum_{j=k+1}^{n} \widehat{a}_{kj} y_j &= \widehat{b}_k \\
0 = \sum_{j=1}^{n} 0 \cdot y_j &= \widehat{b}_{k+1} \\
&\ \ \vdots \\
0 = \sum_{j=1}^{n} 0 \cdot y_j &= \widehat{b}_m.
\end{aligned}
\right.
$$

Notice that the unknowns are denoted by $y_1, \ldots, y_n$ since the original unknowns $x_1, \ldots, x_n$ might have been permuted, and that $0 \le k \le \min(m, n)$. Furthermore, since $[\widehat{A} \,\|\, \widehat{b}]$ has been obtained from $[A' \,\|\, b]$ by row operations, the set of solutions $\widehat{L}$ of the system corresponding to $[\widehat{A} \,\|\, \widehat{b}]$ coincides with the set of solutions L' of the system corresponding to $[A' \,\|\, b]$. The sets $\widehat{L}$ and L can now easily be determined. We have to distinguish between two cases:

Case 1. $k < m$ and there exists i with $k + 1 \le i \le m$ so that $\widehat{b}_i \ne 0$. Then $\widehat{L} = \emptyset$ and hence $L = \emptyset$.

Case 2. Either $[k = m]$ or $[k < m$ and $\widehat{b}_{k+1} = 0, \ldots, \widehat{b}_m = 0]$. Then the system above reduces to the system of equations

$$y_i + \sum_{j=k+1}^{n} \widehat{a}_{ij} y_j = \widehat{b}_i, \quad 1 \leq i \leq k. \tag{1.18}$$

Case 2a. If in addition $k = n$, then the system (1.18) reads $y_i = \widehat{b}_i$ for any $1 \leq i \leq n$. It means that $\widehat{L} = \{(\widehat{b}_1, \ldots, \widehat{b}_n)\}$ and therefore the system with augmented coefficient matrix $[A \, \| \, b]$ we started with, has a unique solution.

Case 2b. If in addition $k < n$, then the system (1.18) reads

$$y_i = \widehat{b}_i - \sum_{j=k+1}^{n} \widehat{a}_{ij} y_j, \quad 1 \leq i \leq k$$

and the unknowns $y_{k+1}, \ldots, y_n$ are *free variables*, also referred to as *parameters* and denoted by $t_{k+1}, \ldots, t_n$. Then $\widehat{L}$ is given by

$$\left\{ \left(\widehat{b}_1 - \sum_{j=k+1}^{n} \widehat{a}_{1j} t_j, \ldots, \widehat{b}_k - \sum_{j=k+1}^{n} \widehat{a}_{kj} t_j, t_{k+1}, \ldots, t_n \right) \middle| (t_{k+1}, \ldots, t_n) \in \mathbb{R}^{n-k} \right\}.$$

Hence the system (1.18) and therefore also the original system with augmented coefficient matrix $[A \, \| \, b]$ has infinitely many solutions. The map

$$F \colon \mathbb{R}^{n-k} \to \mathbb{R}^n, \, (t_{k+1}, \ldots, t_n) \mapsto F(t_{k+1}, \ldots, t_n),$$

with $F(t_{k+1}, \ldots, t_n)$ given by

$$\begin{pmatrix} \widehat{b}_1 \\ \vdots \\ \widehat{b}_k \\ 0 \\ 0 \\ \vdots \\ 0 \end{pmatrix} + t_{k+1} \begin{pmatrix} -\widehat{a}_{1(k+1)} \\ \vdots \\ -\widehat{a}_{k(k+1)} \\ 1 \\ 0 \\ \vdots \\ 0 \end{pmatrix} + t_{k+2} \begin{pmatrix} -\widehat{a}_{1(k+2)} \\ \vdots \\ -\widehat{a}_{k(k+2)} \\ 0 \\ 1 \\ \vdots \\ 0 \end{pmatrix} + \cdots + t_n \begin{pmatrix} -\widehat{a}_{1n} \\ \vdots \\ -\widehat{a}_{kn} \\ 0 \\ 0 \\ \vdots \\ 1 \end{pmatrix} \tag{1.19}$$

is a *parameter representation* of the set of solutions L'.

Let us now illustrate the discussed procedure with a few examples.

Examples

(i) Consider the case of one equation and two unknowns,

$$a_{11}x_1 + a_{12}x_2 = b_1, \quad a_{11} \neq 0.$$

We apply Gaussian elimination. The corresponding augmented coefficient matrix $[A \,\|\, b]$ is given by

$$\begin{bmatrix} a_{11} & a_{12} & \| & b_1 \end{bmatrix}.$$

Since $a_{11} \neq 0$, it is in echelon form. To transform it in reduced echelon form, note that we do not have to permute x_1, x_2. Apply the row operation $(R2)$, $R_1 \rightsquigarrow \frac{1}{a_{11}} R_1$ to get

$$\begin{bmatrix} 1 & \widehat{a}_{12} & \| & \widehat{b}_1 \end{bmatrix}, \qquad \widehat{a}_{12} = \frac{a_{12}}{a_{11}}, \quad \widehat{b}_1 = \frac{b_1}{a_{11}},$$

which is in reduced echelon form. We are in *Case 2b* with $m = 1, n = 2$, and $k = 1$. Hence there is $n - k = 1$ free variable. The set of solutions L coincides with $\widehat{L}$ (since no permutation of the unknowns were necessary) and has the following parameter representation

$$F : \mathbb{R} \to \mathbb{R}^2, \quad t_2 \mapsto \begin{pmatrix} \widehat{b}_1 \\ 0 \end{pmatrix} + t_2 \begin{pmatrix} -\widehat{a}_{21} \\ 1 \end{pmatrix}.$$

It is a straight line in $\mathbb{R}^2$, passing through the point $(\widehat{b}_1, 0)$ and having the direction $(-\widehat{a}_{21}, 1)$.

(ii) Consider the case of one equation and three unknowns,

$$a_{11}x_1 + a_{12}x_2 + a_{13}x_3 = b_1, \quad a_{11} \neq 0.$$

We apply Gaussian elimination. The corresponding augmented coefficient matrix $[A \,\|\, b]$ is given by

$$\begin{bmatrix} a_{11} & a_{12} & a_{13} & \| & b_1 \end{bmatrix}.$$

Since $a_{11} \neq 0$, it is in echelon form. To transform it in reduced echelon form, note that we do not have to permute x_1, x_2, x_3. Apply the row operation $(R2)$, $R_1 \rightsquigarrow \frac{1}{a_{11}} R_1$ to get

$$\begin{bmatrix} 1 & \widehat{a}_{12} & \widehat{a}_{13} & \| & \widehat{b}_1 \end{bmatrix}, \qquad \widehat{a}_{12} = \frac{a_{12}}{a_{11}}, \quad \widehat{a}_{13} = \frac{a_{13}}{a_{11}}, \quad \widehat{b}_1 = \frac{b_1}{a_{11}},$$

which is in reduced echelon form. We are again in *Case 2b*, now with $m = 1, n = 3$, and $k = 1$. Hence there are $n - k = 2$ free variables. The set of solutions L coincides with $\widehat{L}$ and has the following parameter representation

$$F: \mathbb{R}^2 \to \mathbb{R}^3, \quad \begin{pmatrix} t_2 \\ t_3 \end{pmatrix} \mapsto \begin{pmatrix} \widehat{b}_1 \\ 0 \\ 0 \end{pmatrix} + t_2 \begin{pmatrix} -\widehat{a}_{12} \\ 1 \\ 0 \end{pmatrix} + t_3 \begin{pmatrix} -\widehat{a}_{13} \\ 0 \\ 1 \end{pmatrix}.$$

It is a plane in $\mathbb{R}^3$ passing through the point $(\widehat{b}_1, 0, 0)$ and spanned by $(-\widehat{a}_{12}, 1, 0)$ and $(-\widehat{a}_{13}, 0, 1)$.

(iii) Consider the following system

$$\begin{cases} x_1 + x_2 + x_3 = 4 \\ x_1 - x_2 - 2x_3 = 0. \end{cases}$$

We apply Gaussian elimination. The corresponding augmented coefficient matrix $[A \parallel b]$ is given by

$$\begin{bmatrix} 1 & 1 & 1 & \Big\| & 4 \\ 1 & -1 & -2 & \Big\| & 0 \end{bmatrix}.$$

To transform it in echelon form apply the row operation $(R3)$, $R_2 \rightsquigarrow R_2 - R_1$ to get

$$\begin{bmatrix} 1 & 1 & 1 & \Big\| & 4 \\ 0 & -2 & -3 & \Big\| & -4 \end{bmatrix},$$

which is echelon form. To transform in reduced echelon form, apply the row operation $(R2)$, $R_2 \rightsquigarrow -\frac{1}{2} R_2$ to get

$$\begin{bmatrix} 1 & 1 & 1 & \Big\| & 4 \\ 0 & 1 & \sqrt[3]{2} & \Big\| & 2 \end{bmatrix}.$$

Apply the row operation $(R3)$ once more, $R_1 \rightsquigarrow R_1 - R_2$, to obtain

$$\begin{bmatrix} 1 & 0 & -\frac{1}{2} & \Big\| & 2 \\ 0 & 1 & \frac{3}{2} & \Big\| & 2 \end{bmatrix}$$

which is in reduced echelon form. We are in *Case 2b* with $m = 2, n = 3$, and $k = 2$. Hence there is $n - k = 1$ free variable. Since we have not permuted the unknowns, $\widehat{L} = L$ and a parameter representation of L is given by

$$F: \mathbb{R} \to \mathbb{R}^3, \quad t_3 \mapsto \begin{pmatrix} 2 \\ 2 \\ 0 \end{pmatrix} + t_3 \begin{pmatrix} \frac{1}{2} \\ -\frac{3}{2} \\ 1 \end{pmatrix}$$

which is a straight line in $\mathbb{R}^3$, passing through the point $(2, 2, 0)$ and having the direction $(1/2, -3/2, 1)$.

(iv) Consider

$$\begin{cases} x_1 + 2x_2 - x_3 = 1 \\ 2x_1 + x_2 + x_3 = 0 \\ 3x_1 + 0 \cdot x_2 + 3x_3 = -1. \end{cases}$$

We apply Gaussian elimination. The corresponding augmented coefficient matrix $[A \parallel b]$ is given by

$$\left[\begin{array}{ccc|c} 1 & 2 & -1 & 1 \\ 2 & 1 & 1 & 0 \\ 3 & 0 & 3 & -1 \end{array} \right].$$

To transform it in echelon form, apply the row operation $(R3)$, replacing $R_2 \rightsquigarrow R_2 - 2R_1$, $R_3 \rightsquigarrow R_3 - 3R_1$, to obtain

$$\left[\begin{array}{ccc|c} 1 & 2 & -1 & 1 \\ 0 & -3 & 3 & -2 \\ 0 & -6 & 6 & -4 \end{array} \right].$$

Apply $(R3)$ once were, $R_3 \rightsquigarrow R_3 - 2R_2$, to obtain an augmented coefficient matrix in echelon form

$$\left[\begin{array}{ccc|c} 1 & 2 & -1 & 1 \\ 0 & -3 & 3 & -2 \\ 0 & 0 & 0 & 0 \end{array} \right].$$

To transform it in reduced echelon form, apply $(R2)$, $R_2 \rightsquigarrow -\frac{1}{3} R_2$ to get

$$\left[\begin{array}{ccc|c} 1 & 2 & -1 & 1 \\ 0 & 1 & -1 & 2/3 \\ 0 & 0 & 0 & 0 \end{array} \right]$$

and finally we apply $(R3)$ once more, $R_1 \rightsquigarrow R_1 - 2R_2$, to get the following augmented coefficient matrix in reduced echelon form

$$\left[\begin{array}{ccc|c} 1 & 0 & 1 & -1/3 \\ 0 & 1 & -1 & 2/3 \\ 0 & 0 & 0 & 0 \end{array} \right].$$

We are in *Case 2b* with $m = 3$, $n = 3$, and $k = 2$. Hence there us $n - k = 1$ free variable. Furthermore $\widehat{L} = L$. Hence a parameter representation of L is given by

$$F : \mathbb{R} \to \mathbb{R}^3, \quad t_3 \mapsto \begin{pmatrix} -\,{}^1\!/_3 \\ {}^2\!/_3 \\ 0 \end{pmatrix} + t_3 \begin{pmatrix} -1 \\ 1 \\ 1 \end{pmatrix}$$

which is a straight line in $\mathbb{R}^3$ passing through the point $(-\,{}^1\!/_3\,, {}^2\!/_3\,, 0)$ and having the direction $(-1, 1, 1)$.

From our analysis one can deduce the following

Theorem 1.3.5 *If* (S) *is a system of* m *linear equations and* n *unknowns with* $m < n$*, then its set of solutions is either empty or infinite.*

Remark To see that Theorem 1.3.5 holds, one argues as follows: bring the augmented coefficient matrix in reduced row echelon form. Then $k \le \min(m, n) = m < n$ since by assumption $m < n$. Hence *Case 2a* cannot occur and we are either in *Case 1* ($L = \emptyset$) or in *Case 2b* (L infinite).

An important class of linear systems is the one where the number of equations is the same as the number of unknowns, $m = n$.

Definition 1.3.6 A matrix A is called *quadratic* if the number of its rows equals the number of its columns.

Definition 1.3.7 We say that a $n \times n$ matrix $(A) = (a_{ij})_{1 \le i, j \le n}$ is a *diagonal matrix* if $A = \mathrm{diag}(A)$ where $\mathrm{diag}(A) = (d_{ij})_{1 \le i, j \le n}$ is the $n \times n$ matrix with

$$d_{ii} = a_{ii}, \quad 1 \le i \le n, \qquad d_{ij} = 0, \quad i \ne j.$$

It is called the $n \times n$ *identity matrix* if $a_{ii} = 1$ for any $1 \le i \le n$ and $a_{ij} = 0$ for $i \ne j$. We denote it by $\mathrm{Id}_{n \times n}$ or Id_n for short.

Going through the procedure described above for transforming the augmented coefficient matrix $[A \,\|\, b]$ of a given system (S) of n linear equations with n unknowns into reduced echelon form, one sees that (S) has a unique solution if and only if it is possible to bring $[A \,\|\, b]$ *without* permuting the unknowns into the form $[\mathrm{Id}_{n \times n} \,\|\, \widehat{b}]$, yielding the solution $x_1 = \widehat{b}_1, \ldots, x_n = \widehat{b}_n$.

Definition 1.3.8 We say that a $n \times n$ matrix A is *regular* if it can be transformed by the row operations $(R1)$–$(R3)$ to the identity matrix Id_n. Otherwise A is called *singular*.

Theorem 1.3.9 *Assume that (S) is a system of n linear equations and n unknowns with augmented coefficients matrix [A ‖ b], i.e.,*

$$\sum_{j=1}^{n} a_{ij} x_j = b_i, \quad 1 \le i \le n.$$

Then the following holds:

(i) If A is regular, then for any $b' \in \mathbb{R}^n$,

$$\sum_{j=1}^{n} a_{ij} x_j = b'_i, \quad 1 \le i \le n.$$

  has a unique solution.
(ii) If A is singular, then for any $b' \in \mathbb{R}^n$, the system

$$\sum_{j=1}^{n} a_{ij} x_j = b'_i, \quad 1 \le i \le n.$$

  has either no solution at all or infinitely many.

Example Assume that A is a singular $n \times n$ matrix. Then the system with augmented coefficient matrix $[A \parallel 0]$ has infinitely many solutions.

Corollary 1.3.10 *Assume that A is a $n \times n$ matrix, $A = (a_{ij})_{1 \le i, j \le n}$. If there exists $b \in \mathbb{R}^n$ so that*

$$\sum_{j=1}^{n} a_{ij} x_j = b_i, \quad 1 \le i \le n,$$

has a unique solution, then for any $b' \in \mathbb{R}^n$,

$$\sum_{j=1}^{n} a_{ij} x_j = b'_i, \quad 1 \le i \le n,$$

has a unique solution.

Definition 1.3.11 A system of the form

$$\sum_{j=1}^{n} a_{ij} x_j = b_i, \quad 1 \le i \le m$$

is said to be a *homogeneous system of linear equations* if $b = (b_1, \ldots, b_m) = (0, \ldots, 0)$ and *inhomogeneous one* otherwise. Given a inhomogeneous system

$$\sum_{j=1}^{n} a_{ij} x_j = b_i, \quad 1 \le i \le m,$$

the system

$$\sum_{j=1}^{n} a_{ij} x_j = 0, \quad 1 \le i \le m$$

is referred to as the corresponding *homogeneous system.*

Note that a homogeneous system of linear equations has always the zero solution. Theorem 1.3.5 then leads to the following corollary.

Corollary 1.3.12 *A homogeneous system of m linear equations with n unknowns and $m < n$ has infinitely many solutions.*

We summarize the results obtained for a system of n linear equations with n unknowns as follows.

Corollary 1.3.13 *Assume that we are given a system* (S)

$$\sum_{j=1}^{n} a_{ij} x_j = b_i, \quad 1 \le i \le n$$

where $b = (b_1, \ldots, b_n) \in \mathbb{R}^n$. Then the following statements are equivalent:

(i) (S) has a unique solution.
(ii) The homogeneous system corresponding to (S) has only the zero solution.
(iii) $A = (a_{ij})_{1 \le i, j \le n}$ is regular.

In view of Theorem 1.2.2, Corollary 1.3.13 leads to the following characterization of regular 2×2 matrices.

Corollary 1.3.14 *For any 2×2 matrix A, the following statements are equivalent:*

(i) A is regular.
(ii) $\det A \neq 0$.

Finally, we already remark at this point that the set of solutions L of the linear system

$$\sum_{j=1}^{n} a_{ij} x_j = b, \quad 1 \le i \le m,$$

and the set of solutions L_{hom} of the corresponding homogeneous system

$$\sum_{j=1}^{n} a_{ij}x_j = 0, \quad 1 \le i \le m,$$

have the following properties, which can be verified in a straightforward way:

$$(P1) \quad x + x' \in L_{\text{hom}}, \lambda x \in L_{\text{hom}} \qquad x, x' \in L_{\text{hom}}, \lambda \in \mathbb{R}.$$
$$(P2) \quad x - x' \in L_{\text{hom}}, \qquad\qquad\qquad x, x' \in L.$$
$$(P3) \quad x + x' \in L, \qquad\qquad\qquad\quad x \in L, x' \in L_{\text{hom}}.$$

We will come back to these properties after having introduced the notion of a vector space.

Problems

1. Determine the augmented coefficient matrices of the following linear systems and transform them in row echelon form by using Gaussian elimination.

(i) $\begin{cases} x_1 + 2x_2 + x_3 = 0 \\ 2x_1 + 6x_2 + 3x_3 = 4 \\ 2x_2 + 5x_3 = -4 \end{cases}$
(ii) $\begin{cases} x_1 - 3x_2 + x_3 = 1 \\ 2x_1 + x_2 - x_3 = 2 \\ x_1 + 4x_2 - 2x_3 = 1 \end{cases}$

2. Transform the augmented coefficient matrices of the following linear systems into reduced echelon form and find a parameter representation of the sets of its solutions.

(i) $\begin{cases} x_1 - 3x_2 + 4x_3 = 5 \\ x_2 - x_3 = 4 \\ 2x_2 + 4x_3 = 2 \end{cases}$
(ii) $\begin{cases} x_1 + 3x_2 + x_3 + x_4 = 3 \\ 2x_1 - x_2 + x_3 + 2x_4 = 8 \\ x_1 - 5x_2 + x_4 = 5 \end{cases}$

3. Consider the linear system given by the following augmented coefficient matrix

$$\begin{bmatrix} 1 & 1 & 3 & \bigg\| & 2 \\ 1 & 2 & 4 & \bigg\| & 3 \\ 1 & 3 & \alpha & \bigg\| & \beta \end{bmatrix}.$$

(i) For which values of α and β in $\mathbb{R}$ does the system have infinitely many solutions?

(ii) For which values of α and β in $\mathbb{R}$ does the system have no solutions?

4. Determine the set of solutions of the following linear system of n equations and n unknowns.

$$
\begin{cases}
x_1 + 5x_2 & & = 0 \\
\quad x_2 + 5x_3 & & = 0 \\
& \vdots & \\
& x_{n-1} + 5x_n & = 0 \\
5x_1 & \quad + x_n & = 1
\end{cases}
$$

5. Decide whether the following assertions are true or false and justify your answers.

   (i) There exist linear systems with three equations and three unknowns, which have precisely three solutions due to special symmetry properties.
   (ii) Every linear system with two equations and three unknowns has infinitely many solutions.

Chapter 2
Matrices and Related Topics

The aim of this chapter is to discuss the basic operations on matrices, to introduce the notions of linear (in)dependence of elements in $\mathbb{R}^k$, of a basis of $\mathbb{R}^k$ and of coordinates of an element in $\mathbb{R}^k$ with respect to a basis. Furthermore, we extend the notion of the determinant of 2×2 matrices, introduced at the end of Sect. 1.2, to square matrices of arbitrary dimension and characterize invertible square matrices as being those with nonvanishing determinant.

An element $(a_1, \ldots, a_k)$ in $\mathbb{R}^k$ is often referred to as a *vector* (in $\mathbb{R}^k$) and a_j, $1 \leq j \leq k$, as its jth component. The vector in $\mathbb{R}^k$, whose components are all zero, is referred to as the *null vector* in $\mathbb{R}^k$ and is denoted by 0.

2.1 Matrices

The aim of this section is to discuss the basic operations on matrices. We denote by $\mathrm{Mat}_{m \times n}(\mathbb{R})$, or by $\mathbb{R}^{m \times n}$ for short, the set of all real $m \times n$ matrices,

$$A = (a_{ij})_{\substack{1 \leq i \leq m \\ 1 \leq j \leq n}} = \begin{pmatrix} a_{11} & \cdots & a_{1n} \\ \vdots & & \vdots \\ a_{m1} & \cdots & a_{mn} \end{pmatrix}.$$

Definition 2.1.1 Let $A = (a_{ij})_{\substack{1 \leq i \leq m \\ 1 \leq j \leq n}}$, $B = (b_{ij})_{\substack{1 \leq i \leq m \\ 1 \leq j \leq n}}$ be in $\mathbb{R}^{m \times n}$ and let $\lambda \in \mathbb{R}$.

(i) By $A + B$ we denote the $m \times n$ matrix given by

$$(a_{ij} + b_{ij})_{\substack{1 \leq i \leq m \\ 1 \leq j \leq n}} \in \mathbb{R}^{m \times n}.$$

The matrix $A + B$ is referred to as the *sum of the matrices A and B*.

© The Author(s), under exclusive license to Springer Nature Switzerland AG 2023
M. Benz, T. Kappeler, *Linear Algebra for the Sciences*, La Matematica
per il 3+2 151, https://doi.org/10.1007/978-3-031-27220-2_2

(ii) By λA we denote the $m \times n$ matrix

$$(\lambda a_{ij})_{\substack{1 \le i \le m \\ 1 \le j \le n}} \in \mathbb{R}^{m \times n}.$$

The matrix λA is referred to as the *scalar multiple* of A by the factor λ.

Note that for the sum of two matrices A and B to be defined, they have to have the same dimensions, i.e., the same number of rows and the same number of columns. So, e.g., the two matrices

$$\begin{pmatrix} 1 & 2 \\ 3 & 4 \end{pmatrix} \qquad \text{and} \qquad \begin{pmatrix} 6 & 7 \end{pmatrix}$$

cannot be added.

The definition of the multiplication of matrices is more complicated. It is motivated by the interpretation of a $m \times n$ matrix as a linear map from $\mathbb{R}^n$ to $\mathbb{R}^m$, which we will discuss in Sect. 4.2 in detail. We will see that the multiplication of matrices corresponds to the composition of the corresponding linear maps. At this point however it is only important to know that the definition of the multiplication of matrices is very well motivated.

Definition 2.1.2 Let $A = (a_{i\ell})_{\substack{1 \le i \le m \\ 1 \le \ell \le n}} \in \mathbb{R}^{m \times n}$ and $B = (b_{\ell j})_{\substack{1 \le \ell \le n \\ 1 \le j \le k}} \in \mathbb{R}^{n \times k}$. Then the product of A and B, denoted by $A \cdot B$, or AB for short, is the $m \times k$ matrix with coefficients given by

$$(A \cdot B)_{ij} := \sum_{\ell=1}^{n} a_{i\ell} b_{\ell j}.$$

Remark

(i) In the case, $m = 1$ and $n = 1$, the matrices $A, B \in \mathbb{R}^{m \times n}$ are 1×1 matrices, $A = (a_{11})$ and $B = (b_{11})$, and the product $AB = (a_{11}b_{11})$ is the product of a_{11} with b_{11}.

(ii) Note that in order for the multiplication AB of two matrices A and B to be defined, A must have the same number of columns as B has rows. Furthermore, the coefficient $(A \cdot B)_{ij}$ of the matrix product AB can be viewed as the matrix product of the ith row $R_i(A) \in \mathbb{R}^{1 \times n}$ of A and the jth column $C_j(B) \in \mathbb{R}^{n \times 1}$ of B,

$$(A \cdot B)_{ij} = R_i(A) \cdot C_j(B), \quad R_i(A) := \begin{pmatrix} a_{i1} & \cdots & a_{in} \end{pmatrix}, \quad C_j(B) := \begin{pmatrix} b_{1j} \\ \vdots \\ b_{nj} \end{pmatrix}.$$

Examples

(i) Let $A = \begin{pmatrix} 1 & 2 \\ 3 & 4 \end{pmatrix}$, $B = \begin{pmatrix} 1 \\ 2 \end{pmatrix}$. Then $A \in \mathbb{R}^{2 \times 2}$, $B \in \mathbb{R}^{2 \times 1}$, and AB is well defined,

$$AB = \begin{pmatrix} 1 \cdot 1 + 2 \cdot 2 \\ 3 \cdot 1 + 4 \cdot 2 \end{pmatrix} = \begin{pmatrix} 5 \\ 11 \end{pmatrix} \in \mathbb{R}^{2 \times 1}.$$

Note that the matrix product BA is *not* defined.

(ii) Let $A = \begin{pmatrix} 1 & 2 \\ 3 & 4 \end{pmatrix}$, $B = \begin{pmatrix} 1 & 0 \\ 3 & 1 \end{pmatrix}$. Then $A, B \in \mathbb{R}^{2 \times 2}$ and

$$AB = \begin{pmatrix} 1 \cdot 1 + 2 \cdot 3 & 1 \cdot 0 + 2 \cdot 1 \\ 3 \cdot 1 + 4 \cdot 3 & 3 \cdot 0 + 4 \cdot 1 \end{pmatrix} = \begin{pmatrix} 7 & 2 \\ 15 & 4 \end{pmatrix} \in \mathbb{R}^{2 \times 2}.$$

Note that BA is well defined and can be computed in a similar way.

(iii) Let $A = \begin{pmatrix} 1 & 2 \\ 3 & 4 \end{pmatrix}$, $B = \begin{pmatrix} 1 & 0 & 1 \\ 3 & 1 & 0 \end{pmatrix}$. Then $A \in \mathbb{R}^{2 \times 2}$, $B \in \mathbb{R}^{2 \times 3}$ and

$$AB = \begin{pmatrix} 1 \cdot 1 + 2 \cdot 3 & 1 \cdot 0 + 2 \cdot 1 & 1 \cdot 1 + 2 \cdot 0 \\ 3 \cdot 1 + 4 \cdot 3 & 3 \cdot 0 + 4 \cdot 1 & 3 \cdot 1 + 4 \cdot 0 \end{pmatrix} = \begin{pmatrix} 7 & 2 & 1 \\ 15 & 4 & 3 \end{pmatrix} \in \mathbb{R}^{2 \times 3}.$$

(iv) Let $A = \begin{pmatrix} 1 & 2 \end{pmatrix}$, $B = \begin{pmatrix} 3 \\ 4 \end{pmatrix}$. Then $A \in \mathbb{R}^{1 \times 2}$, $B \in \mathbb{R}^{2 \times 1}$. Hence both AB and BA are well defined and

$$AB = 3 + 8 = 11 \in \mathbb{R}^{1 \times 1} \quad (\simeq \mathbb{R})$$

and

$$BA = \begin{pmatrix} 3 \cdot 1 & 3 \cdot 2 \\ 4 \cdot 1 & 4 \cdot 2 \end{pmatrix} = \begin{pmatrix} 3 & 6 \\ 4 & 8 \end{pmatrix} \in \mathbb{R}^{2 \times 2}.$$

The following theorem states elementary properties of matrix multiplication. We recall that $\mathrm{Id}_{n \times n}$ denotes the $n \times n$ identity matrix.

Theorem 2.1.3 *The following holds:*

(i) Matrix multiplication is associative,

$$(AB)C = A(BC), \quad A \in \mathbb{R}^{m \times n}, B \in \mathbb{R}^{n \times k}, C \in \mathbb{R}^{k \times \ell}.$$

(ii) Matrix multiplication is distributive,

$$(A + B)C = AC + BC, \quad A, B \in \mathbb{R}^{m \times n}, C \in \mathbb{R}^{n \times k},$$

$$A(B + C) = AB + AC, \quad A \in \mathbb{R}^{m \times n}, B, C \in \mathbb{R}^{n \times k}.$$

(iii) Multiplication with the identity matrix $\mathrm{Id}_{n \times n}$ *satisfies*

$$A \cdot \mathrm{Id}_{n \times n} = A, \quad A \in \mathbb{R}^{m \times n}, \qquad \mathrm{Id}_{n \times n} \cdot B = B, \quad B \in \mathbb{R}^{n \times k}.$$

Remark To get acquainted with matrix multiplication, let us verify Theorem (i) of Theorem 2.1.3 in the case $m = 2$, $n = 2$, and $k = 2$. For any $A, B, C \in \mathbb{R}^{2 \times 2}$, the identity $A(BC) = (AB)C$ is verified as follows: let $D := BC$, $E := AB$. It is to show that $AD = EC$. Indeed, for any $1 \leq i, j \leq 2$,

$$(AD)_{ij} = \sum_{k=1}^{2} a_{ik} d_{kj} = \sum_{k=1}^{2} a_{ik} \left(\sum_{\ell=1}^{2} b_{k\ell} c_{\ell j} \right) = \sum_{k=1}^{2} \sum_{\ell=1}^{2} a_{ik} b_{k\ell} c_{\ell j}$$

and

$$(EC)_{ij} = \sum_{\ell=1}^{2} e_{i\ell} c_{\ell j} = \sum_{\ell=1}^{2} \left(\sum_{k=1}^{2} a_{ik} b_{k\ell} \right) c_{\ell j} = \sum_{\ell=1}^{2} \sum_{k=1}^{2} a_{ik} b_{k\ell} c_{\ell j}.$$

This shows that $AD = EC$.

Note that in the case A and B are square matrices of the same dimensions, i.e., A, B in $\mathbb{R}^{n \times n}$, the products AB and BA are well defined and both are elements in $\mathbb{R}^{n \times n}$. In particular, AA is well defined. It is demoted by A^2. More generally, for any $n \in \mathbb{N}$, we denote by A^n the product $A \cdots A$ with n factors.

It is important to be aware that matrix multiplication of square matrices in $\mathbb{R}^{n \times n}$ with $n \geq 2$ is *not commutative*, i.e., in general, for $A, B \in \mathbb{R}^{n \times n}$, one has $AB \neq BA$. As an example consider $A = \begin{pmatrix} 1 & 0 \\ 0 & 2 \end{pmatrix}$ and $B = \begin{pmatrix} 1 & 0 \\ 1 & 1 \end{pmatrix}$. Then $AB \neq BA$ since

$$AB = \begin{pmatrix} 1 & 0 \\ 0 & 2 \end{pmatrix} \begin{pmatrix} 1 & 0 \\ 1 & 1 \end{pmatrix} = \begin{pmatrix} 1 & 0 \\ 2 & 2 \end{pmatrix},$$

whereas

$$BA = \begin{pmatrix} 1 & 0 \\ 1 & 1 \end{pmatrix} \begin{pmatrix} 1 & 0 \\ 0 & 2 \end{pmatrix} = \begin{pmatrix} 1 & 0 \\ 1 & 2 \end{pmatrix}.$$

As a consequence, in general $(AB)^2 \neq A^2 B^2$.

However matrices in certain special classes commute with each other. The set of diagonal $n \times n$ matrices is such a class. Indeed if $A, B \in \mathbb{R}^{n \times n}$ are both diagonal matrices, $A = \text{diag}(A)$, $B = \text{diag}(B)$, then AB is also a diagonal matrix, $AB = \text{diag}(AB)$, with

$$(AB)_{ii} = a_{ii} b_{ii}, \quad 1 \le i \le n,$$

implying that $AB = BA$.

Definition 2.1.4 Let $A, B \in \mathbb{R}^{n \times n}$.

(i) A and B *commute* if $AB = BA$. They *anti-commute* if $AB = -BA$.
(ii) A is *invertible* if there exists $C \in \mathbb{R}^{n \times n}$ so that $AC = \text{Id}_{n \times n}$ and $CA = \text{Id}_{n \times n}$.

Remark

(i) One easily verifies that for any given $A \in \mathbb{R}^{n \times n}$, there exists at most one matrix $B \in \mathbb{R}^{n \times n}$ so that $AB = \text{Id}_{n \times n}$ and $BA = \text{Id}_{n \times n}$. Indeed assume that $C \in \mathbb{R}^{n \times n}$ satisfies $AC = \text{Id}_{n \times n}$ and $CA = \text{Id}_{n \times n}$. Then

$$C = C \cdot \text{Id}_{n \times n} = C(AB) = (CA)B = \text{Id}_{n \times n} \cdot B = B.$$

Hence if $A \in \mathbb{R}^{n \times n}$ is invertible, there exists a unique matrix $B \in \mathbb{R}^{n \times n}$ so that $AB = \text{Id}_{n \times n}$ and $BA = \text{Id}_{n \times n}$. This matrix is denoted by A^{-1} and is called the *inverse* of A. Note that the notion of the inverse of a matrix is only defined for square matrices.
(ii) A matrix $A = (a_{11}) \in \mathbb{R}^{1 \times 1}$ is invertible if and only of $a_{11} \ne 0$ and in such a case, $(A^{-1})_{11} = 1/a_{11}$.

Examples Decide which of the following 2×2 matrices are invertible and which are not.

(i) The matrix $\begin{pmatrix} 0 & 0 \\ 0 & 0 \end{pmatrix}$, referred to as *null matrix*, is not invertible since for any $B \in \mathbb{R}^{2 \times 2}$

$$\begin{pmatrix} 0 & 0 \\ 0 & 0 \end{pmatrix} B = \begin{pmatrix} 0 & 0 \\ 0 & 0 \end{pmatrix} \begin{pmatrix} b_{11} & b_{12} \\ b_{21} & b_{22} \end{pmatrix} = \begin{pmatrix} 0 & 0 \\ 0 & 0 \end{pmatrix} \ne \text{Id}_{2 \times 2}.$$

(ii) The matrix $\begin{pmatrix} 0 & 1 \\ 0 & 0 \end{pmatrix}$ is not invertible since for any $B \in \mathbb{R}^{2 \times 2}$

$$\begin{pmatrix} 0 & 1 \\ 0 & 0 \end{pmatrix} B = \begin{pmatrix} 0 & 1 \\ 0 & 0 \end{pmatrix} \begin{pmatrix} b_1 & b_2 \\ b_3 & b_4 \end{pmatrix} = \begin{pmatrix} b_3 & b_4 \\ 0 & 0 \end{pmatrix} \ne \text{Id}_{2 \times 2}.$$

(iii) The identity matrix $\text{Id}_{2 \times 2}$ is invertible and $(\text{Id}_{2 \times 2})^{-1} = \text{Id}_{2 \times 2}$.

(iv) The diagonal matrix $A = \begin{pmatrix} 3 & 0 \\ 0 & 4 \end{pmatrix}$ is invertible and

$$A^{-1} = \begin{pmatrix} {}^1\!/_3 & 0 \\ 0 & {}^1\!/_4 \end{pmatrix}.$$

(v) The matrix $A = \begin{pmatrix} 1 & 2 \\ 3 & 4 \end{pmatrix}$ is invertible. One can easily verify that

$$A^{-1} = -\frac{1}{2} \begin{pmatrix} 4 & -2 \\ -3 & 1 \end{pmatrix}.$$

How can the inverse of an invertible 2×2 matrix $A = \begin{pmatrix} a & b \\ c & d \end{pmatrix}$ be computed? In the case where $\det(A) \neq 0$, it turns out that A is invertible and its inverse A^{-1} is given by

$$A^{-1} = \frac{1}{\det(A)} \begin{pmatrix} d & -b \\ -c & a \end{pmatrix} = \begin{pmatrix} \frac{d}{\det(A)} & \frac{-b}{\det(A)} \\ \frac{-c}{\det(A)} & \frac{a}{\det(A)} \end{pmatrix}. \tag{2.1}$$

Indeed, one has

$$A^{-1}A = \frac{1}{\det(A)} \begin{pmatrix} d & -b \\ -c & a \end{pmatrix} \begin{pmatrix} a & b \\ c & d \end{pmatrix}$$

$$= \frac{1}{\det(A)} \begin{pmatrix} da - bc & db - bd \\ -ca + ac & -cb + ad \end{pmatrix} = \mathrm{Id}_{2 \times 2}.$$

Similarly one verifies that $AA^{-1} = \mathrm{Id}_{2 \times 2}$.

Before we describe a procedure to find the inverse of an invertible $n \times n$ matrix, let us state some general results on invertible $n \times n$ matrices. First let us introduce

$$\mathrm{GL}_{\mathbb{R}}(n) := \{ A \in \mathbb{R}^{n \times n} \mid A \text{ is invertible} \}. \tag{2.2}$$

We remark that GL stands for 'general linear'.

Theorem 2.1.5 *The following holds:*

(i) *For any $A, B \in \mathrm{GL}_{\mathbb{R}}(n)$, one has $AB \in \mathrm{GL}_{\mathbb{R}}(n)$ and*

$$(AB)^{-1} = B^{-1}A^{-1}.$$

As a consequence, for any $k \in \mathbb{N}$, $A^k \in \mathrm{GL}_{\mathbb{R}}(n)$ and $(A^k)^{-1} = (A^{-1})^k$. We denote $(A^{-1})^k$ by A^{-k} and define $A^0 := \mathrm{Id}_{n \times n}$.

(ii) For any $A \in \mathrm{GL}_{\mathbb{R}}(n)$, $A^{-1} \in \mathrm{GL}_{\mathbb{R}}(n)$ and

$$(A^{-1})^{-1} = A.$$

To get acquainted with the notion of the inverse of an invertible matrix, let us verify Theorem 2.1.5(i): first note that if $A, B \in \mathrm{GL}_{\mathbb{R}}(n)$, then A^{-1}, B^{-1} are well defined and so is $B^{-1}A^{-1}$. To see that $B^{-1}A^{-1}$ is the inverse of AB we compute

$$(B^{-1}A^{-1})(AB) = B^{-1}(A^{-1}A)B = B^{-1} \cdot \mathrm{Id}_{n \times n} \cdot B = B^{-1}B = \mathrm{Id}_{n \times n}$$

and similarly

$$(AB)B^{-1}A^{-1} = A(BB^{-1})A^{-1} = A \cdot \mathrm{Id}_{n \times n} \cdot A^{-1} = AA^{-1} = \mathrm{Id}_{n \times n}.$$

Hence by the definition of the inverse we have that AB is invertible and $(AB)^{-1}$ is given by $B^{-1}A^{-1}$. To see that Theorem 2.1.5(ii) holds we argue similarly. Note that in general, $A^{-1}B^{-1}$ is not the inverse of AB, but of BA. Hence in case A and B do not commute, neither do A^{-1} and B^{-1}.

Questions of interest with regard to the invertibility of a square matrix A are the following ones:

$(Q1)$ How can we decide if A is invertible?
$(Q2)$ In case A is invertible, how can we find its inverse?

It turns out that the two questions are closely related and can be answered by the following procedure: consider the system (S) of n linear equations with n unknowns

$$\begin{cases} a_{11}x_1 + \cdots + a_{1n}x_n = b_1 \\ \qquad \vdots \\ a_{n1}x_1 + \cdots + a_{nn}x_n = b_n \end{cases}$$

where $A = (a_{ij})_{1 \le i, j \le n}$ and $b = (b_1, \ldots, b_n) \in \mathbb{R}^n$. We want to write this system in matrix notation. For this purpose we consider b as a $n \times 1$ matrix and similarly, we do so for x,

$$b = \begin{pmatrix} b_1 \\ \vdots \\ b_n \end{pmatrix}, \qquad x = \begin{pmatrix} x_1 \\ \vdots \\ x_n \end{pmatrix}.$$

Since A is a $n \times n$ matrix, the matrix multiplication of A and x is well defined and $Ax \in \mathbb{R}^{n \times 1}$. Note that

$$(Ax)_j = \sum_{k=1}^{n} a_{jk} x_k = a_{j1} x_1 + \cdots + a_{jn} x_n.$$

Hence the above linear system (S), when written in matrix notation, takes the form

$$Ax = b.$$

Let us assume that A is invertible. Then the matrix multiplication of A^{-1} and Ax is well defined and

$$A^{-1}(Ax) = (A^{-1}A)x = \mathrm{Id}_{n \times n}\, x = x.$$

Hence multiplying left and right hand side of $Ax = b$ with A^{-1}, we get

$$x = A^{-1}b.$$

It follows that for any $b \in \mathbb{R}^n$, the linear system $Ax = b$ has a unique solution, given by $A^{-1}b$. Introduce the following vectors in $\mathbb{R}^n$,

$$e^{(1)} := (1, 0, \cdots, 0), \quad e^{(2)} := (0, 1, 0, \cdots, 0), \quad \ldots, \quad e^{(n)} := (0, \cdots, 0, 1).$$

If $b = e^{(1)}$ with $e^{(1)}$ viewed as $n \times 1$ matrix, then $x = A^{-1}e^{(1)}$ is the *first column* of A^{-1}. More generally, if $b = e^{(j)}$, $1 \le j \le n$, with $e^{(j)}$ viewed as $n \times 1$ matrix, then $A^{-1}e^{(j)}$ is the *jth column* of A^{-1}. Summarizing, we have seen that if A^{-1} exists, then for any $b \in \mathbb{R}^n$, $Ax = b$ has a unique solution and A is regular. Furthermore, we can determine A^{-1} by solving the following systems of linear equations

$$Ax = e^{(1)}, \qquad Ax = e^{(2)}, \qquad \cdots, \qquad Ax = e^{(n)}.$$

Conversely, assume that A is regular. Then the latter equations have unique solutions, which we denote by $x^{(1)}, \cdots, x^{(n)}$, and one verifies that the matrix B, whose columns are given by $x^{(1)}, \cdots, x^{(n)}$, satisfies $AB = \mathrm{Id}_{n \times n}$. One can prove that this implies that B equals A^{-1} (cf. Corollary 2.2.6).

Our considerations suggest to compute the inverse of A by forming the following version of the augmented coefficient matrix

$$\begin{bmatrix} a_{11} & \cdots & a_{1n} & 1 & 0 & \cdots & 0 \\ \vdots & & \vdots & 0 & 1 & \cdots & 0 \\ \vdots & & \vdots & \vdots & \vdots & \vdots & \vdots \\ a_{n1} & \cdots & a_{nn} & 0 & 0 & \cdots & 1 \end{bmatrix}$$

and use Gaussian elimination to determine the solutions $x^{(1)}, \ldots, x^{(n)}$.

We summarize our findings as follows.

Theorem 2.1.6 *The square matrix A is invertible if and only if A is regular. In case A is invertible, the linear system $Ax = b$ has the (unique) solution $x = A^{-1}b$.*

We recall that by Definition 1.3.8, a $n \times n$ matrix A is said to be regular if it can be transformed into the identity matrix $\mathrm{Id}_{n \times n}$ by the row operations $(R1)$–$(R3)$. In such a case, the above version of the augmented coefficient matrix

$$
\begin{bmatrix}
a_{11} & \cdots & a_{1n} & 1 & 0 & \cdots & 0 \\
\vdots & & \vdots & 0 & 1 & \cdots & 0 \\
\vdots & & \vdots & \vdots & \vdots & \vdots & \vdots \\
a_{n1} & \cdots & a_{nn} & 0 & 0 & \cdots & 1
\end{bmatrix}
$$

gets transformed into

$$
\begin{bmatrix}
1 & \cdots & 0 & b_{11} & \cdots & b_{1n} \\
\vdots & & \vdots & \vdots & & \vdots \\
0 & \cdots & 1 & b_{n1} & \cdots & b_{nn}
\end{bmatrix}
$$

and A^{-1} is given by the matrix $(b_{ij})_{1 \leq i, j \leq n}$.

Examples For each of the matrices A below, decide if A is invertible and in case it is, determine its inverse.

(i) For $A = \begin{pmatrix} 1 & 1 \\ -1 & 1 \end{pmatrix}$ consider the augmented coefficient matrix

$$
\begin{bmatrix}
1 & 1 & 1 & 0 \\
-1 & 1 & 0 & 1
\end{bmatrix}
$$

and compute its reduced echelon form.

(a) By the row operation $R_2 \rightsquigarrow R_2 + R_1$, one gets

$$
\begin{bmatrix}
1 & 1 & 1 & 0 \\
0 & 2 & 1 & 1
\end{bmatrix}.
$$

(b) Then apply the row operations $R_2 \rightsquigarrow \frac{1}{2} R_2$,

$$
\begin{bmatrix}
1 & 1 & 1 & 0 \\
0 & 1 & \frac{1}{2} & \frac{1}{2}
\end{bmatrix}.
$$

(c) Finally, the row operation $R_1 \rightsquigarrow R_1 - R_2$ yields

$$\left[\begin{array}{cc||cc} 1 & 0 & 1/2 & -1/2 \\ 0 & 1 & 1/2 & 1/2 \end{array}\right],$$

thus

$$A^{-1} = \begin{pmatrix} 1/2 & -1/2 \\ 1/2 & 1/2 \end{pmatrix} = \frac{1}{2}\begin{pmatrix} 1 & -1 \\ 1 & 1 \end{pmatrix}.$$

Note that the result coincides with the one obtained by the formula

$$\frac{1}{\det(A)}\begin{pmatrix} d & -b \\ -c & a \end{pmatrix}.$$

(ii) For $A = \begin{pmatrix} 1 & -2 & 2 \\ 1 & 1 & -1 \\ 2 & 3 & 1 \end{pmatrix}$ consider the augmented coefficient matrix

$$\left[\begin{array}{ccc||ccc} 1 & -2 & 2 & 1 & 0 & 0 \\ 1 & 1 & -1 & 0 & 1 & 0 \\ 2 & 3 & 1 & 0 & 0 & 1 \end{array}\right]$$

and compute its reduced echelon form.

(a) By the row operations $R_1 \rightsquigarrow R_2 - R_1$, $R_3 \rightsquigarrow R_3 - 2R_1$ one gets

$$\left[\begin{array}{ccc||ccc} 1 & -2 & 2 & 1 & 0 & 0 \\ 0 & 3 & -3 & -1 & 1 & 0 \\ 0 & 7 & -3 & -2 & 0 & 1 \end{array}\right].$$

(b) Then apply the row operation $R_3 \rightsquigarrow R_3 - 7/3\, R_2$,

$$\left[\begin{array}{ccc||ccc} 1 & -2 & 2 & 1 & 0 & 0 \\ 0 & 3 & -3 & -1 & 1 & 0 \\ 0 & 0 & 4 & 1/3 & -7/3 & 1 \end{array}\right].$$

(c) In a next step, apply the row operations $R_2 \rightsquigarrow 1/3\, R_2$, $R_3 \rightsquigarrow 1/4\, R_3$,

$$\left[\begin{array}{ccc||ccc} 1 & -2 & 2 & 1 & 0 & 0 \\ 0 & 1 & -1 & -1/3 & 1/3 & 0 \\ 0 & 0 & 1 & 1/12 & -7/12 & 1/4 \end{array}\right].$$

(d) Finally, by first applying the row operation $R_1 \rightsquigarrow R_1 + 2R_2$,

$$\begin{bmatrix} 1 & 0 & 0 & \Big\| & \frac{1}{3} & \frac{2}{3} & 0 \\ 0 & 1 & -1 & \Big\| & -\frac{1}{3} & \frac{1}{3} & 0 \\ 0 & 0 & 1 & \Big\| & \frac{1}{12} & -\frac{7}{12} & \frac{1}{4} \end{bmatrix},$$

(e) and then the row operations $R_2 \rightsquigarrow R_2 + R_3$, one arrives at

$$\begin{bmatrix} 1 & 0 & 0 & \Big\| & \frac{1}{3} & \frac{2}{3} & 0 \\ 0 & 1 & 0 & \Big\| & -\frac{1}{4} & -\frac{1}{4} & \frac{1}{4} \\ 0 & 0 & 1 & \Big\| & \frac{1}{12} & -\frac{7}{12} & \frac{1}{4} \end{bmatrix}.$$

Hence

$$A^{-1} = \begin{pmatrix} \frac{1}{3} & \frac{2}{3} & 0 \\ -\frac{1}{4} & -\frac{1}{4} & \frac{1}{4} \\ \frac{1}{12} & -\frac{7}{12} & \frac{1}{4} \end{pmatrix}.$$

(iii) For $A = \begin{pmatrix} 2 & 3 \\ -4 & -6 \end{pmatrix}$ consider the augmented coefficients matrix

$$\begin{bmatrix} 2 & 3 & \Big\| & 1 & 0 \\ -4 & -6 & \Big\| & 0 & 1 \end{bmatrix}$$

and try to compute its reduced row echelon form.

(a) Apply the row operation $R_2 \rightsquigarrow R_2 + 2R_1$ to get

$$\begin{bmatrix} 2 & 3 & \Big\| & 1 & 0 \\ 0 & 0 & \Big\| & 2 & 1 \end{bmatrix}.$$

It follows that A is not regular and hence according to Theorem 2.1.6, A is not invertible.

We finish this section by introducing the notion of the transpose of a matrix.

Definition 2.1.7 Given $A \in \mathbb{R}^{m \times n}$, we denote by A^{T} the $n \times m$ matrix for which the ith row is given by the jth column of A and call it the *transpose* of A. More formally,

$$(A^{\mathrm{T}})_{ij} = a_{ji}, \quad 1 \le i \le n, 1 \le j \le m.$$

Examples

(i) $A = \begin{pmatrix} 1 & 2 \\ 3 & 4 \end{pmatrix} \in \mathbb{R}^{2\times2} \quad \rightsquigarrow \quad A^{\mathrm{T}} = \begin{pmatrix} 1 & 3 \\ 2 & 4 \end{pmatrix} \in \mathbb{R}^{2\times2}.$

(ii) $A = \begin{pmatrix} 1 & 4 \\ 2 & 5 \\ 3 & 6 \end{pmatrix} \in \mathbb{R}^{3\times2} \quad \rightsquigarrow \quad A^{\mathrm{T}} = \begin{pmatrix} 1 & 2 & 3 \\ 4 & 5 & 6 \end{pmatrix} \in \mathbb{R}^{2\times3}.$

(iii) $A = \begin{pmatrix} 1 \\ 2 \\ 3 \end{pmatrix} \in \mathbb{R}^{3\times1} \quad \rightsquigarrow \quad A^{\mathrm{T}} = \begin{pmatrix} 1 \\ 2 \\ 3 \end{pmatrix} \in \mathbb{R}^{1\times3}.$

Definition 2.1.8 A square matrix $A = (a_{ij})_{1\le i,j\le n} \in \mathbb{R}^{n\times n}$ is said to be *symmetric* if $A = A^{\mathrm{T}}$ or, written coefficient wise,

$$a_{ij} = a_{ji} \quad 1 \le i, j \le n.$$

Examples

(i) $A = \begin{pmatrix} 1 & 2 \\ 2 & 3 \end{pmatrix}$ is symmetric.

(ii) $A = \begin{pmatrix} 1 & 2 \\ 1 & 3 \end{pmatrix}$ is not symmetric.

(iii) If $A \in \mathbb{R}^{n\times n}$ is a diagonal matrix, then A is symmetric. (Recall that A is a diagonal matrix if $A = \mathrm{diag}(A)$.)

Theorem 2.1.9

(i) *For any $A \in \mathbb{R}^{m\times n}$, $B \in \mathbb{R}^{n\times k}$,*

$$(AB)^{\mathrm{T}} = B^{\mathrm{T}}A^{\mathrm{T}}.$$

(ii) *For any $A \in \mathrm{GL}_{\mathbb{R}}(n)$, also $A^{\mathrm{T}} \in \mathrm{GL}_{\mathbb{R}}(n)$ and*

$$(A^{\mathrm{T}})^{-1} = (A^{-1})^{\mathrm{T}}.$$

To get more acquainted with the notion of the transpose of a matrix, let us verify the statements of Theorem 2.1.9: given $A \in \mathbb{R}^{m\times n}$, $B \in \mathbb{R}^{n\times k}$, one has $AB \in \mathbb{R}^{m\times k}$ and for any $1 \le i \le m$, $1 \le j \le k$,

$$(AB)_{ij} = \sum_{\ell=1}^{n} a_{i\ell}b_{\ell j} = \sum_{\ell=1}^{n}(A^{\mathrm{T}})_{\ell i}(B^{\mathrm{T}})_{j\ell} = \sum_{\ell=1}^{n}(B^{\mathrm{T}})_{j\ell}(A^{\mathrm{T}})_{\ell i} = (B^{\mathrm{T}}A^{\mathrm{T}})_{ji}$$

and on the other hand, by the definition of the transpose matrix, $((AB)^{\mathrm{T}})_{ji} = (AB)_{ij}$. Combining the two identities yields $(AB)^{\mathrm{T}} = B^{\mathrm{T}}A^{\mathrm{T}}$. To see that for any

$A \in \mathrm{GL}_{\mathbb{R}}(n)$ also $A^{\mathrm{T}} \in \mathrm{GL}_{\mathbb{R}}(n)$, the candidate for the inverse of A^{T} is the matrix $(A^{-1})^{\mathrm{T}}$. Indeed, by Theorem 2.1.9(i), one has

$$(A^{\mathrm{T}})(A^{-1})^{\mathrm{T}} = (A^{-1}A)^{\mathrm{T}} = \mathrm{Id}_{n \times n}^{\mathrm{T}} = \mathrm{Id}_{n \times n}$$

and

$$(A^{-1})^{\mathrm{T}}A^{\mathrm{T}} = (AA^{-1})^{\mathrm{T}} = \mathrm{Id}_{n \times n}^{\mathrm{T}} = \mathrm{Id}_{n \times n}.$$

Problems

1. Let A, B, C be the following matrices

$$A = \begin{pmatrix} 2 & -1 & 2 \\ 4 & -2 & 4 \end{pmatrix}, \qquad B = \begin{pmatrix} -1 & 0 \\ 2 & 2 \\ 2 & 1 \end{pmatrix}, \qquad C = \begin{pmatrix} -1 & 2 \\ 0 & 2 \end{pmatrix}.$$

   (i) Determine which product $Q \cdot P$ are defined for $Q, P \in \{A, B, C\}$ and which are not (the matrices Q and P do not have to be different).
   (ii) Compute AB and BA.
   (iii) Compute $3C^5 + 2C^2$.
   (iv) Compute ABC.

2. Determine which of the following matrices are regular and if so, determine their inverses.

   (i) $A = \begin{pmatrix} 1 & 2 & -2 \\ 0 & -1 & 1 \\ 2 & 3 & 0 \end{pmatrix}$

   (ii) $B = \begin{pmatrix} 1 & 2 & 2 \\ 0 & 2 & -1 \\ -1 & 0 & -3 \end{pmatrix}$

3. (i) Determine all real numbers α, β for which the 2×2 matrix

$$A := \begin{pmatrix} \alpha & \beta \\ \beta & \alpha \end{pmatrix}$$

   is invertible and compute for those numbers the inverse of A.
   (ii) Determine all real numbers a, b, c, d, e, f for which the matrix

$$B := \begin{pmatrix} a & d & e \\ 0 & b & f \\ 0 & 0 & c \end{pmatrix}$$

   is invertible and compute for those numbers the inverse of B.

4. (i) Find symmetric 2×2 matrices A, B so that the product AB is not symmetric.

   (ii) Verify: for any 4×4 matrices of the form

$$
A = \begin{pmatrix} A_1 & A_2 \\ 0 & A_3 \end{pmatrix}, \qquad B = \begin{pmatrix} B_1 & B_2 \\ 0 & B_3 \end{pmatrix}
$$

   where A_1, A_2, A_3 and B_1, B_2, B_3 are 2×2 matrices, the 4×4 matrix AB is given by

$$
AB = \begin{pmatrix} A_1 B_1 & A_1 B_2 + A_2 B_3 \\ 0 & A_3 B_3 \end{pmatrix}.
$$

5. Decide whether the following assertions are true or false and justify your answers.

   (i) For arbitrary matrices A, B in $\mathbb{R}^{2 \times 2}$,

$$
(A + B)^2 = A^2 + 2AB + B^2.
$$

   (ii) Let A be the 2×2 matrix $A = \begin{pmatrix} 1 & 2 \\ 3 & 5 \end{pmatrix}$. Then for any $k \in \mathbb{N}$, A^k is invertible and for any $n, m \in \mathbb{Z}$,

$$
A^{n+m} = A^n A^m.
$$

   (Recall that $A^0 = \mathrm{Id}_{2 \times 2}$ and for any $k \in \mathbb{N}$, A^{-k} is defined as $A^{-k} = (A^{-1})^k$.)

2.2 Linear Dependence, Bases, Coordinates

In this section we introduce the important notions of linear independence / linear dependence of elements in $\mathbb{R}^k$, of a basis of $\mathbb{R}^k$, and of coordinates of an element in $\mathbb{R}^k$ with respect to a basis. In Chap. 4, we will discuss these notions in the more general framework of vector spaces, of which $\mathbb{R}^k$ is an example. An element a of $\mathbb{R}^k$ is referred to as a *vector* and is written as $a = (a_1, \ldots, a_k)$ where $a_j, 1 \le j \le k$, are called the components of a. Depending on the context, it is convenient to view a alternatively as a $1 \times k$ matrix $(a_1 \cdots a_k) \in \mathbb{R}^{1 \times k}$ or as a $k \times 1$ matrix

$$
a = \begin{pmatrix} a_1 \\ \vdots \\ a_k \end{pmatrix} \in R^{k \times 1}.
$$

Definition 2.2.1 Assume that $a^{(1)}, \ldots, a^{(n)}$ are vectors in $\mathbb{R}^k$ where $n \geq 2$ and $k \geq 1$. A vector $b \in \mathbb{R}^k$ is said to be a *linear combination* of the vectors $a^{(1)}, \ldots, a^{(n)}$ if there exist real numbers $\alpha_1, \ldots, \alpha_n$ so that

$$b = \alpha_1 a^{(1)} + \cdots + \alpha_n a^{(n)} \qquad \text{or} \qquad b = \sum_{j=1}^{n} \alpha_j a^{(j)}.$$

The notion of linear combination in $\mathbb{R}$ is trivial, involving the addition and multiplication of real numbers. The following examples illustrate the notion of linear combination in $\mathbb{R}^2$.

Examples

(i) $n = 2$. Consider the vectors $a^{(1)} = (1, 2)$ and $a^{(2)} = (2, 1) \in \mathbb{R}^2$. The vector $b = (b_1, b_2) = (1, 5)$ is a linear combination of $a^{(1)}$ and $a^{(2)}$, $b = 3a^{(1)} - a^{(2)}$. Indeed,

$$3a^{(1)} - a^{(2)} = 3(1, 2) - (1, 2) = (3, 6) - (2, 1) = (1, 5) = b.$$

(ii) $n = 3$. Consider the vectors $a^{(1)} = (1, 2), a^{(2)} = (2, 1),$ and $a^{(3)} = (2, 2) \in \mathbb{R}^2$. The vector $b = (b_1, b_2) = (1, 5)$ is a linear combination of $a^{(1)}, a^{(2)},$ and $a^{(3)}$, $b = 2a^{(1)} - 2a^{(2)} + \frac{3}{2} a^{(3)}$. Indeed,

$$2a^{(1)} - 2a^{(2)} + \frac{3}{2} a^{(3)} = 2(1, 2) - 2(2, 1) + \frac{3}{2} (2, 2)$$

$$= (2, 4) - (4, 2) + (3, 3) = (1, 5).$$

Note that according to (i), one also has $b = 3a^{(1)} - a^{(2)} + 0a^{(3)}$. Hence the representation of b as a linear combination of $a^{(1)}, a^{(2)},$ and $a^{(3)}$ is not unique.

Definition 2.2.2

(i) Assume that $a^{(1)}, \ldots, a^{(n)}$ are vectors in $\mathbb{R}^k$ where $n \geq 2$ and $k \geq 1$. They are said to be *linearly dependent* if there exists $1 \leq i \leq n$ so that $a^{(i)}$ is a linear combination of $a^{(j)}, j \neq i$, i.e., if there exist $\alpha_j \in \mathbb{R}, j \neq i$, so that

$$a^{(i)} = \sum_{\substack{j \neq i \\ 1 \leq j \leq n}} \alpha_j a^{(j)}.$$

Equivalently, $a^{(1)}, \ldots, a^{(n)}$ are linearly dependent if there exist $\beta_1, \ldots, \beta_n \in \mathbb{R}$, not all zero, so that

$$0 = \sum_{j=1}^{n} \beta_j a^{(j)}.$$

We say in such a case that 0 is a nontrivial linear combination of $a^{(1)}, \ldots, a^{(n)}$.

(ii) The vectors $a^{(1)}, \ldots, a^{(n)}$ are said to be *linearly independent* if they are not
linearly dependent. Equivalently, $a^{(1)}, \ldots, a^{(n)}$ are linearly independent if for
any $\beta_1, \ldots, \beta_n \in \mathbb{R}$, the identity $\sum_{j=1}^{n} \beta_j a^{(j)} = 0$ implies that $\beta_j = 0$ for any
$1 \leq j \leq n$.

According to Definition 2.2.2, any set of numbers $a^{(1)}, \ldots, a^{(n)}$, $n \geq 2$, in $\mathbb{R}$
is linearly dependent. Furthermore, any two vectors $a^{(1)}$ and $a^{(2)}$ in $\mathbb{R}^k$, $k \geq 2$,
are linearly dependent if and only if $a^{(1)}$ is a scalar multiple of $a^{(2)}$ or $a^{(2)}$ is
a scalar multiple of $a^{(1)}$. The following examples illustrate the notion of linear
(in)dependence in $\mathbb{R}^2$.

Examples

(i) $n = 2$. The two vectors $a^{(1)} = (1, 2)$, $a^{(2)} = (2, 1)$ are linearly independent
in $\mathbb{R}^2$. Indeed, since $(1, 2)$ is not a scalar multiple of $(2, 1)$ and $(2, 1)$ is not a
scalar multiple of $(1, 2)$, $a^{(1)}$ and $a^{(2)}$ are linearly independent.
To verify that $a^{(1)}$ and $a^{(2)}$ are linearly independent, one can use the alternative
definition, given in Definition 2.2.2. In this case one needs to verify that for any
β_1, β_2 in $\mathbb{R}$, $\beta_1 a^{(1)} + \beta_2 a^{(2)} = 0$ implies that $\beta_1 = 0$ and $\beta_2 = 0$. Indeed, the
equation $\beta_1 a^{(1)} + \beta_2 a^{(2)} = (0, 0)$ can be written in matrix notation as

$$A \begin{pmatrix} \beta_1 \\ \beta_2 \end{pmatrix} = \begin{pmatrix} 0 \\ 0 \end{pmatrix}, \qquad A := \begin{pmatrix} 1 & 2 \\ 2 & 1 \end{pmatrix}. \tag{2.3}$$

Since $\det A = 1 \cdot 1 - 2 \cdot 2 = -3 \neq 0$, the homogeneous linear system (2.3) has
only the trivial solution $\beta_1 = 0$, $\beta_2 = 0$.

(ii) The three vectors $a^{(1)} = (1, 2)$, $a^{(2)} = (2, 1)$, and $a^{(3)} = (2, 2) \in \mathbb{R}^2$ are
linearly dependent. Indeed,

$$a^{(3)} = \frac{2}{3} a^{(1)} + \frac{2}{3} a^{(2)}. \tag{2.4}$$

To verify that $a^{(1)}$, $a^{(2)}$, and $a^{(3)}$ are linearly dependent by using the alternative
definition, given in Definition 2.2.2, it suffices to note that by (2.4) one has
$\frac{2}{3} a^{(1)} + \frac{2}{3} a^{(2)} - a^{(3)} = 0$, meaning that 0 can be represented as a nontrivial
linear combination of the vectors $a^{(1)}$, $a^{(2)}$, and $a^{(3)}$.

Figures 2.1, 2.2, 2.3, and Fig. 2.4 show the notion of linear (in)dependence and linear
combination in $\mathbb{R}^2$.

Important questions with regard to the notions of linear combination and linear
(in)dependence are the following ones:

(Q1) How can we decide whether given vectors $a^{(1)}, \ldots, a^{(n)}$ in $\mathbb{R}^k$ are linearly
independent?

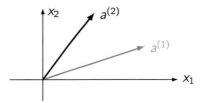

Fig. 2.1 Linearly independent vectors in $\mathbb{R}^2$. Note that the two arrows representing the vectors point in different directions

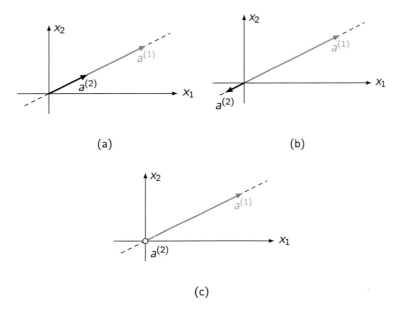

(a)   (b)

(c)

Fig. 2.2 Linearly dependent vectors in $\mathbb{R}^2$. **(a)** The linear dependence is shown by the fact that the two arrows representing the vectors point in the same direction or... **(b)** that they point in opposite directions or... **(c)** that one vector is the null vector

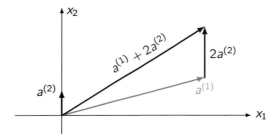

Fig. 2.3 A linear combination of vectors in $\mathbb{R}^2$. The vector $a^{(1)} + 2a^{(2)}$ is the sum of vector $a^{(1)}$ and twice vector $a^{(2)}$

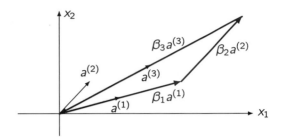

Fig. 2.4 The null vector as a linear combination of $a^{(1)}, a^{(2)}, a^{(3)} \in \mathbb{R}^2$. Note that because the sum of $\beta_1 a^{(1)} + \beta_2 a^{(2)} + \beta_3 a^{(3)}$ yields the null vector, the arrows representing this sum form a closed loop with start and end at the origin

(Q2) Given vectors $b, a^{(1)}, \ldots, a^{(n)}$ in $\mathbb{R}^k$, how can we decide whether b is a linear combination of $a^{(1)}, \ldots, a^{(n)}$? In case, b can be represented as a linear combination of $a^{(1)}, \ldots, a^{(n)}$, how can we find $\alpha_1, \ldots, \alpha_n \in \mathbb{R}$ so that

$$b = \sum_{j=1}^{n} \alpha_j a^{(j)}?$$

Are the numbers $\alpha_1, \ldots, \alpha_n$ uniquely determined?

It turns out that these questions are closely related with each other and can be rephrased in terms of systems of linear equations: assume that $b, a^{(1)}, \ldots, a^{(n)}$ are vectors in $\mathbb{R}^k$. We view them as $k \times 1$ matrices and write

$$b = \begin{pmatrix} b_1 \\ \vdots \\ b_k \end{pmatrix}, \quad a^{(1)} = \begin{pmatrix} a_1^{(1)} \\ \vdots \\ a_k^{(1)} \end{pmatrix} = \begin{pmatrix} a_{11} \\ \vdots \\ a_{k1} \end{pmatrix}, \quad \ldots, \quad a^{(n)} = \begin{pmatrix} a_1^{(n)} \\ \vdots \\ a_k^{(n)} \end{pmatrix} = \begin{pmatrix} a_{1n} \\ \vdots \\ a_{kn} \end{pmatrix}.$$

Denote by A the $k \times n$ matrix with columns $C_1(A) = a^{(1)}, \ldots, C_n(A) = a^{(n)}$,

$$A = \begin{pmatrix} a_{11} & \ldots & a_{1n} \\ \vdots & & \vdots \\ a_{k1} & \ldots & a_{kn} \end{pmatrix} = \begin{pmatrix} a_1^{(1)} & \ldots & a_1^{(n)} \\ \vdots & & \vdots \\ a_k^{(1)} & \ldots & a_k^{(n)} \end{pmatrix}.$$

Recall that b is a linear combination of the vectors $a^{(1)}, \ldots, a^{(n)}$ if there exist real numbers $\alpha_1, \ldots, \alpha_n$ so that

$$b = \sum_{j=1}^{n} \alpha_j a^{(j)}.$$

Written componentwise, the latter identity reads

$$
\begin{cases}
b_1 = \alpha_1 a_1^{(1)} + \cdots + \alpha_n a_1^{(n)} = \displaystyle\sum_{j=1}^{n} a_{1j}\alpha_j \\[2em]
\quad\vdots \\[1em]
b_k = \alpha_1 a_k^{(1)} + \cdots + \alpha_n a_k^{(n)} = \displaystyle\sum_{j=1}^{n} a_{kj}\alpha_j
\end{cases}
$$

or in matrix notation, $b = A\alpha$, where

$$
\alpha = \begin{pmatrix} \alpha_1 \\ \vdots \\ \alpha_n \end{pmatrix} \in \mathbb{R}^{n \times 1}.
$$

Note that $A \in \mathbb{R}^{k \times n}$ and hence the matrix multiplication of A by α is well defined. Hence we can express the questions, whether b is a linear combination of $a^{(1)}, \ldots, a^{(n)}$, and whether $a^{(1)}, \ldots, a^{(n)}$ are linearly (in)dependent, in terms of matrices as follows:

(LC) b is a *linear combination* of $a^{(1)}, \ldots, a^{(n)}$ if and only if the linear system $Ax = b$ has a solution $x \in \mathbb{R}^{n \times 1}$.

(LD) $a^{(1)}, \ldots, a^{(n)}$ are *linearly dependent* if and only if the linear homogeneous system $Ax = 0$ has a solution $x \in \mathbb{R}^{n \times 1}$ with $x \neq 0$.

(LI) $a^{(1)}, \ldots, a^{(n)}$ are *linearly independent* if and only if the linear homogeneous system $Ax = 0$ has only the trivial solution $x = 0$.

The following example illustrates how to apply (LC) in the case where $k = 4$ and $n = 3$.

Example Let $b = (5, 5, -1, -2) \in \mathbb{R}^4$ and

$$
a^{(1)} = (1, 1, 0, 1), \qquad a^{(2)} = (0, 2, -1, -1), \qquad a^{(3)} = (-2, -1, 0, 1).
$$

Then b is a linear combination of $a^{(1)}, a^{(2)}, a^{(3)}$. *Verification.* We look for a solution $x \in \mathbb{R}^3$ of the linear system $Ax = b$ where

$$
A = \begin{pmatrix} a_1^{(1)} & a_1^{(2)} & a_1^{(3)} \\ \vdots & \vdots & \vdots \\ a_4^{(1)} & a_4^{(2)} & a_4^{(3)} \end{pmatrix} = \begin{pmatrix} 1 & 0 & -2 \\ 1 & 2 & -1 \\ 0 & -1 & 0 \\ 1 & -1 & 1 \end{pmatrix}, \qquad x = \begin{pmatrix} x_1 \\ x_2 \\ x_3 \end{pmatrix}, \qquad b = \begin{pmatrix} 5 \\ 5 \\ -1 \\ -2 \end{pmatrix}.
$$

To this is end, we transform the augmented coefficient matrix

$$\left[\begin{array}{ccc|c} 1 & 0 & -2 & 5 \\ 1 & 2 & -1 & 5 \\ 0 & -1 & 0 & -1 \\ 1 & -1 & 1 & -2 \end{array}\right]$$

in row echelon form.

Step 1. Apply the row operations $R_2 \rightsquigarrow R_2 - R_1$, $R_4 \rightsquigarrow R_4 - R_1$,

$$\left[\begin{array}{ccc|c} 1 & 0 & -2 & 5 \\ 0 & 2 & 1 & 0 \\ 0 & -1 & 0 & -1 \\ 0 & -1 & 3 & -7 \end{array}\right].$$

Step 2. Apply the row operation $R_{2\rightsquigarrow3}$,

$$\left[\begin{array}{ccc|c} 1 & 0 & -2 & 5 \\ 0 & -1 & 0 & -1 \\ 0 & 2 & 1 & 0 \\ 0 & -1 & 3 & -7 \end{array}\right].$$

Step 3. Apply the row operations $R_3 \rightsquigarrow R_3 + 2R_2$, $R_4 \rightsquigarrow R_4 - R_2$,

$$\left[\begin{array}{ccc|c} 1 & 0 & -2 & 5 \\ 0 & -1 & 0 & -1 \\ 0 & 0 & 1 & -2 \\ 0 & 0 & 3 & -6 \end{array}\right].$$

Step 4. Apply the row operation $R_4 \rightsquigarrow R_4 - 3R_3$,

$$\left[\begin{array}{ccc|c} 1 & 0 & -2 & 5 \\ 0 & -1 & 0 & -1 \\ 0 & 0 & 1 & -2 \\ 0 & 0 & 0 & 0 \end{array}\right].$$

Therefore, $x_3 = -2$, $x_2 = 1$, $x_1 = 5 + 2x_3 = 1$ and hence

$$b = 1 \cdot a^{(1)} + 1 \cdot a^{(2)} + (-2) \cdot a^{(3)} = a^{(1)} + a^{(2)} - 2a^{(3)}.$$

In the important case where $k = n$, we have, in view of the definition of a regular matrix, the following

Theorem 2.2.3 *Assume that $a^{(1)}, \ldots, a^{(n)}$ are vectors in $\mathbb{R}^n$. Then the following two statements are equivalent:*

(i) $a^{(1)}, \ldots, a^{(n)}$ are linearly independent.
(ii) The $n \times n$ matrix A is regular where

$$
A = \begin{pmatrix} a_1^{(1)} & \ldots & a_1^{(n)} \\ \vdots & & \vdots \\ a_n^{(1)} & \ldots & a_n^{(n)} \end{pmatrix}.
$$

Definition 2.2.4 The vectors $a^{(1)}, \ldots, a^{(n)}$ in $\mathbb{R}^k$ are called a *basis* of $\mathbb{R}^k$ if every vector $b \in \mathbb{R}^k$ can be represented in a unique way as linear combination of $a^{(1)}, \ldots, a^{(n)}$, i.e., if for every $b \in \mathbb{R}^k$ there exist uniquely determined real numbers $\alpha_1, \ldots, \alpha_n$ such that

$$
b = \sum_{j=1}^n \alpha_j a^{(j)}.
$$

The numbers $\alpha_1, \ldots, \alpha_n$ are called the *coordinates* of b with respect to the basis $a^{(1)}, \ldots, a^{(n)}$. For a basis, consisting of the vectors $a^{(1)}, \ldots, a^{(n)}$, we will often use the notation $[a^{(1)}, \ldots, a^{(n)}]$ or $[a]$ for short.

Theorem 2.2.5

(i) Any basis of $\mathbb{R}^k$ consists of k vectors.
(ii) If the vectors $a^{(1)}, \ldots, a^{(k)} \in \mathbb{R}^k$ are linearly independent, then they form a basis in $\mathbb{R}^k$.
(iii) If $a^{(1)}, \ldots, a^{(n)}$ are linearly independent vectors in $\mathbb{R}^k$ with $1 \leq n < k$, then there exist vectors $a^{(n+1)}, \ldots, a^{(k)} \in \mathbb{R}^k$ so that $a^{(1)}, \ldots, a^{(n)}, a^{(n+1)}, \ldots, a^{(k)}$ form a basis of $\mathbb{R}^k$. In words: a collection of linearly independent vectors of $\mathbb{R}^k$ can always be completed to a basis of $\mathbb{R}^k$.

Examples

(i) The vectors

$$
e^{(1)} = (1, 0, \ldots, 0), \quad e^{(2)} = (0, 1, 0, \ldots, 0), \quad \ldots, \quad e^{(k)} = (0, \ldots, 0, 1),
$$

form a basis of $\mathbb{R}^k$, referred to as the *standard basis* of $\mathbb{R}^k$. Indeed, any vector $b = (b_1, \ldots, b_k) \in \mathbb{R}^k$ can be written in a unique way as

$$
b = \sum_{j=1}^k b_j e^{(j)}.
$$

The components $b_1, \ldots, b_k$ of b are the coordinates of b with respect to the standard basis $[e] = [e^{(1)}, \ldots, e^{(k)}]$ of $\mathbb{R}^k$.

(ii) The vectors $a^{(1)} = (1, 1)$, $a^{(2)} = (2, 1)$ form a basis of $\mathbb{R}^2$. Indeed, according to Theorem 2.2.5(ii), it suffices to verify that $a^{(1)}, a^{(2)}$ are linearly independent in $\mathbb{R}^2$. By Theorem 2.2.3, this is the case if and only if

$$A = \begin{pmatrix} a_1^{(1)} & a_1^{(2)} \\ a_2^{(1)} & a_2^{(2)} \end{pmatrix} = \begin{pmatrix} 1 & 2 \\ 1 & 1 \end{pmatrix}$$

is regular. By Gaussian elimination,

$$\begin{pmatrix} 1 & 2 \\ 1 & 1 \end{pmatrix} \overset{R_2 \rightsquigarrow R_2 - R_1}{\rightsquigarrow} \begin{pmatrix} 1 & 2 \\ 0 & -1 \end{pmatrix} \overset{R_2 \rightsquigarrow -R_2}{\rightsquigarrow} \begin{pmatrix} 1 & 2 \\ 0 & 1 \end{pmatrix} \overset{R_1 \rightsquigarrow R_1 - 2R_1}{\rightsquigarrow} \begin{pmatrix} 1 & 0 \\ 0 & 1 \end{pmatrix}$$

and hence A is regular. Alternatively, to see that A is regular, one can show that $\det A \neq 0$. Indeed, one has $\det A = 1 \cdot 1 - 2 \cdot 1 = -1 \neq 0$.

(iii) Let us compute the coordinates of the vector $b = (1, 3) \in \mathbb{R}^2$ with respect to the basis $[a] = [a^{(1)}, a^{(2)}]$ of $\mathbb{R}^2$ of Item (ii). To this end, we solve the linear system $Ax = b$ where A is given as in Item (ii). By Gaussian elimination,

$$\begin{bmatrix} 1 & 2 & \| & 1 \\ 1 & 1 & \| & 3 \end{bmatrix} \overset{R_2 \rightsquigarrow R_2 - R_1}{\rightsquigarrow} \begin{bmatrix} 1 & 2 & \| & 1 \\ 0 & -1 & \| & 2 \end{bmatrix}$$

yielding $x_2 = -2$, $x_1 = 1 - 2x_2 = 5$, and hence

$$b = 5a^{(1)} - 2a^{(2)}. \tag{2.5}$$

Note that by (2.5), the coordinates of b with respect to the basis $[a^{(1)}, a^{(2)}]$ of $\mathbb{R}^2$ are 5 and -2, whereas the ones with respect to the standard basis of $\mathbb{R}^2$ are 1 and 3. For a geometric interpretation, see Fig. 2.5. An important question is how the coefficients $5, -2$ and $1, 3$ are related to each other. We consider this question in a more general context. Assume that

$$[a] := [a^{(1)}, \ldots, a^{(n)}] \qquad \text{and} \qquad [b] := [b^{(1)}, \ldots, b^{(n)}]$$

are bases of $\mathbb{R}^n$ and consider a vector $u \in \mathbb{R}^n$. Then

$$u = \sum_{j=1}^{n} \alpha_j a^{(j)}, \qquad u = \sum_{j=1}^{n} \beta_j b^{(j)}$$

where $\alpha_1, \ldots, \alpha_n$ are the coordinates of u with respect to $[a]$ and $\beta_1, \ldots, \beta_n$ the ones of u with respect to $[b]$. We would like to have a method of computing β_j,

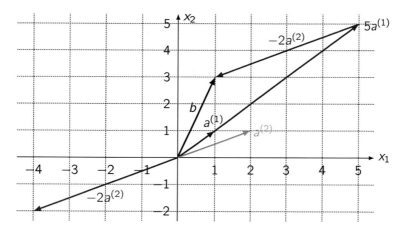

Fig. 2.5 A geometric interpretation of Example (iii). Note that vector b has coordinates 1 and 3 in the coordinate system using the standard basis of $\mathbb{R}^2$, but can be expressed by $5a^{(1)} - 2a^{(2)}$ and hence has the coordinates 5 and -2 with respect to the basis $a^{(1)}, a^{(2)}$

$1 \leq j \leq n$ from α_j, $1 \leq j \leq n$. To this end we have to express the vectors $a^{(j)}$ as linear combination of the vectors $b^{(1)}, \dots, b^{(n)}$:

$$a^{(j)} = \sum_{i=1}^{n} t_{ij} b^{(i)}, \qquad T := \begin{pmatrix} t_{11} & \cdots & t_{1n} \\ \vdots & & \vdots \\ t_{n1} & \cdots & t_{nn} \end{pmatrix}.$$

Then

$$\sum_{i=1}^{n} \beta_i b^{(i)} = u = \sum_{j=1}^{n} \alpha_j a^{(j)} = \sum_{j=1}^{n} \alpha_j \sum_{i=1}^{n} t_{ij} b^{(i)}$$

or

$$\sum_{i=1}^{n} \beta_i b^{(i)} = \sum_{i=1}^{n} \left(\sum_{j=1}^{n} \alpha_j t_{ij} \right) b^{(i)}.$$

Since the coordinates of u with respect to the basis $[b]$ are uniquely determined one concludes that $\beta_i = \sum_{j=1}^{n} t_{ij} \alpha_j$ for any $1 \leq i \leq n$. In matrix notation, we thus obtain the relation

$$\beta = T\alpha, \qquad \alpha = \begin{pmatrix} \alpha_1 \\ \vdots \\ \alpha_n \end{pmatrix}, \qquad \beta = \begin{pmatrix} \beta_1 \\ \vdots \\ \beta_n \end{pmatrix}. \qquad (2.6)$$

It turns out that a convenient notation for T is the following one

$$T = \text{Id}_{[a] \to [b]}. \tag{2.7}$$

The jth column of T is the vector of coordinates of $a^{(j)}$ with respect to the basis $[b]$. Similarly, we express $b^{(j)}$ as a linear combination of $a^{(1)}, \ldots, a^{(n)}$,

$$b^{(j)} = \sum_{i=1}^{n} s_{ij} a^{(i)}, \qquad S = \begin{pmatrix} s_{11} & \cdots & s_{1n} \\ \vdots & & \vdots \\ s_{n1} & \cdots & s_{nn} \end{pmatrix}.$$

We then obtain

$$\alpha = S\beta, \qquad S = \text{Id}_{[b] \to [a]}. \tag{2.8}$$

We will motivate the notation $\text{Id}_{[a] \to [b]}$ for T in a much broader context in Sect. 4.2, once we have introduced the notion of vector spaces and linear maps. At this point, we just record that $\text{Id}_{[a] \to [b]}$ can be viewed as the matrix representation of the identity operator on $\mathbb{R}^n$ with respect to the bases $[a]$ and $[b]$.

Theorem 2.2.6 *Assume that $[a]$ and $[b]$ are bases of $\mathbb{R}^n$. Then*

(i) S and T are regular $n \times n$ matrices, hence invertible.
(ii) $T = S^{-1}$.

Note that Theorem 2.2.6 can be deduced from (2.6) and (2.8). Indeed

$$\beta = T\alpha \quad \overset{(2.8)}{\leadsto} \quad \beta = T S\beta$$
$$\alpha = S\beta \quad \overset{(2.6)}{\leadsto} \quad \alpha = ST\alpha.$$

It then follows that $TS = \text{Id}_{n \times n}$ and $ST = \text{Id}_{n \times n}$, i.e., $T = S^{-1}$.

Example Consider the standard basis $[e] = [e^{(1)}, e^{(2)}, e^{(3)}]$ of $\mathbb{R}^3$ and the basis $[b] = [b^{(1)}, b^{(2)}, b^{(3)}]$ of $\mathbb{R}^3$, given by

$$b^{(1)} = (1, 1, 1), \qquad b^{(2)} = (0, 1, 2), \qquad b^{(3)} = (2, 1, -1).$$

Let us compute $S = \text{Id}_{[b] \to [e]}$. The first column of S is the vector of coordinates of $b^{(1)}$ with respect to the standard basis $[e]$,

$$b^{(1)} = 1 \cdot e^{(1)} + 1 \cdot e^{(2)} + 1 \cdot e^{(3)},$$

and similarly,

$$b^{(2)} = 0 \cdot e^{(1)} + 1 \cdot e^{(2)} + 2 \cdot e^{(3)}, \qquad b^{(3)} = 2 \cdot e^{(1)} + 1 \cdot e^{(2)} - 1 \cdot e^{(3)}.$$

Hence

$$S = \begin{pmatrix} 1 & 0 & 2 \\ 1 & 1 & 1 \\ 1 & 2 & -1 \end{pmatrix}.$$

By Gaussian elimination, we then compute $T = \mathrm{Id}_{[e] \to [b]} = S^{-1}$,

$$T = \begin{pmatrix} 3 & -4 & 2 \\ -2 & 3 & -1 \\ -1 & 2 & -1 \end{pmatrix}.$$

The coordinates $\beta_1, \beta_2, \beta_3$ of $u = (1, 2, 3) \in \mathbb{R}^3$ with respect to the basis $[b]$, $u = \sum_{j=1}^{3} \beta_j b^{(j)}$, can then be computed as follows:

Step 1. Compute the coordinates of u with respect to the basis $[e]$,

$$\alpha_1 = 1, \alpha_2 = 2, \alpha_3 = 3.$$

Step 2.

$$\beta = \mathrm{Id}_{[e] \to [b]} \alpha = T\alpha = \begin{pmatrix} 3 & -4 & 2 \\ -2 & 3 & -1 \\ -1 & 2 & -1 \end{pmatrix} \begin{pmatrix} 1 \\ 2 \\ 3 \end{pmatrix} = \begin{pmatrix} 1 \\ 1 \\ 0 \end{pmatrix}.$$

Hence $u = 1 \cdot b^{(1)} + 1 \cdot b^{(2)} + 0 \cdot b^{(3)}$.

The formula $\beta = \mathrm{Id}_{[e] \to [b]} \alpha$ can be expressed in words as follows,

$$\begin{cases} \beta = \text{'new' coordinates,} & \alpha = \text{'old' coordinates;} \\ [b] = \text{'new' basis,} & [e] = \text{'old' basis.} \end{cases}$$

We will come back to this topic when we discuss the notion of a linear map and its matrix representation with respect to bases.

Problems

1. (i) Decide whether $a^{(1)} = (2, 5)$, $a^{(2)} = (5, 2)$ are linearly dependent in $\mathbb{R}^2$.
   (ii) If possible, represent $b = (1, 9)$ as a linear combination of $a^{(1)} = (1, 1)$ and $a^{(2)} = (3, -1)$.
2. Decide whether the following vectors in $\mathbb{R}^3$ are linearly dependent or not.

   (i) $a^{(1)} = (1, 2, 1)$, $a^{(2)} = (-1, 1, 3)$.

(ii) $a^{(1)} = (1, 1, -1)$, $a^{(2)} = (0, 4, -3)$, $a^{(3)} = (1, 0, 3)$.

3. (i) Let $a^{(1)} := (1, 1, 1)$, $a^{(2)} := (1, 1, 2)$. Find a vector $a^{(3)} \in \mathbb{R}^3$ so that $a^{(1)}$, $a^{(2)}$, and $a^{(3)}$ form a basis of $\mathbb{R}^3$.

   (ii) Let $[a] = [a^{(1)}, a^{(2)}]$ be the basis of $\mathbb{R}^2$, given by

$$a^{(1)} = (1, -1), \qquad a^{(2)} = (2, 1).$$

   Compute $\mathrm{Id}_{[e] \to [a]}$ and $\mathrm{Id}_{[a] \to [e]}$.

4. Consider the basis $[a] = [a^{(1)}, a^{(2)}, a^{(3)}]$ of $\mathbb{R}^3$, given by $a^{(1)} = (1, 1, 0)$, $a^{(2)} = (1, 0, 1)$ and $a^{(3)} = (0, 1, 1)$, and denote by $[e] = [e^{(1)}, e^{(2)}, e^{(3)}]$ the standard basis of $\mathbb{R}^3$.

   (i) Compute $S := \mathrm{Id}_{[a] \to [e]}$ and $T := \mathrm{Id}_{[e] \to [a]}$.

   (ii) Compute the coordinates $\alpha_1, \alpha_2, \alpha_3$ of the vector $b = (1, 2, 3)$ with respect to the basis $[a]$, $b = \alpha_1 a^{(1)} + \alpha_2 a^{(2)} + \alpha_3 a^{(3)}$, and determine the coefficients $\beta_1, \beta_2, \beta_3$ of the vector $a^{(1)} + 2a^{(2)} + 3a^{(3)}$ with respect to the standard basis $[e]$.

5. Decide whether the following assertions are true or false and justify your answers.

   (i) Let $n, m \geq 2$. If one of the vectors $a^{(1)}, \ldots, a^{(n)} \in \mathbb{R}^m$ is the null vector, then $a^{(1)}, \ldots, a^{(n)}$ are linearly dependent.

   (ii) Assume that $a^{(1)}, \ldots, a^{(n)}$ are vectors in $\mathbb{R}^2$. If $n \geq 3$, then any vector $b \in \mathbb{R}^2$ can be written as a linear combination of $a^{(1)}, \ldots, a^{(n)}$.

2.3 Determinants

In Sect. 1.2, we introduced the notion of the determinant of a 2×2 matrix,

$$\det(A) = a_{11} a_{22} - a_{12} a_{21}, \qquad A = \begin{pmatrix} a_{11} & a_{12} \\ a_{21} & a_{22} \end{pmatrix}$$

and recorded the following important properties (cf. Theorem 1.2.2):

(i) For any $b \in \mathbb{R}^2$, $Ax = b$ has a unique solution if and only if $\det(A) \neq 0$.

(ii) Cramer's rule for solving $Ax = b$, $b \in \mathbb{R}^2$. Let

$$C_1 \equiv C_1(A) := \begin{pmatrix} a_{11} \\ a_{21} \end{pmatrix}, \qquad C_2 \equiv C_2(A) := \begin{pmatrix} a_{12} \\ a_{22} \end{pmatrix}, \qquad b = \begin{pmatrix} b_1 \\ b_2 \end{pmatrix}.$$

Then

$$x_1 = \frac{\det\left(b \ C_2\right)}{\det\left(C_1 \ C_2\right)}, \qquad x_2 = \frac{\det\left(C_1 \ b\right)}{\det\left(C_1 \ C_2\right)}.$$

We now want to address the question whether the notion of determinant can be extended to $n \times n$ matrices with properties similar to the ones of the determinant of 2×2 matrices. The answer is yes! Among the many equivalent ways of defining the determinant of $n \times n$ matrices, we choose a recursive definition, which defines the determinant of a $n \times n$ matrix in terms of determinants of certain $(n-1) \times (n-1)$ matrices. First we need to introduce some more notations. For $A \in \mathbb{R}^{n \times n}$ and $1 \leq i, j \leq n$, we denote by $A^{(i,j)} \in \mathbb{R}^{(n-1) \times (n-1)}$ the $(n-1) \times (n-1)$ matrix, obtained from A by deleting the ith row and the jth column.

Example For

$$A = \begin{pmatrix} 1 & 2 & 3 \\ 4 & 5 & 6 \\ 7 & 8 & 9 \end{pmatrix} \in \mathbb{R}^{3 \times 3}$$

one has

$$A^{(1,1)} = \begin{pmatrix} 5 & 6 \\ 8 & 9 \end{pmatrix}, \qquad A^{(1,3)} = \begin{pmatrix} 4 & 5 \\ 7 & 8 \end{pmatrix}, \qquad A^{(1,2)} = \begin{pmatrix} 4 & 6 \\ 7 & 9 \end{pmatrix},$$

$$A^{(2,1)} = \begin{pmatrix} 2 & 3 \\ 8 & 9 \end{pmatrix}, \qquad A^{(2,2)} = \begin{pmatrix} 1 & 3 \\ 7 & 9 \end{pmatrix}, \qquad A^{(2,3)} = \begin{pmatrix} 1 & 2 \\ 7 & 8 \end{pmatrix}.$$

To motivate the inductive definition of the determinant of a $n \times n$ matrix, we first consider the cases $n = 1, n = 2$. One has

$$n = 1: \quad A = (a_{11}) \in \mathbb{R}^{1 \times 1} \quad \rightsquigarrow \quad \det(A) = a_{11}.$$

$$n = 2: \quad A = \begin{pmatrix} a_{11} & a_{12} \\ a_{21} & a_{22} \end{pmatrix} \quad \rightsquigarrow \quad \det(A) = a_{11}a_{22} - a_{12}a_{21}$$

which can be written as

$$\det(A) = a_{11} \cdot \det(A^{(1,1)}) - a_{12} \cdot \det(A^{(1,2)})$$

$$= (-1)^{1+1}a_{11} \cdot \det(A^{(1,1)}) + (-1)^{1+2}a_{12} \cdot \det(A^{(1,2)})$$

$$= \sum_{j=1}^{2} (-1)^{1+j} a_{1j} \cdot \det(A^{(1,j)}).$$

Definition 2.2.1 For any $A \in \mathbb{R}^{n \times n}$ with $n \geq 3$, we define the *determinant* of A as

$$\det(A) = \sum_{j=1}^{n} (-1)^{1+j} a_{1j} \det(A^{(1,j)}). \tag{2.9}$$

Since $A^{(1,j)}$ is a $(n-1) \times (n-1)$ matrix for any $1 \leq j \leq n$, this is indeed a recursive definition. We refer to (2.9) as the expansion of $\det(A)$ with respect to the first row of A.

Example The determinant of

$$A = \begin{pmatrix} 1 & 2 & 3 \\ 4 & 2 & 1 \\ 1 & 0 & 1 \end{pmatrix} \in \mathbb{R}^{3 \times 3}$$

can be computed as follows. By Definition 2.2.1

$$\det(A) = (-1)^{1+1} 1 \cdot \det(A^{(1,1)}) + (-1)^{1+2} 2 \cdot \det(A^{(1,2)}) + (-1)^{1+3} 3 \cdot \det(A^{(1,3)}).$$

Since

$$A^{(1,1)} = \begin{pmatrix} 2 & 1 \\ 0 & 1 \end{pmatrix}, \qquad A^{(1,2)} = \begin{pmatrix} 4 & 1 \\ 1 & 1 \end{pmatrix}, \qquad A^{(1,3)} = \begin{pmatrix} 4 & 2 \\ 1 & 0 \end{pmatrix},$$

one gets

$$\det(A) = (2 \cdot 1 - 0) - 2(4 \cdot 1 - 1 \cdot 1) + 3(4 \cdot 0 - 2 \cdot 1) = 2 - 6 - 6 = -10.$$

Let us state some important properties of the determinant.

Theorem 2.2.2 *For any $A \in \mathbb{R}^{n \times n}$ and $1 \leq k \leq n$, the following holds:*

(i) Expansion of $\det(A)$ with respect to the kth row of A.

$$\det(A) = \sum_{j=1}^{n} (-1)^{k+j} a_{kj} \det(A^{(k,j)}).$$

(ii) Expansion of $\det(A)$ with respect to the kth column of A.

$$\det(A) = \sum_{j=1}^{n} (-1)^{j+k} a_{jk} \det(A^{(j,k)}).$$

(iii) The determinant of the transpose of A.

$$\det(A^{\mathrm{T}}) = \det(A).$$

Note that Item (iii) of the latter theorem follows from Item (ii), since $(A^{\mathrm{T}})_{1j} = a_{j1}$ and $(A^{\mathrm{T}})^{(1,j)} = A^{(j,1)}$.

To state the next theorem let us introduce some more notations. For $A \in \mathbb{R}^{n \times n}$, denote by $C_1 \equiv C_1(A), \ldots, C_n \equiv C_n(A)$ its columns and by $R_1 \equiv R_1(A), \ldots, R_n \equiv R_n(A)$ its rows. We then have

$$A = (C_1 \cdots C_n), \qquad A = \begin{pmatrix} R_1 \\ \vdots \\ R_n \end{pmatrix}.$$

Theorem 2.2.3 *For any $A \in \mathbb{R}^{n \times n}$, the following identities hold:*

(i) For any $1 \le j < i \le n$,

$$\det(C_1 \cdots C_j \cdots C_i \cdots C_n) = -\det(C_1 \cdots C_i \cdots C_j \cdots C_n).$$

By Theorem 2.2.2(iii), it then follows that a corresponding result holds for the rows of A,

$$\det \begin{pmatrix} R_1 \\ \vdots \\ R_j \\ \vdots \\ R_i \\ \vdots \\ R_n \end{pmatrix} = -\det \begin{pmatrix} R_1 \\ \vdots \\ R_i \\ \vdots \\ R_j \\ \vdots \\ R_n \end{pmatrix}.$$

(ii) For any $\lambda \in \mathbb{R}$ and $1 \le i \le n$,

$$\det(C_1 \cdots \lambda C_i \cdots C_n) = \lambda \cdot \det(C_1 \cdots C_i \cdots C_n)$$

and for any $b \in \mathbb{R}^{n \times 1}$,

$$\det(C_1 \cdots (C_i + b) \cdots C_n) = \det(C_1 \cdots C_i \cdots C_n) + \det(C_1 \cdots b \cdots C_n).$$

By Theorem 2.2.2(iii), it then follows that analogous statements hold for the rows of A.

(iii) For any $1 \le i, j \le n$ and $i \ne j$, $\lambda \in \mathbb{R}$,

$$\det(C_1 \cdots (C_i + \lambda C_j) \cdots C_n) = \det(C_1 \cdots C_i \cdots C_n).$$

By Theorem 2.2.2(iii), it then follows that an analogous statement holds for the rows of A.

(iv) If A is upper triangular, namely $a_{ij} = 0, i > j$, then

$$\det(A) = a_{11}a_{22}\cdots a_{nn} = \prod_{j=1}^{n} a_{jj}.$$

In particular, $\det(\mathrm{Id}_{n \times n}) = 1$ *and* $\det(0_{n \times n}) = 0$ *where* $0_{n \times n}$ *denotes the* $n \times n$ *matrix, all whose coefficients are zero.*

Remark Expressed in words, Theorem 2.2.3(ii) says that for any $1 \leq i \leq n$, $\det(A)$ is linear with respect to its ith column and Theorem 2.2.3(iii) says that for any $1 \leq i \leq n$, $\det(A)$ is linear with respect to its ith row (cf. Chap. 4 for the notion of linear maps).

In view of the rules for computing $\det(A)$, stated in Theorem 2.2.3, one can compute $\det(A)$ with the help of the row operation $(R1)$–$(R3)$.

Example Let

$$A = \begin{pmatrix} 1 & 2 & 3 \\ 4 & 2 & 1 \\ 1 & 0 & 1 \end{pmatrix} \in \mathbb{R}^{3 \times 3}.$$

Then $\det(A)$ can be computed as follows.

Step 1. Applying the row operations $R_2 \rightsquigarrow R_2 - 4R_1$, $R_3 \rightsquigarrow R_3 - R_1$ to A does not change its determinant (cf. Theorem 2.2.3(iii)),

$$\det(A) = \det \begin{pmatrix} 1 & 2 & 3 \\ 0 & -6 & -11 \\ 0 & -2 & -2 \end{pmatrix}.$$

Step 2. Next apply the row operation $R_3 \rightsquigarrow R_3 - \frac{1}{3} R_2$ to get

$$\det(A) = \det \begin{pmatrix} 1 & 2 & 3 \\ 0 & -6 & -11 \\ 0 & 0 & \frac{5}{3} \end{pmatrix},$$

and hence (cf. Theorem 2.2.3(iv))

$$\det(A) = \det \begin{pmatrix} 1 & 2 & 3 \\ 0 & -6 & -11 \\ 0 & 0 & \frac{5}{3} \end{pmatrix} = -10.$$

We finish this section by stating the following important properties of determinants.

Theorem 2.2.4 *For any $A \in \mathbb{R}^{n \times n}$, A is regular if and only if $\det(A) \neq 0$.*

Theorem 2.2.5 *For any $A, B \in \mathbb{R}^{n \times n}$, the following holds:*

(i) $\det(AB) = \det(A)\det(B)$.
(ii) A is invertible if and only if $\det(A) \neq 0$ and in case $\det(A) \neq 0$,

$$\det(A^{-1}) = \frac{1}{\det(A)}.$$

Theorem 2.2.5 implies the following

Corollary 2.2.6 *Assume that for a given $A \in \mathbb{R}^{n \times n}$, there exists $B \in \mathbb{R}^{n \times n}$ so that $AB = \mathrm{Id}_{n \times n}$. Then A is invertible and $A^{-1} = B$.*

Theorem 2.2.7 (Cramer's Rule) *Assume that $A \in \mathbb{R}^{n \times n}$ is regular. Then for any $b \in \mathbb{R}^{n \times 1}$, the unique solution of $Ax = b$ is given by $x = A^{-1}b$ and for any $1 \leq j \leq n$, the jth component x_j of x by*

$$x_j = \frac{\det(A_{C_j(A) \rightsquigarrow b})}{\det(A)}$$

where $A_{C_j(A) \rightsquigarrow b}$ is the $n \times n$ matrix, obtained from A by replacing the jth column $C_j(A)$ of A by b.

Problems

1. Decide whether the following vectors in $\mathbb{R}^3$ form a basis of $\mathbb{R}^3$ and if so, represent $b = (1, 0, 1)$ as a linear combination of the basis vectors.

   (i) $a^{(1)} = (1, 0, 0)$, $a^{(2)} = (0, 4, -1)$, $a^{(3)} = (2, 2, -3)$,
   (ii) $a^{(1)} = (2, -4, 5)$, $a^{(2)} = (1, 5, 6)$, $a^{(3)} = (1, 1, 1)$.

2. Compute the determinants of the following 3×3 matrices

   (i) $A = \begin{pmatrix} -1 & 2 & 3 \\ 4 & 5 & 6 \\ 7 & 8 & 9 \end{pmatrix}$,

   (ii) $B = \begin{pmatrix} 1 & 2 & 3 \\ 4 & 5 & 6 \\ 7 & 8 & 9 \end{pmatrix}$.

3. (i) Compute the determinant of the 3×3 matrix

$$A = \begin{pmatrix} 1 & 0 & 1 \\ -1 & 1 & 0 \\ 1 & 1 & 1 \end{pmatrix}^{25}.$$

(ii) Determine all numbers $a \in \mathbb{R}$ for which the determinant of the 2×2 matrix

$$B = \begin{pmatrix} 4 & 3 \\ 1 & 0 \end{pmatrix} + a \begin{pmatrix} -2 & 1 \\ -1 & -1 \end{pmatrix}$$

vanishes.

4. Verify that for any basis $[a^{(1)}, a^{(2)}, a^{(3)}]$ of $\mathbb{R}^3$, $[-a^{(1)}, 2a^{(2)}, a^{(1)}+a^{(3)}]$ is also a basis.

5. Decide whether the following assertions are true or false and justify your answers.

(i) $\det(\lambda A) = \lambda \det(A)$ for any $\lambda \in \mathbb{R}$, $A \in \mathbb{R}^{n \times n}$, $n \geq 1$.

(ii) Let $A \in \mathbb{R}^{n \times n}$, $n \geq 1$, have the property that $\det(A^k) = 0$ for some $k \in \mathbb{N}$. Then $\det(A) = 0$.

Chapter 3
Complex Numbers

So far we have worked with real numbers and used that they are ordered and can be added and multiplied, tacitly assuming that addition and multiplication satisfy the classical computational rules, i.e., that these operations are commutative, associative, ... It turns out that for many reasons, it is necessary to consider an extension of the set $\mathbb{R}$ of real numbers. These more general numbers are referred to as *complex numbers* and the set of them is denoted by $\mathbb{C}$. They can be added and multiplied. One important feature of complex numbers is that for any equation of the form

$$x^n + a_{n-1}x^{n-1} + \cdots + a_1 x + a_0 = 0$$

with $n \geq 1$ and $a_{n-1}, \ldots, a_0$ arbitrary real numbers, there exists at least one solution in $\mathbb{C}$. In particular, the equation

$$x^2 + 1 = 0,$$

admits a solution in $\mathbb{C}$, denoted by i. Additionally, $-\mathrm{i} := (-1)\,\mathrm{i}$ is a second solution.

Before introducing the complex numbers in a formal way, let us give a brief overview on the steps of gradually extending the set of natural numbers to larger systems of numbers. The *natural numbers* $1, 2, 3, \ldots$ appear in the process of counting and have been studied for thousands of years. The set of these numbers is denoted by $\mathbb{N}$. Natural numbers are ordered and can be added and multiplied.

A first extension of $\mathbb{N}$ is necessary to solve an equation of the form

$$x + a = b, \quad a, b \in \mathbb{N}, \quad a \geq b.$$

Note that such equations frequently come up in the business of accounting. Since $a \geq b$, this equation has no solution in $\mathbb{N}$ and it is necessary to introduce negative

© The Author(s), under exclusive license to Springer Nature Switzerland AG 2023
M. Benz, T. Kappeler, *Linear Algebra for the Sciences*, La Matematica
per il 3+2 151, https://doi.org/10.1007/978-3-031-27220-2_3

numbers and zero. The set

$$\ldots, -2, -1, 0, 1, 2, \ldots$$

is denoted by $\mathbb{Z}$ and its elements are referred to as *integers*. For each $a \in \mathbb{N}$, $x + a = a$ has the unique solution $0 \in \mathbb{Z}$. More generally, for any $a, b \in \mathbb{Z}$, $x + a = b$ has the unique solution

$$x = b + (-a) \in \mathbb{Z}.$$

Integers are ordered and can be added, subtracted, and multiplied. To solve equations of the form

$$ax = b, \quad a, b \in \mathbb{Z}, \quad a \neq 0,$$

one needs to extend $\mathbb{Z}$ and introduce the set of *rational numbers*

$$\mathbb{Q} := \left\{ \frac{p}{q} \mid p \in \mathbb{Z}, q \in \mathbb{N}; p, q \text{ relatively prime} \right\}.$$

Rational numbers are ordered. They can be added, subtracted, multiplied, and be divided by nonzero rational numbers.

Note that the equation

$$x^2 = 2$$

has no solution in $\mathbb{Q}$. (To see this, argue by assuming that it does and show that this leads to a contradiction.) We remark that x can be interpreted as the length of the hypotenuse of a rectangular triangle whose smaller sides have both length 1. Considerations of this type led to the extension of $\mathbb{Q}$ to the set $\mathbb{R}$ of *real numbers*. Elements in $\mathbb{R}$, which are not in $\mathbb{Q}$, are referred to as *irrational numbers*. Irrational numbers can be further distinguished. Irrational numbers, which are roots of polynomials with rational numbers as coefficients, are referred to as *algebraic*, whereas numbers, which do not have this property such as π and e, are called *transcendental numbers*.) Real numbers are ordered. They can be added, subtracted, multiplied, and divided by nonzero real numbers. But as mentioned above, $\mathbb{R}$ is not algebraically closed, i.e., there are polynomials with real coefficients which have no roots in $\mathbb{R}$.

3.1   Complex Numbers: Definition and Operations

A complex number is an element $(a, b) \in \mathbb{R}^2$ which we conveniently write as $z = a + \mathrm{i}\, b$ where the letter i stands for *imaginary*. The real numbers a and b are referred

Fig. 3.1 The complex plane
with the real and the
imaginary axis

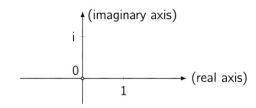

to as *real* and *imaginary parts* of z and denoted as follows

$$a = \mathrm{Re}(z), \qquad b = \mathrm{Im}(z).$$

The set of all the complex numbers is denoted by $\mathbb{C}$ and sometimes referred to as the
complex plane (Fig. 3.1). We write $z = 0$ if $\mathrm{Re}(z) = 0$ and $\mathrm{Im}(z) = 0$. A complex
number in $\mathbb{C} \setminus \{0\}$ is referred to as a *nonzero complex number*. If $z = \mathrm{i}b, b \in \mathbb{R}$,
then z is called a *purely imaginary number*.

It turns out that addition and multiplication of complex numbers can be defined
in such a way that they satisfy the same computational rules as real numbers.

Addition   The addition of complex numbers is defined by the addition of vectors in
$\mathbb{R}^2$. Recall that vectors $(a, b), (a', b') \in \mathbb{R}^2$ are added componentwise,

$$(a, b) + (a', b') = (a + a', b + b').$$

Accordingly, the sum of two complex numbers $z = a + \mathrm{i}b$, $z' = a' + \mathrm{i}b'$ is defined
as

$$z + z' = (a + \mathrm{i}b) + (a' + \mathrm{i}b') := (a + a') + \mathrm{i}(b + b').$$

Note that for $z' = 0$ one has

$$z + 0 = (a + 0) + \mathrm{i}(b + 0) = z.$$

Example   The sum of the complex numbers $z = 2 + 4\mathrm{i}$, $z' = 3 + \mathrm{i}$ is computed as

$$z + z' = (2 + 3) + \mathrm{i}(4 + 1) = 5 + 5\mathrm{i}.$$

Multiplication   The key idea to define a multiplication between complex numbers
is to interpret i as a solution of $x^2 + 1 = 0$, i.e.,

$$i^2 = -1.$$

Assuming that the terms involved commute with each other, the multiplication of $z = a + i b$ and $z' = a' + i b'$ can be computed formally as follows,

$$zz' = (a + i b)(a' + i b') := aa' + i ba' + i ab' + i^2 bb'.$$

Using that $i^2 = -1$ and collecting terms containing i and those which do not, one gets $zz' = (aa' - bb') + i(ab' + ba')$. Hence we define the multiplication zz' of two complex numbers, $z = a + i b$, $z' = a' + i b'$ as

$$zz' := (aa' - bb') + i(ab' + ba').$$

For reasons of clarity we sometimes write $z \cdot z'$ instead of zz'.

Example For $z = 2 + 4 i$, $z' = 3 + i$, one computes

$$zz' = (2 \cdot 3 - 4 \cdot 1) + i(2 \cdot 1 + 4 \cdot 3) = 2 + i \, 14.$$

We have the following special cases:

(i) If $z = a \in \mathbb{R}$, then

$$zz' = az' = (aa') + i(ab')$$

corresponds to the scalar multiplication of the vector $(a', b') \in \mathbb{R}^2$ by the real number $a \in \mathbb{R}$. In particular $1 \cdot z' = z'$.

(ii) If $z = i b$, $b \in \mathbb{R}$, then

$$zz' = i b(a' + i b') = b(-b' + i a') = -bb' + i \, ba'.$$

Geometrically, multiplication by $i b$ can be viewed as the composition of a rotation in $\mathbb{R}^2$ by $\pi/2$ (in counterclockwise orientation) with the scalar multiplication by b.

(iii) If $z = 0$, then $0 \cdot z' = 0$.

One can verify that the standard computational rules are satisfied: the operations of addition and multiplication are associative and commutative and the distributive laws hold.

Absolute Value, Polar Representation The absolute value of a complex number $z = a + i b$ is defined as the length of the vector $(a, b) \in \mathbb{R}^2$ and denoted by $|z|$,

$$|z| = \sqrt{a^2 + b^2}.$$

In particular $|z| = 0$, if and only if $z = 0$. It leads to the polar representation of a complex number $z \neq 0$. Indeed

$$(a, b) = \sqrt{a^2 + b^2} \, (\cos \varphi, \sin \varphi) = (\sqrt{a^2 + b^2} \cos \varphi, \sqrt{a^2 + b^2} \sin \varphi) \qquad (3.1)$$

where φ is the oriented angle (determined modulo 2π) between the x-axis and the vector (a, b). Hence, we have by the definition of multiplication that

$$z = |z| \cos \varphi + i |z| \sin \varphi = |z|(\cos \varphi + i \sin \varphi).$$

To shorten our notation, we introduce

$$e(\varphi) := \cos \varphi + i \sin \varphi,$$

yielding the polar representation

$$z = |z| e(\varphi).$$

Note that $|e(\varphi)| = 1$. For reasons of clarity, we can write $z = |z| \cdot e(\varphi)$. Note that the angle φ is only determined modulo 2π, i.e., φ might be replaced by $\varphi + 2\pi k$ for an arbitrary integer $k \in \mathbb{Z}$.

Examples

(i) Polar representation of $z = -2$:

$$|z| = 2 \quad \rightsquigarrow \quad z = 2(\cos \pi + i \sin \pi) = 2 \cdot e(\pi).$$

(ii) Polar representation of $z = 1 + i$ (Fig. 3.2):

$$|z| = \sqrt{1 + 1} = \sqrt{2} \quad \rightsquigarrow \quad z = \sqrt{2}\left(\cos(\pi/4) + i \sin(\pi/4)\right) = \sqrt{2} \cdot e(\pi/4).$$

(iii) Polar representation of $z = 1 - i$ (Fig. 3.3):

$$|z| = \sqrt{2} \quad \rightsquigarrow \quad z = \sqrt{2} \cdot e(-\pi/4) = \sqrt{2} \cdot e(7\pi/4).$$

Fig. 3.2 A graphical illustration of the polar representation of the number $z = 1 + i$

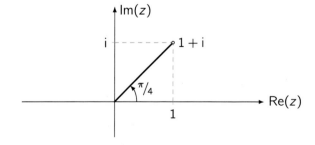

Fig. 3.3 A graphical
illustration of the polar
representation of the number
$z = 1 - i$

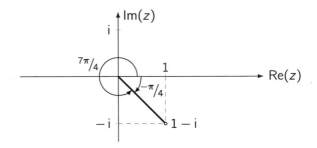

The polar representation is particularly useful for the multiplication and the division of non zero complex numbers: let z, z' be nonzero complex numbers with polar representation $z = |z|e(\varphi)$, $z' = |z'|e(\varphi')$. Then

$$zz' = |z||z'|(\cos\varphi + i\sin\varphi)(\cos\varphi' + i\sin\varphi')$$
$$= |z||z'|\big((\cos\varphi\cos\varphi' - \sin\varphi\sin\varphi') + i(\cos\varphi\sin\varphi' + \sin\varphi\cos\varphi')\big).$$

Since by the trigonometric addition theorems

$$\cos\varphi\cos\varphi' - \sin\varphi\sin\varphi' = \cos(\varphi+\varphi'), \qquad \cos\varphi\sin\varphi' + \sin\varphi\cos\varphi' = \sin(\varphi+\varphi'),$$

one obtains

$$zz' = |z||z'|e(\varphi + \varphi').$$

In particular, one has $|zz'| = |z||z'|$.

Given $z \neq 0$, the formula above can be used to determine the inverse of z, denoted by $\frac{1}{z}$. It is the complex number z' characterized by

$$1 = zz' = |z||z'|e(\varphi + \varphi').$$

Hence

$$|z'| = \frac{1}{|z|}, \qquad \varphi' = -\varphi \pmod{2\pi},$$

i.e.,

$$\frac{1}{z} = \frac{1}{|z|}e(-\varphi).$$

More generally, the formula for the multiplication of two nonzero complex numbers in polar representation can be applied to compute their quotient. More precisely, let z, z' be nonzero complex numbers with polar representations

$z = |z|e(\varphi)$ and $z' = |z'|e(\varphi')$, respectively. Then

$$\frac{z'}{z} := z' \cdot \frac{1}{z} = \frac{|z'|}{|z|} e(\varphi' - \varphi).$$

Conjugation The complex conjugate of a complex number $z = a + ib$ is the complex number

$$\bar{z} := a - ib.$$

Note that $\bar{0} = 0$ and that for any $z \neq 0$ with polar representation $z = |z|e(\varphi)$, one has

$$\bar{z} = |z|e(-\varphi).$$

For a geometrical interpretation of the effects of a complex conjugation of the complex number z, see Fig. 3.4.

Note that for any $z \in \mathbb{C}$,

$$z\bar{z} = (a + ib)(a - ib) = a^2 + b^2 = |z|^2.$$

This can be used to compute the real and imaginary parts of the quotient z'/z of the nonzero complex numbers z' and z. Indeed, let $z = a + ib$ and $z' = a' + ib'$. Then we multiply nominator and denominator of z'/z by $\bar{z}$ to obtain

$$\frac{z'}{z} = \frac{z'\bar{z}}{z\bar{z}} = \frac{(a' + ib')(a - ib)}{a^2 + b^2}$$

$$= \frac{a'a + b'b}{a^2 + b^2} + i\frac{b'a - a'b}{a^2 + b^2}.$$

Example Compute real and imaginary part of the quotient $(2 + 3i)/(1 - i)$. Since $\overline{1 - i} = 1 + i$, one has

$$\frac{2 + 3i}{1 - i} \cdot \frac{1 + i}{1 + i} = \frac{(2 + 3i)(1 + i)}{1 + 1} = \frac{2 - 3 + i(3 + 2)}{2} = -\frac{1}{2} + i\frac{5}{2}.$$

Fig. 3.4 Geometrically, the map $z \mapsto \bar{z}$ corresponds to the reflection in $\mathbb{R}^2$ at the x-axis, $(a, b) \mapsto (a, -b)$

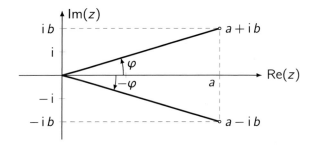

Note that in this example, the computations in polar coordinates would be more complicated as the angle in the polar representation of $2 + 3\,\mathrm{i}$ is not integer valued.

The following identities can be easily verified,

$$\overline{z + z'} = \overline{z} + \overline{z'}, \qquad \overline{zz'} = \overline{z} \cdot \overline{z'}.$$

Furthermore, $z = \overline{z}$ if and only if z is real, i.e., $\mathrm{Im}(z) = 0$.

Powers and Roots The powers z^n, $n = 1, 2 \ldots$ of a complex number $z \neq 0$ can easily be computed by using the polar representation of z, $z = |z|e(\varphi)$. For $n = 2$ one gets

$$z^2 = |z|e(\varphi)|z|e(\varphi) = |z|^2 e(2\varphi).$$

Arguing by induction, one obtains for any $n \geq 2$,

$$z^n = \big(|z|e(\varphi)\big)^n = |z|^n \big(\cos(n\varphi) + \mathrm{i}\sin(n\varphi)\big) = |z|^n e(n\varphi).$$

Example Compute real and imaginary parts of $(1 + \mathrm{i})^{12}$. Note that $|1 + \mathrm{i}| = \sqrt{2}$ and thus $1 + \mathrm{i} = \sqrt{2}\,e(\pi/4)$. Hence

$$(1 + \mathrm{i})^{12} = 2^{\frac{12}{2}} e(12\pi/4) = 64e(3\pi) = -64.$$

Hence $\mathrm{Re}((1 + \mathrm{i})^{12}) = -64$ and $\mathrm{Im}((1 + \mathrm{i})^{12}) = 0$.

The same method works to compute for any nonzero complex number z and any natural number $n \geq 1$,

$$z^{-n} := \Big(\frac{1}{z}\Big)^n = \frac{1}{|z|^n}\, e(-n\varphi).$$

Example Compute real and imaginary part of $(1 + \mathrm{i})^{-8}$.
Since $1 + \mathrm{i} = \sqrt{2}\,e(\pi/4)$, one has $\frac{1}{1+\mathrm{i}} = \frac{1}{\sqrt{2}}\,e(-\pi/4)$ and therefore

$$(1 + \mathrm{i})^{-8} = \frac{1}{2^{\frac{8}{2}}}\, e(-8\pi/4) = \frac{1}{16},$$

yielding $\mathrm{Re}((1 + \mathrm{i})^{-8}) = \frac{1}{16}$ and $\mathrm{Im}((1 + \mathrm{i})^{-8}) = 0$.

Next we want to find all complex numbers w, satisfying the equation

$$w^n = z$$

where z is a given complex number and $n \geq 2$ a natural number. A solution of the equation $w^n = z$ is called a nth root of z. In the case $n = 2$, w is referred as a *square root* of z, rather than as the second root of z.

If $z = 0$, then the only solution is $w = 0$. If $z \neq 0$, consider its polar representation, $z = |z| e(\varphi)$. A solution w of $w^n = z$ then satisfies $w \neq 0$ and hence has also a polar representation $w = \rho e(\psi)$. Hence we want to solve

$$\rho^n e(n\psi) = |z| e(\varphi).$$

Clearly one has

$$\rho = |z|^{\frac{1}{n}}, \qquad n\psi = \varphi \ (\mathrm{mod} \ 2\pi).$$

There are n solutions of the latter equation, given by

$$\psi_k = \frac{\varphi}{n} + k \frac{2\pi}{n} \ (\mathrm{mod} \ 2\pi), \quad 0 \leq k \leq n - 1.$$

Hence if $z \neq 0$, $w^n = z$ has n solutions, given by

$$w_k = |z|^{\frac{1}{n}} e\left(\frac{\varphi}{n} + k \frac{2\pi}{n}\right), \quad 0 \leq k \leq n - 1.$$

In the case $z = 1$, $w_0, \ldots, w_{n-1}$ are called the nth roots of unity. As polar representation of $z = 1$, one can choose $z = e(0)$ and hence one has $w_0 = 1$.

Examples

(i) Compute the 6th roots of the unity, $w^6 = 1$.
   Note that $w_0 = 1$, $w_1 = e\left(\frac{2\pi}{6}\right) = e\left(\frac{\pi}{3}\right)$ and

$$w_2 = e\left(2 \cdot \frac{2\pi}{6}\right) = e\left(\frac{2\pi}{3}\right), \qquad w_3 = e\left(3 \cdot \frac{2\pi}{6}\right) = e(\pi) = -1,$$

$$w_4 = e\left(4 \cdot \frac{2\pi}{6}\right) = e\left(\frac{4\pi}{3}\right), \qquad w_5 = e\left(5 \cdot \frac{2\pi}{6}\right) = e\left(\frac{5\pi}{3}\right).$$

(ii) Compute the solutions of $w^3 = 2(\sqrt{3} + i)$.
   Note that $2(\sqrt{3} + i) = 4e(\pi/6)$. Thus $w_0 = 4^{1/3} e(\pi/18)$ and

$$w_1 = 4^{\frac{1}{3}} e\left(\frac{\pi}{18} + \frac{2\pi}{3}\right), \qquad w_2 = 4^{\frac{1}{3}} e\left(\frac{\pi}{18} + 2 \cdot \frac{2\pi}{3}\right).$$

Remark The expression $e(\varphi) = \cos\varphi + i\sin\varphi$ can be computed via the complex exponential function, given by the series

$$e^z = \sum_{n=0}^{\infty} \frac{1}{n!} z^n = 1 + z + \frac{1}{2!} z^2 + \dots, \quad z \in \mathbb{C}. \tag{3.2}$$

Euler showed that

$$e^{ix} = \cos x + i\sin x \qquad \text{(Euler's formula)}.$$

The complex exponential function has similar properties as the real exponential function e^x. In particular, the law of the exponents holds,

$$e^{z+z'} = e^z e^{z'},$$

implying that for any $z = a + ib$

$$e^z = e^{a+ib} = e^a e^{ib} = e^a(\cos b + i\sin b).$$

It means that

$$|e^z| = e^a,$$

and that $b \pmod{2\pi}$ is the argument of e^z. By Euler's formula, one has

$$\cos x = \frac{e^{ix} + e^{-ix}}{2}, \qquad \sin x = \frac{e^{ix} - e^{-ix}}{2i}.$$

We remark that the trigonometric addition theorems are a straightforward consequence of the law of exponents. Indeed, by Euler's formula,

$$e^{i(\varphi+\psi)} = \cos(\varphi + \psi) + i\sin(\varphi + \psi),$$

and by the law of the exponents,

$$e^{i(\varphi+\psi)} = e^{i\varphi} e^{i\psi} = (\cos\varphi + i\sin\varphi)(\cos\psi + i\sin\psi)$$
$$= \cos\varphi\cos\psi - \sin\varphi\sin\psi + i(\sin\varphi\cos\psi + \cos\varphi\sin\psi).$$

Comparing the two expressions leads to

$$\cos(\varphi+\psi) = \cos\varphi\cos\psi - \sin\varphi\sin\psi, \qquad \sin(\varphi+\psi) = \sin\varphi\cos\psi + \sin\psi\cos\varphi.$$

Problems

1. Compute the real and the imaginary part of the following complex numbers.

(i) $\dfrac{1+i}{2+3i}$ (ii) $(2+3i)^2$ (iii) $\dfrac{1}{(1-i)^3}$ (iv) $\dfrac{1+\frac{1-i}{1+i}}{1+\frac{1}{1+2i}}$

2. (i) Compute the polar coordinates r, φ of the complex number $z = \sqrt{3} + i$ and find all possible values of $z^{1/3}$.
   (ii) Compute the polar coordinates r, φ of the complex number $z = 1 + i$ and find all possible values of $z^{1/5}$.
   (iii) Find all solutions of $z^4 = 16$.

3. (i) Compute $\left|\dfrac{2-3i}{3+4i}\right|$.
   (ii) Compute real and imaginary part of the complex number $\sum_{n=1}^{16} i^n$.
   (iii) Express $\sin^3 \varphi$ in terms of sin and cos of multiples of the angle φ.

4. Sketch the following subsets of the complex plane $\mathbb{C}$.

(i) $M_1 = \{z \in \mathbb{C} \mid |z - 1 + 2i| \geq |z + 1|\}$
(ii) $M_2 = \{z \in \mathbb{C} \mid |z + i| \geq 2;\ |z - 2| \leq 1\}$

5. Decide whether the following assertions are true or false and justify your answers.

(i) There are complex numbers $z_1 \neq 0$ and $z_2 \neq 0$ so that $z_1 z_2 = 0$.
(ii) The identity $i^0 = i$ holds.
(iii) The identity $i = e^{-i\pi/2}$ holds.

3.2   The Fundamental Theorem of Algebra

The goal of this section is to discuss complex valued functions, $p\colon \mathbb{C} \to \mathbb{C}$, given by polynomials. In Chap. 5, such polynomials come into play when computing the eigenvalues of a matrix.

By definition, a function $p\colon \mathbb{C} \to \mathbb{C}$ is said to be a *polynomial* in the complex variable z, if it can be written in the form

$$p(z) = a_n z^n + a_{n-1} z^{n-1} + \cdots + a_1 z + a_0$$

where n is a nonnegative integer and $a_0, a_1, \ldots, a_n$ are complex numbers. Written in compact form using the symbol $\sum$, p reads

$$p(z) = \sum_{j=0}^{n} a_j z^j.$$

The complex numbers $a_0, a_1, \ldots, a_n$ are referred to as the *coefficients* of p. If $a_n \neq 0$, p is said to be a *polynomial of degree n*. Note that a polynomial p of degree 0 is the constant function $p = a_0 \neq 0$. In case $n = 0$ and $a_0 = 0$, p is said to be the *zero polynomial*. It is customary to define the degree of the zero polynomial to be $-\infty$. In case all the coefficients are real, p is said to be a *polynomial with real coefficients*.

Of special interest are the roots of a polynomial p, also referred to as the *zeroes* of p. By definition, a (complex) root of a polynomial p of degree $n \geq 1$ is a complex number w so that $p(w) = 0$.

Lemma 3.2.1 *Assume that $p(z) = \sum_{j=0}^{n} a_j z^j$ and $q(z) = \sum_{\ell=0}^{m} b_\ell z^\ell$ are polynomials with complex coefficients of degree n and m, respectively. Then:*

(i) *The sum $p(z) + q(z)$ is a polynomial. In case $n > m$, it is polynomial of degree n,*

$$p(z) + q(z) = \sum_{j=m+1}^{n} a_j z^j + \sum_{j=0}^{m} (a_j + b_j) z^j.$$

In case $n = m$, $p(z) + q(z) = \sum_{j=0}^{n} (a_j + b_j) z^j$ is a polynomial of degree $\leq n$.

(ii) *The product $p(z) \cdot q(z)$ of p and q is a polynomial of degree $m + n$,*

$$p(z) \cdot q(z) = \sum_{k=0}^{n+m} c_k z^k, \quad c_k = \sum_{j+\ell=k} a_j b_\ell.$$

One says that p and q are factors of $p(z) \cdot q(z)$.

(iii) *A complex number $w \in \mathbb{C}$ is a root of $p(z) \cdot q(z)$ if and only if $p(w) = 0$ or $q(w) = 0$. (Note that it is possible that $p(w) = 0$ and $q(w) = 0$.)*

Lemma 3.2.1 has the following

Corollary 3.2.2 *Let $n \geq 1$ and $a_n \neq 0$, z_k, $1 \leq k \leq n$, be arbitrary complex numbers. Then the product $a_n \prod_{k=1}^{n} (z - z_k)$ is a polynomial of degree n. It has precisely n roots, when counted with multiplicities. They are given by z_k, $1 \leq k \leq n$. For any $1 \leq k \leq n$, $z - z_k$ is a polynomial of degree one and is referred to as a linear factor of $a_n \prod_{k=1}^{n} (z - z_k)$.*

Examples

(i) The function $p \colon \mathbb{C} \to \mathbb{C}$, given by $p(z) = (z - 1)(z + i)$ is a polynomial of degree two. Indeed, when multiplied out, one gets

$$p(z) = z^2 - z + i z + i = z^2 + (-1 + i)z - i.$$

The degree of p is two and the coefficients are given by $a_2 = 1$, $a_1 = -1 + i$, and $a_0 = -i$. Note $(z - 1)$ and $(z + i)$ are polynomials of degree one and hence linear factors of p. The roots of p are 1 and $-i$.

(ii) Let $p(z) = (z - 1)(z + i)$ and $q(z) = -(z - 1)(z - i)$. Then $p(z) + q(z)$ is a polynomial of degree one. Indeed

$$p(z) + q(z) = z^2 + (-1 + i)z - i - \left(z^2 + (-1 - i)z + i\right) = 2iz - 2i,$$

or $p(z) + q(z) = 2i(z - 1)$. Hence $p(z) + q(z)$ is a polynomial of degree one and its only root is 1.

A version of the fundamental theorem of algebra says that any polynomial with complex coefficients of degree $n \geq 1$ of the form $z^n + \cdots$ is a product of n polynomials of degree 1. More precisely one has

Theorem 3.2.3 *Any polynomial $p(z) = \sum_{k=0}^{n} a_k z^k$ of degree $n \geq 1$ with complex coefficients $a_0, a_1, \ldots, a_n$ can be uniquely written as a product of the constant a_n and of n polynomials of degree one,*

$$p(z) = a_n(z - z_1) \cdots (z - z_n) = a_n \prod_{k=1}^{n} (z - z_k). \qquad (3.3)$$

Remark

(i) The complex numbers $z_1, \ldots, z_n$ in (3.3) do not need to be different from each other. Alternatively we can write

$$p(z) = a_n(z - \zeta_1)^{m_1} \cdots (z - \zeta_\ell)^{m_\ell}$$

where $\zeta_1, \ldots, \zeta_\ell$ are the distinct complex numbers among $z_1, \ldots, z_n$ and for any $1 \leq j \leq \ell$, $m_j \in \mathbb{N}$ is referred to as the *multiplicity* of the root ζ_j. One has $\sum_{j=1}^{\ell} m_j = n$.

(ii) The result, corresponding to the one of Theorem 3.2.3, does not hold for polynomials p with *real* coefficients. More precisely, a polynomial p of degree $n \geq 2$ with real coefficients might not have n real zeroes. As an example we mention the polynomial $p(x) = x^2 + 1$ of degree two, with real coefficients $a_2 = 1$, $a_1 = 0$ and $a_0 = 1$. Its graph does not intersect the x-axis and hence p has no real zeroes (Fig. 3.5).

However, as it should be, p has two complex roots, $z_1 = i$, $z_2 = -i$ and p factors over the complex numbers,

$$p(z) = (z - i)(z + i).$$

Note that the two complex roots are complex conjugates of each other (see Lemma 3.2.4 below for a general result about roots of polynomials with real coefficients). The fact that the version of Theorem 3.2.3 for polynomials p with

Fig. 3.5 No intersection with
the x-axis implies that there
are no real zeroes for
$p(x) = x^2 + 1$

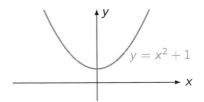

$y = x^2 + 1$

real coefficients does not hold, is one of the main reasons why we need to work with complex numbers in the sequel: the eigenvalues of a real $n \times n$ matrix ($n \geq 2$), might be complex numbers.

The next result concerns polynomials with real coefficients. Assume that $p(z) = \sum_{k=0}^{n} a_k z^k$ is a polynomial of degree $n \geq 1$ with real coefficients a_k, $0 \leq k \leq n$. Assume that $w \in \mathbb{C}$ is a root of p, $0 = p(w) = \sum_{k=0}^{n} a_k w^k$. Taking the complex conjugate of the left and right hand sides of the latter identity yields

$$0 = \overline{\sum_{k=0}^{n} a_k w^k} = \sum_{k=0}^{n} \overline{a_k w^k} = \sum_{k=0}^{n} \overline{a_k}(\overline{w})^k.$$

Since by assumption $\overline{a_k} = a_k$, it then follows

$$0 = \sum_{k=0}^{n} a_k (\overline{w})^k = p(\overline{w}),$$

i.e., $\overline{w}$ is also a root of p. We record our findings as follows.

Lemma 3.2.4 *Let p be a polynomial of degree $n \geq 1$ with real coefficients. Then the complex conjugate $\overline{w}$ of any root $w \in \mathbb{C}$ of p, is also a root of p. Furthermore, w and $\overline{w}$ have the same multiplicity.*

Example (Polynomials of Degree Two with Complex Coefficients) Consider $p(z) = az^2 + bz + c$ with $a, b, c \in \mathbb{C}$, $a \neq 0$. To find the roots of p, we proceed by completing the square. One obtains

$$p(z) = a\left(z^2 + \frac{b}{a}z + \frac{b^2}{4a^2}\right) + c - \frac{b^2}{4a} = a\left(z + \frac{b}{2a}\right)^2 + c - \frac{b^2}{4a}.$$

To find z_1, z_2 with $p(z_i) = 0$, one needs to solve

$$a\left(z + \frac{b}{2a}\right)^2 = \frac{b^2}{4a} - c$$

or

$$\left(z + \frac{b}{2a}\right)^2 = \frac{b^2}{4a^2} - \frac{c}{a}.$$

We need to distinguish two cases:

Case 1. $\frac{b^2}{4a^2} - \frac{c}{a} = 0$. Then

$$\left(z + \frac{b}{2a}\right)^2 = 0,$$

implying that

$$z_1 = z_2 = -\frac{b}{2a}$$

and

$$p(z) = a\left(z + \frac{b}{2a}\right)^2.$$

Case 2. $\frac{b^2}{4a^2} - \frac{c}{a} \neq 0$. In this case, the equation

$$w^2 = \frac{b^2}{4a^2} - \frac{c}{a} = \frac{1}{4a^2}(b^2 - 4ac)$$

has two distinct complex solutions w_1 and w_2 where $w_2 = -w_1$ and $w_1 \neq 0$ is given by

$$w_1 = \frac{1}{2a}\sqrt{b^2 - 4ac}$$

with a specific choice of the sign of the square root (cf. Sect. 3.1). Hence the two roots of p are

$$z_1 = -\frac{b}{2a} + w_1, \qquad z_2 = -\frac{b}{2a} - w_1,$$

and we have

$$p(z) = a\left(z - \left(-\frac{b}{2a} + w_1\right)\right)\left(z - \left(-\frac{b}{2a} - w_1\right)\right).$$

Example (Polynomials of Degree Two with Real Coefficients) Consider the polynomial $p(z) = az^2 + bz + c$ with real coefficients $a, b, c \in \mathbb{R}$ and $a \neq 0$. We argue in a similar way as in previous example and write by completing the square

$$p(z) = a\left(z + \frac{b}{2a}\right)^2 + c - \frac{b^2}{4a}.$$

Case 1. $\frac{b^2}{4a^2} - \frac{c}{a} = 0$. Then

$$\left(z + \frac{b}{2a}\right)^2 = 0$$

and we conclude that

$$z_1 = z_2 = -\frac{b}{2a} \in \mathbb{R}.$$

The polynomial $p(z) = az^2 + bz + c$ can be written as a product of real linear factors,

$$p(z) = a\left(z + \frac{b}{2a}\right)^2.$$

Case 2. $\frac{b^2}{4a^2} - \frac{c}{a} > 0$. Then

$$w_1 = \frac{\sqrt{b^2 - 4ac}}{2a}, \qquad w_2 = -\frac{\sqrt{b^2 - 4ac}}{2a} < 0,$$

are real numbers and so are the roots

$$z_1 = -\frac{b}{2a} + \frac{1}{2a}\sqrt{b^2 - 4ac}, \qquad z_2 = -\frac{b}{2a} - \frac{1}{2a}\sqrt{b^2 - 4ac}.$$

Here and in the future, $\sqrt{b^2 - 4ac} > 0$ denotes the positive square root of $b^2 - 4ac$. In this case, $p(x)$ can again be written as the product of two real linear factors,

$$p(z) = a\left(z - \left(-\frac{b}{2a} + \frac{1}{2a}\sqrt{b^2 - 4ac}\right)\right)\left(z - \left(-\frac{b}{2a} - \frac{1}{2a}\sqrt{b^2 - 4ac}\right)\right).$$

Case 3. $b^2 - 4ac < 0$. Then

$$w_1 = i\frac{\sqrt{4ac - b^2}}{2a}, \qquad w_2 = -i\frac{\sqrt{4ac - b^2}}{2a},$$

are purely imaginary numbers and the roots z_1, z_2 are now the complex numbers

$$z_1 = -\frac{b}{2a} + i\frac{1}{2a}\sqrt{4ac - b^2}, \qquad z_2 = -\frac{b}{2a} - i\frac{1}{2a}\sqrt{4ac - b^2}.$$

Note that z_2 is the complex conjugate of z_1, $z_2 = \bar{z}_1$. As a consequence

$$p(z) = a\left(z - \left(-\frac{b}{2a} + \frac{i}{2a}\sqrt{4ac - b^2}\right)\right)\left(z - \left(-\frac{b}{2a} - \frac{i}{2a}\sqrt{4ac - b^2}\right)\right).$$

Remark For polynomials of degree three, there are formulas for the three roots, due to Cardano, but they are rather complicated. For polynomials of degree $n \geq 5$, it can be proved that the roots can no longer be written as an expression of roots of complex numbers.

Examples

(i) Represent $p(z) = z^4 + z^3 - z - 1$ as a product of linear factors.
By inspection, one sees that $z_1 = 1$, $z_2 = -1$ are roots of p. We get the representation

$$p(z) = (z - 1)(z + 1)(z^2 + z + 1).$$

The polynomial $q(z) = z^2 + z + 1$ has the roots

$$z_3 = -\frac{1}{2} + i\frac{\sqrt{3}}{2}, \qquad z_4 = \bar{z}_3 = -\frac{1}{2} - i\frac{\sqrt{3}}{2}.$$

(ii) Represent $p(z) = z^6 - 1$ as a product of linear factors. By inspection, one sees that

$$z^6 - 1 = (z^3 - 1)(z^3 + 1)$$

and

$$z^3 - 1 = (z - 1)(z^2 + z + 1), \qquad z^3 + 1 = (z + 1)(z^2 - z + 1).$$

Hence $z_1 = 1$, $z_2 = -1$ are two real roots. As in Item (i), the roots of $z^2 + z + 1$ are

$$z_3 = -\frac{1}{2} + i\frac{\sqrt{3}}{2}, \qquad z_4 = \bar{z}_3 = -\frac{1}{2} - i\frac{\sqrt{3}}{2},$$

whereas the roots of $z^2 - z + 1$ are

$$z_5 = \frac{1}{2} + i\frac{\sqrt{3}}{2}, \qquad z_6 = \frac{1}{2} - i\frac{\sqrt{3}}{2}.$$

Problems

1. Find the roots of the following polynomials and write the latter as a product of linear factors.

   (i) $p(z) = 2z^2 - 4z - 5$
   (ii) $p(z) = z^2 + 2z + 3$

2. Let $p(z) = z^3 - 5z^2 + z - 5$.

   (i) Verify that i is a root of $p(z)$.
   (ii) Write $p(z)$ as a product of linear factors.

3. (i) Find all roots of $p(z) = z^4 + 2z^2 - 5$ and write $p(z)$ as a product of linear factors.
   (ii) Find all roots of $p(z) = z^5 + 3z^3 + z$ and write $p(z)$ as a product of linear factors.

4. Consider the function

$$f: \mathbb{R} \to \mathbb{R}, \quad x \mapsto x^3 + 5x^2 + x - 1.$$

   (i) Compute the values of f at $x = -1$, $x = 0$ and $x = 1$.
   (ii) Conclude that f has three real roots.

5. Decide whether the following assertions are true or false and justify your answers.

   (i) Any polynomial of degree 5 with real coefficients has at least three real roots.
   (ii) Any polynomial $p(z)$ of degree $n \geq 1$ with real coefficients can be written as a product of polynomials with real coefficients, each of which has degree one or two.

3.3 Linear Systems with Complex Coefficients

The results about systems of linear equations with real coefficients of Chap. 1 and the ones about matrices and related topics of Chap. 2 extend in a straightforward way to a setup where the real numbers are replaced by the complex numbers. The goal of this section is to introduce the main definitions and results.

Given any $n \in \mathbb{N}$, we denote by $\mathbb{C}^n$ the Cartesian product of n copies of $\mathbb{C}$, i.e., $\mathbb{C}^n = \mathbb{C} \times \mathbb{C} \times \cdots \times \mathbb{C}$ (n copies of $\mathbb{C}$). Elements in $\mathbb{C}^n$ are denoted by $(z_1, \ldots, z_n)$, $z_j \in \mathbb{C}$. They are often referred to as *(complex) vectors*. Complex vectors can be added and one can define a (complex) scalar multiple of a complex vector similarly as in the case of real vectors,

$$(z_1, \ldots, z_n) + (z'_1, \ldots, z'_n) = (z_1 + z'_1, \ldots, z_n + z'_n),$$

$$\lambda(z_1, \ldots, z_n) = (\lambda z_1, \ldots, \lambda z_n), \quad \lambda \in \mathbb{C}.$$

For notational convenience, the zero vector $(0, \ldots, 0)$ is often denoted by 0.

Definition 3.3.1 An $m \times n$ matrix A with complex coefficients is an array of complex numbers of the form

$$A = (a_{ij})_{\substack{1 \le i \le m \\ 1 \le j \le n}}, \quad a_{ij} \in \mathbb{C}.$$

The set of all $m \times n$ matrices is denoted by $\mathbb{C}^{m \times n}$ or $\mathrm{Mat}_{m \times n}(\mathbb{C})$.

A system of m linear equations with complex coefficients and n unknowns $z_1, \ldots, z_n$ is a system of equations of the form

$$\sum_{j=1}^{n} a_{ij} z_j = b_i, \quad 1 \le i \le m$$

where $a_{ij} \in \mathbb{C}$ ($1 \le i \le m, 1 \le j \le n$) and $b_i \in \mathbb{C}$ ($1 \le i \le m$). It is referred to as a *complex linear system*. In matrix notation, it is given by $Az = b$ where

$$z = \begin{pmatrix} z_1 \\ \vdots \\ z_n \end{pmatrix} \in \mathbb{C}^{n \times 1}, \quad b = \begin{pmatrix} b_1 \\ \vdots \\ b_m \end{pmatrix} \in \mathbb{C}^{m \times 1}, \quad A = (a_{ij})_{\substack{1 \le i \le m \\ 1 \le j \le n}} \in \mathbb{C}^{m \times n},$$

and Az denotes matrix multiplication of the $m \times n$ matrix A with the $n \times 1$ matrix z, resulting in the $m \times 1$ matrix Az with coefficients

$$\sum_{j=1}^{n} a_{ij} z_j, \quad 1 \le i \le m.$$

As in the case of real linear systems, complex linear systems can be solved by Gaussian elimination, using the row operations $(R1)$, $(R2)$, and $(R3)$, now involving complex numbers instead of real ones. We say that a $n \times n$ matrix $A \in \mathbb{C}^{n \times n}$ is *regular* if A can be transformed into the $n \times n$ identity matrix $\mathrm{Id}_{n \times n}$ by the row operations $(R1)$, $(R2)$, and $(R3)$.

Definition 3.3.2 The *general linear group* over $\mathbb{C}$ is defined as

$$\mathrm{GL}_{\mathbb{C}}(n) := \{A \in \mathbb{C}^{n \times n} \mid A \text{ is regular}\}.$$

The *determinant* of a complex $n \times n$ matrix is defined in the same way as the one of a real $n \times n$ matrix and it can be shown that for any $A \in \mathbb{C}^{n \times n}$, A is regular if and only if $\det(A) \neq 0$.

Definition 3.3.3 We say that $b \in \mathbb{C}^k$ is a $\mathbb{C}$-linear combination (or linear combination for short) of complex vectors $a^{(1)}, \ldots, a^{(n)} \in \mathbb{C}^k$ if there exist $\lambda_1, \ldots, \lambda_n \in \mathbb{C}$ such that

$$b = \sum_{j=1}^{n} \lambda_j a^{(j)}.$$

Definition 3.3.4 The vectors $a^{(1)}, \ldots, a^{(n)} \in \mathbb{C}^k$ are said to be $\mathbb{C}$-*linearly independent* (or *linearly independent* for short) if for any $\lambda_1, \ldots, \lambda_n \in \mathbb{C}$ with

$$\sum_{j=1}^{n} \lambda_j a^{(j)} = 0,$$

it follows that $\lambda_1 = 0, \ldots, \lambda_n = 0$. Otherwise $a^{(1)}, \ldots, a^{(n)}$ are said to be $\mathbb{C}$-*linearly dependent* (or *linearly dependent* for short).

Definition 3.3.5 The vectors $a^{(1)}, \ldots, a^{(n)} \in \mathbb{C}^k$ form a *basis* of $\mathbb{C}^k$ if the following holds:

(i) $a^{(1)}, \ldots, a^{(n)}$ are $\mathbb{C}$-linearly independent;
(ii) every vector $b \in \mathbb{C}^k$ can be represented as a $\mathbb{C}$-linear combination of $a^{(1)}, \ldots, a^{(n)}$.

If $a^{(1)}, \ldots, a^{(n)} \in \mathbb{C}^k$ form a basis of $\mathbb{C}^k$, one has $n = k$ and every element in $\mathbb{C}^k$ can be written in a *unique way* as a $\mathbb{C}$-linear combination of $a^{(1)}, \ldots, a^{(k)}$. A basis $a^{(1)}, \ldots, a^{(k)}$ of $\mathbb{C}^k$ is denoted by $[a] = [a^{(1)}, \ldots, a^{(k)}]$.

Example The vectors $e^{(1)} = (1, 0)$, $e^{(2)} = (0, 1)$ in $\mathbb{R}^2$ are $\mathbb{R}$-linearly independent in $\mathbb{R}^2$. Viewed as complex vectors in $\mathbb{C}^2$, $e^{(1)}$, $e^{(2)}$, are $\mathbb{C}$-linearly independent and hence form basis of $\mathbb{C}^2$. It is referred as the *standard basis* of $\mathbb{C}^2$. Any vector $b = (b_1, b_2) \in \mathbb{C}^2$ can be written as a $\mathbb{C}$-linear combination of $e^{(1)}$ and $e^{(2)}$ as follows,

$$b = b_1 e^{(1)} + b_2 e^{(2)}.$$

Problems

1. Find the set of solutions of the following complex linear systems.

(i) $\begin{cases} z_1 - \mathrm{i} z_2 = 2 \\ (-1 + \mathrm{i})z_1 + (2 + \mathrm{i})z_2 = 0 \end{cases}$

(ii) $\begin{cases} z_1 + i z_2 - (1+i)z_3 = 0 \\ i z_1 + z_2 + (1+i)z_3 = 0 \\ (1+2i)z_1 + (1+i)z_2 + 2z_3 = 0 \end{cases}$

2. (i) Compute the determinant of the following complex 3×3 matrix.

$$A = \begin{pmatrix} 1 & i & 1+i \\ 0 & -1+i & 2 \\ i & 2 & 1+2i \end{pmatrix}$$

(ii) Find all complex numbers $z \in \mathbb{C}$ with the property that $\det(B) = 0$ where

$$B = \begin{pmatrix} 1+2i & 3+4i \\ z & 1-2i \end{pmatrix}^{25}.$$

3. (i) Compute AA^T and $A^T A$ where A^T is the transpose of A and $A \in \mathbb{C}^{2 \times 3}$ is given by

$$A = \begin{pmatrix} 1 & i & 2 \\ i & -2 & i \end{pmatrix}.$$

(ii) Compute the inverse of the following complex 2×2 matrix

$$A = \begin{pmatrix} 2i & 1 \\ 1+i & 1-i \end{pmatrix}.$$

4. Decide whether the following vectors in $\mathbb{C}^3$ are $\mathbb{C}$-linearly independent or $\mathbb{C}$-linearly dependent.

(i) $a^{(1)} = (1, 2+i, i)$, $a^{(2)} = (-1+3i, 1+i, 3+i)$
(ii) $b^{(1)} = (1+i, 1-i, -1+i)$, $b^{(2)} = (0, 4-2i, -3+5i)$, $b^{(3)} = (1+i, 0, 3)$

5. Decide whether the following assertions are true or false and justify your answers.

(i) There exists vectors $a^{(1)}$, $a^{(2)}$ in $\mathbb{C}^2$ with the following two properties:
   (P1) For any $\alpha_1, \alpha_2 \in \mathbb{R}$ with $\alpha_1 a^{(1)} + \alpha_2 a^{(2)} = 0$, it follows that $\alpha_1 = 0$ and $\alpha_2 = 0$ (i.e. $a^{(1)}$, $a^{(2)}$ are $\mathbb{R}$-linearly independent).
   (P2) There exist $\beta_1, \beta_2 \in \mathbb{C} \setminus \{0\}$ so that $\beta_1 a^{(1)} + \beta_2 a^{(2)} = 0$ (i.e. $a^{(1)}$, $a^{(2)}$ are $\mathbb{C}$-linearly dependent).
(ii) Any basis $[a] = [a^{(1)}, \ldots, a^{(n)}]$ of $\mathbb{R}^n$ gives rise to a basis of $\mathbb{C}^n$, when $a^{(1)}, \ldots, a^{(n)}$ are considered as vectors in $\mathbb{C}^n$.
(iii) The vectors $a^{(1)} = (i, 1, 0)$, $a^{(2)} = (i, -1, 0)$ are $\mathbb{C}$-linearly independent in $\mathbb{C}^3$ and there exists $a^{(3)} \in \mathbb{C}^3$ so that $[a] = [a^{(1)}, a^{(2)}, a^{(3)}]$ is a basis of $\mathbb{C}^3$.

Chapter 4
Vector Spaces and Linear Maps

In this chapter we introduce the notion of a *vector space* over $\mathbb{R}$ or $\mathbb{C}$ and of maps between such spaces which respect to their structure. Maps of this type are referred to as *linear maps*.

4.1 Vector Spaces

Before we introduce the notion of a vector space in a formal way, let us review two prominent examples of such spaces which we have already considered in the previous chapters.

Recall that we denoted by $\mathbb{R}^n$ the Cartesian product of n copies of the set of real numbers $\mathbb{R}$. Elements of $\mathbb{R}^n$ are denoted by $a = (a_1, \ldots, a_n)$ where $a_j \in \mathbb{R}$, $1 \le j \le n$, and referred to as *vectors*. Such vectors can be added and they can be multiplied by a real number componentwise:

Vector addition. For $a = (a_1, \ldots, a_n)$ and $b = (b_1, \ldots, b_n)$ in $\mathbb{R}^n$,

$$a + b := (a_1 + b_1, \ldots, a_n + b_n).$$

Multiplication by a scalar. For any $\lambda \in \mathbb{R}$ and $a = (a_1, \ldots, a_n) \in \mathbb{R}^n$,

$$\lambda a \equiv \lambda \cdot a = (\lambda a_1, \ldots, \lambda a_n).$$

Similarly, we considered the Cartesian product $\mathbb{C}^n$ of n copies of the set of complex numbers $\mathbb{C}$. Elements of $\mathbb{C}^n$ are denoted by $a = (a_1, \ldots, a_n)$, $a_j \in \mathbb{C}$, $1 \le j \le n$, and are also referred to as *vectors* or *complex vectors*. Such vectors can be added,

$$a + b = (a_1 + b_1, \ldots, a_n + b_n),$$

M. Benz, T. Kappeler, *Linear Algebra for the Sciences*, La Matematica
per il 3+2 151, https://doi.org/10.1007/978-3-031-27220-2_4

and they can be multiplied by a complex scalar $\lambda \in \mathbb{C}$,

$$\lambda a = \lambda \cdot a = (\lambda a_1, \ldots, \lambda a_n).$$

We now define the notion of a $\mathbb{K}$-vector space where $\mathbb{K}$ denotes either $\mathbb{R}$ or $\mathbb{C}$.

Definition 4.1.1 A $\mathbb{K}$-*vector space* is a non empty set V, endowed with two operations $+$ (addition) and $\cdot$ (multiplication with a scalar $\lambda \in \mathbb{K}$),

$$+: V \times V \to V, \quad (a, b) \mapsto a + b, \qquad \cdot: \mathbb{K} \times V \to V, \quad (\lambda, a) \mapsto \lambda \cdot a,$$

with the following properties:
$(VS1)$ $(V, +)$ is an *abelian group*:

(i) $+$ is associative: $(a + b) + c = a + (b + c)$ for any $a, b, c \in V$.
(ii) $+$ is commutative: $a + b = b + a$ for any $a, b \in V$.
(iii) There is an element $0 \in V$ so that $0 + a = a + 0 = a$ for any $a \in V$. The element is uniquely determined and referred to as zero or zero element of V.
(iv) For any element $a \in V$, there exists a unique element $b \in V$ so that $a + b = 0$. The element b is referred to as the *inverse* of a and denoted by $-a$.

$(VS2)$ The two distributive laws hold, i.e., for any $\lambda, \mu \in \mathbb{K}$ and $a, b \in V$,

(i) $(\lambda + \mu) \cdot a = \lambda \cdot a + \mu \cdot a$,
(ii) $\lambda \cdot (a + b) = \lambda \cdot a + \lambda \cdot b$.

$(VS3)$ The scalar multiplication satisfies the following properties:

(i) It is associative, i.e., for any $\lambda, \mu \in \mathbb{K}$ and $a \in V$, $\lambda \cdot (\mu \cdot a) = (\lambda \mu) \cdot a$.
(ii) For any $a \in V$, $1 \cdot a = a$.

Elements of a $\mathbb{K}$-vector space are often referred to as *vectors*.

Remark Various useful identities can be derived from $(VS1)$–$(VS3)$. Let us discuss the following two examples:

(i) For any $a \in V$, $0 \cdot a = 0$.
    To verify this identity, note that $0 \cdot a = (0 + 0) \cdot a \overset{(v)}{=} 0 \cdot a + 0 \cdot a$. Add on both sides of the latter equality $-(0 \cdot a)$ to conclude that $0 \overset{(iv)}{=} 0 \cdot a + (-(0 \cdot a)) = 0 \cdot a + 0 \cdot a + (-(0 \cdot a)) \overset{(i)}{=} 0 \cdot a$.
(ii) For any $a \in V$, $(-1) \cdot a = -a$.
    To verify this identity, note that $a + (-1) \cdot a \overset{(viii)}{=} 1 \cdot a + (-1) \cdot a \overset{(v)}{=} (1 + (-1)) \cdot a = 0 \cdot a \overset{\text{Remark (i)}}{=} 0$.

Examples

(i) For any $n \in \mathbb{Z}_{\geq 1}$, $(\mathbb{R}^n, +, \cdot)$ with addition $+$ and scalar multiplication $\cdot$ described as above is a $\mathbb{R}$-vector space. Similarly $(\mathbb{C}^n, +, \cdot)$ is a $\mathbb{C}$-vector space. In summary, for any $n \in \mathbb{Z}_{\geq 1}$, $(\mathbb{K}^n, +, \cdot)$ is a $\mathbb{K}$-vector space.

(ii) For any $m, n \in \mathbb{Z}_{\geq 1}$, let $V = \mathbb{K}^{m \times n}$ be the set of $m \times n$ matrices A with coefficients in $\mathbb{K}$,

$$A = (a_{ij})_{\substack{1 \leq i \leq m \\ 1 \leq j \leq n}}, \quad a_{ij} \in \mathbb{K},$$

and define addition of matrices and multiplication of a matrix by a scalar $\lambda \in \mathbb{K}$ as in Sect. 2.1 in the case of $m \times n$ matrices with real coefficients,

$$+ \colon \mathbb{K}^{m \times n} \times \mathbb{K}^{m \times n} \to \mathbb{K}^{m \times n}, \quad \left((a_{ij})_{\substack{1 \leq i \leq m \\ 1 \leq j \leq n}}, (b_{ij})_{\substack{1 \leq i \leq m \\ 1 \leq j \leq n}}\right) \mapsto (a_{ij} + b_{ij})_{\substack{1 \leq i \leq m \\ 1 \leq j \leq n}},$$

$$\cdot \colon \mathbb{R} \times \mathbb{K}^{m \times n} \to \mathbb{K}^{m \times n}, \quad \left(\lambda, (a_{ij})_{\substack{1 \leq i \leq m \\ 1 \leq j \leq n}}\right) \mapsto (\lambda a_{ij})_{\substack{1 \leq i \leq m \\ 1 \leq j \leq n}}.$$

Then $(\mathbb{K}^{m \times n}, +, \cdot)$ is a $\mathbb{K}$-vector space.

(iii) Let $M[0, 1] := \{ f \mid [0, 1] \to \mathbb{R} \}$. Define addition on $M[0, 1]$ by

$$+ \colon M[0, 1] \times M[0, 1] \to M[0, 1], \quad (f, g) \mapsto f + g$$

where

$$f + g \colon [0, 1] \to \mathbb{R}, \quad t \mapsto (f + g)(t) = f(t) + g(t),$$

and scalar multiplication

$$\cdot \colon \mathbb{R} \times M[0, 1] \to M[0, 1], \quad (\lambda, f) \mapsto \lambda \cdot f$$

where

$$\lambda \cdot f \colon [0, 1] \to \mathbb{R}, \quad t \mapsto (\lambda f)(t) := \lambda \cdot f(t).$$

Then $(M[0, 1], +, \cdot)$ is a $\mathbb{R}$-vector space. In a similar way, one can define addition and multiplication by a scalar on $M(\mathbb{K}) := \{ f \colon \mathbb{K} \to \mathbb{K} \}$ so that $(M(\mathbb{K}), +, \cdot)$ is a $\mathbb{K}$-vector space.

(iv) For any integer $n \geq 1$, let $\mathcal{P}_n$ denote the set of all polynomials with real coefficients of degree at most n. An element $p \in \mathcal{P}_n$ can be viewed as a function

$$p \colon \mathbb{R} \to \mathbb{R}, \quad t \mapsto a_n t^n + a_{n-1} t^{n-1} + \cdots + a_1 t + a_0$$

where $a_n, \ldots, a_0$ are the coefficients of p and are assumed to be real numbers. Define addition and scalar multiplication as in (3) above. Then $(\mathcal{P}_n, +, \cdot)$ is a $\mathbb{R}$-vector space.

Definition 4.1.2 If $V \equiv (V, +, \cdot)$ is a $\mathbb{K}$-vector space and $W \subseteq V$ a nonempty subset of V, then W is said to be a $\mathbb{K}$-*subspace* (or *subspace* for short) of V if the following holds:

$(SS1)$ W is closed with respect to addition: $a + b \in W, a, b \in W$.
$(SS2)$ W is closed with respect to scalar multiplication: $\lambda a \in W, a \in W, \lambda \in \mathbb{K}$.

A subset of V of the form $a + W$ where $a \in V$ and W a subspace of V, is referred as an *affine subspace* of V.

Note that W with addition $+$ and scalar multiplication $\cdot$ from V forms a $\mathbb{K}$-vector space.

Examples

(i) Assume that $A \in \mathbb{R}^{m \times n}$ and let $L := \{x \in \mathbb{R}^{n \times 1} \mid Ax = 0\}$. Then L is a subspace of $\mathbb{R}^{n \times 1}$. (Recall that $\mathbb{R}^{n \times 1}$ can be identified with $\mathbb{R}^n$). To verify this assertion we have to show that L is non empty and $(SS1)$, $(SS2)$ hold. Indeed, $0 \in \mathbb{R}^{n \times 1}$ is an element of L since $A0 = 0$. Furthermore, for any $x, y \in L$, $\lambda \in \mathbb{R}$, one has

$$A(x + y) = Ax + Ay = 0 + 0 = 0, \qquad A(\lambda x) = \lambda Ax = \lambda \cdot 0 = 0.$$

(ii) Recall that $M(\mathbb{R}) := \{f : \mathbb{R} \to \mathbb{R}\}$ and $\mathcal{P}_n$ are $\mathbb{R}$-vector spaces. In fact, we verified in the previous example that $\mathcal{P}_n$ is a subspace of $\big(M(\mathbb{R}), +, \cdot\big)$.

(iii) The set of all subspaces of $\mathbb{R}^2$ can be described as follows: $\{0\}$ and $\mathbb{R}^2$ are the trivial subspaces of $\mathbb{R}^2$, whereas the nontrivial subspaces of $\mathbb{R}^2$ are given by the one parameter family of straight lines, $W_\theta := \{\lambda(\cos\theta, \sin\theta) \mid \lambda \in \mathbb{R}, 0 \le \theta < \pi\}$.

The notions of linear combination of vectors in $\mathbb{R}^n$ (Definition 2.2.1) and $\mathbb{C}^n$ (Definition 3.3.3) as well as the notion of linearly dependent/linearly independent vectors (cf. Definition 2.2.2, Definition 3.3.4) extends in a natural way to $\mathbb{K}$-vector spaces.

Definition 4.1.3 We say that a vector b in a $\mathbb{K}$-vector space V is a $\mathbb{K}$-*linear combination* (or *linear combination* for short) of vectors $a^{(1)}, \ldots, a^{(n)} \in V$ if there exist $\lambda_1, \ldots, \lambda_n \in \mathbb{K}$ so that

$$b = \sum_{j=1}^n \lambda_j a^{(j)}.$$

Definition 4.1.4 The vectors $a^{(1)}, \ldots, a^{(n)}$ in a $\mathbb{K}$-vector space V are said to be $\mathbb{K}$-*linearly independent* (or *linearly independent* for short) if for any $\lambda_1, \ldots, \lambda_n \in \mathbb{K}$ with

$$\sum_{j=1}^{n} \lambda_j a^{(j)} = 0,$$

it follows that $\lambda_1 = 0, \ldots, \lambda_n = 0$. Otherwise $a^{(1)}, \ldots, a^{(n)}$ are said to be $\mathbb{K}$-*linearly dependent* (or *linearly dependent* for short).

Similarly, the notion of a basis of $\mathbb{R}^n$ (Definition 2.2.4) and the one of $\mathbb{C}^n$ (Definition 3.3.5) extends in a natural way to $\mathbb{K}$-vector spaces.

Definition 4.1.5 Assume that V is a $\mathbb{K}$-vector space and $v^{(1)}, \ldots, v^{(n)}$ are elements of V. Then $[v] = [v^{(1)}, \ldots, v^{(n)}]$ is a *basis* of V if the following holds:

($B1$)   Any vector $a \in V$ can be represented as a $\mathbb{K}$-linear combination of $v^{(1)}, \ldots, v^{(n)}$, i.e. there exist $\lambda_1, \ldots, \lambda_n \in \mathbb{K}$ so that $a = \sum_{j=1}^{n} \lambda_j v^{(j)}$.
($B2$)   The vectors $v^{(1)}, \ldots, v^{(n)}$ are $\mathbb{K}$-linearly independent.

Remark Equivalently, $[v]$ is a basis if the following holds:

(B)   Any $a \in V$ can be written in a *unique* way as a $\mathbb{K}$-linear combination, $a = \sum_{j=1}^{n} \lambda_j v^{(j)}$.

The numbers $\lambda_1, \ldots, \lambda_n \in \mathbb{K}$ are called the coordinates of a with respect to the basis $[v]$.

The properties of bases of $\mathbb{R}^n$ and $\mathbb{C}^n$, discussed in Sect. 2.2 and respectively, Sect. 3.3, also hold for bases of $\mathbb{K}$-vector spaces.

Theorem 4.1.6 *Assume that V is a $\mathbb{K}$-vector space and $[v] = [v^{(1)}, \ldots, v^{(n)}]$ is a basis of V. Then every basis of V has precisely n elements.*

Definition 4.1.7

  (i) Assume that V is a $\mathbb{K}$-vector space and that $[v]$ is a basis of V with $n \geq 1$ elements, $[v] = [v^{(1)}, \ldots, v^{(n)}]$. Then V is said to be a $\mathbb{K}$-*vector space of dimension n*.
 (ii) A $\mathbb{K}$-vector space is said to be *finite dimensional* if it admits a basis with finitely many elements. The dimension of a finite dimensional vector space is denoted by $\dim_{\mathbb{K}}(V)$ or $\dim(V)$ for short.
(iii) If $V = \{0\}$, then $\dim(V) = 0$.

Examples

  (i) For any $n \geq 1$, $V = \mathbb{R}^n$ is a $\mathbb{R}$-vector basis of dimension n since the standard basis $[e]$ of $\mathbb{R}^n$ has n elements,

$$e^{(1)} = (1, 0, \ldots, 0), \quad e^{(2)} = (0, 1, 0, \ldots, 0), \quad \ldots, \quad e^{(n)} = (0, \ldots, 0, 1).$$

(ii) For any $n \geq 1$, the space $\mathcal{P}_n \equiv \mathcal{P}_n(\mathbb{C}) = \{p(z) = \sum_{k=0}^{n} a_k z^k \mid a_0, \ldots,$
$a_n \in \mathbb{C}\}$ is a $\mathbb{C}$-vector space of dimension $n + 1$, since the polynomials
$p^{(0)}, \ldots, p^{(n)}$, given by

$$p^{(0)}(z) := 1, \qquad p^{(1)}(z) := z, \qquad \ldots, \qquad p^{(n)}(z) := z^n.$$

form a basis of $\mathcal{P}_n$.

Theorem 4.1.8 *Assume that V is a $\mathbb{K}$-vector space of dimension $n \geq 1$. Then:*

(i) *Every subspace W of V is a finite dimensional $\mathbb{K}$-vector space and $\dim(W) \leq \dim(V)$.*
(ii) *If the vectors $v^{(1)}, \ldots, v^{(m)}$ span all of V,*

$$V = \Big\{ \sum_{j=1}^{m} \lambda_j v^{(j)} \mid \lambda_1, \ldots, \lambda_m \in \mathbb{K} \Big\},$$

then the following holds:

(ii1) *$m \geq n$;*
(ii2) *$v^{(1)}, \ldots, v^{(m)}$ is a basis of V if and only if $m = n$;*
(ii3) *if $m > n$, then one can choose n elements among $v^{(1)}, \ldots, v^{(m)}$, which form a basis of V.*

(iii) *If the vectors $v^{(1)}, \ldots, v^{(m)}$ in V are $\mathbb{K}$-linearly independent then one has:*

(iii1) *$m \leq n$;*
(iii2) *$v^{(1)}, \ldots, v^{(m)}$ is a basis of V if and only if $m = n$;*
(iii3) *if $m < n$, then there exist elements $w^{(1)}, \ldots, w^{(n-m)}$ in V so that $v^{(1)}, \ldots, v^m, w^{(1)}, \ldots, w^{(n-m)}$ form a basis of V.*

Let us discuss an example.

Example Consider the following vectors in $\mathbb{R}^{3 \times 1}$,

$$v^{(1)} = \begin{pmatrix} 1 \\ 1 \\ 1 \end{pmatrix}, \qquad v^{(2)} = \begin{pmatrix} 0 \\ 2 \\ 1 \end{pmatrix}, \qquad v^{(3)} = \begin{pmatrix} 1 \\ -1 \\ 0 \end{pmatrix}, \qquad v^{(4)} = \begin{pmatrix} 1 \\ 2 \\ 3 \end{pmatrix}.$$

(i) Verify that $v^{(1)}, \ldots, v^{(4)}$ span all of $\mathbb{R}^{3 \times 1}$.
   It means that every element $b \in \mathbb{R}^{3 \times 1}$ can be represented as a linear combination of $v^{(1)}, \ldots, v^{(4)}$, i.e., $b = \sum_{j=1}^{4} \lambda_j v^{(j)}$ where $\lambda_1, \ldots, \lambda_4 \in \mathbb{R}$. Expressed in terms of linear systems, it is to show that for any $b \in \mathbb{R}^{3 \times 1}$, the linear system $\sum_{j=1}^{4} x_j v^{(j)} = b$ has at least one solution. We form the

corresponding augmented coefficient matrix

$$\left[\begin{array}{cccc|c} 1 & 0 & 1 & 1 & b_1 \\ 1 & 2 & -1 & 2 & b_2 \\ 1 & 1 & 0 & 3 & b_3 \end{array}\right]$$

and use Gaussian elimination to transform it in row echelon form.

Step 1. $R_2 \rightsquigarrow R_2 - R_1$, $R_3 \rightsquigarrow R_3 - R_1$,

$$\left[\begin{array}{cccc|c} 1 & 0 & 1 & 1 & b_1 \\ 0 & 2 & -2 & 1 & b_2 - b_1 \\ 0 & 1 & -1 & 2 & b_3 - b_1 \end{array}\right].$$

Step 2. $R_3 \rightsquigarrow R_3 - \frac{1}{2} R_2$

$$\left[\begin{array}{cccc|c} 1 & 0 & 1 & 1 & b_1 \\ 0 & 2 & -2 & 1 & b_2 - b_1 \\ 0 & 0 & 0 & \frac{3}{2} & b_3 - \frac{1}{2} b_1 - \frac{1}{2} b_2 \end{array}\right].$$

We conclude that $b = \sum_{j=1}^{4} \lambda_j v^{(j)}$ with $\frac{3}{2} \lambda_4 = b_3 - \frac{1}{2} b_1 - \frac{1}{2} b_2$, $\lambda_3 = 0$, $2\lambda_2 = -\lambda_4 + b_2 - b_1$, and $\lambda_1 = -\lambda_4 + b_1$.

(ii) By Theorem 4.1.8(ii3) we can find a basis of $\mathbb{R}^{3 \times 1}$ by choosing three of the four vectors $v^{(1)}, \ldots, v^{(4)}$. How to proceed? Note that $\dim(\mathbb{R}^{3 \times 1}) = 3$, and hence $v^{(1)}, \ldots, v^{(4)}$ are linearly dependent. We want to find a vector among $v^{(1)}, \ldots, v^{(4)}$ which can be expressed as a linear combination of the remaining three vectors. It means to find a non trivial solution of the homogeneous system $\sum_{j=1}^{4} x_j v^{(j)} = 0$. According to the above scheme with $b = 0$ we get

$$\left[\begin{array}{cccc|c} 1 & 0 & 1 & 1 & 0 \\ 0 & 2 & -2 & 1 & 0 \\ 0 & 0 & 0 & \frac{3}{2} & 0 \end{array}\right].$$

With the solution x, given by $x_4 = 0$, $x_3 = 1$, $x_2 = 1$, and $x_1 = -x_3 = -1$, one obtains $-v^{(1)} + v^{(2)} + v^{(3)} = 0$ or

$$v^{(1)} = v^{(2)} + v^{(3)}.$$

Since $v^{(1)}, \ldots, v^{(4)}$ span $\mathbb{R}^{3 \times 1}$ and $\dim(\mathbb{R}^{3 \times 1}) = 3$, the vectors $v^{(2)}, v^{(3)}, v^{(4)}$ form a basis of $\mathbb{R}^{3 \times 1}$.

As an illustration of the notion of a subspace of a vector space and the one of the dimension of a finite dimensional vector space, we introduce the notion of the rank of a matrix. For any $m \times n$ matrix $A \in \mathbb{K}^{m \times n}$, denote by $C_1(A), \ldots, C_n(A) \in \mathbb{K}^{m \times 1}$

its columns and define the set

$$C_A := \Big\{ \sum_{j=1}^{n} \lambda_j C_j(A) \mid \lambda_1, \ldots, \lambda_n \in \mathbb{K} \Big\}.$$

Then C_A is a $\mathbb{K}$-subspace of $\mathbb{K}^{m \times 1}$ and hence a finite dimensional $\mathbb{K}$-vector space.

Definition 4.1.9 The rank of A, denoted by rank(A), is defined as

$$\text{rank}(A) := \dim(C_A).$$

Note that rank$(A) \le m$ since $C_A \subseteq \mathbb{K}^{m \times 1}$ and rank$(A) \le n$ since A has n columns, implying that rank$(A) \le \min\{m, n\}$.

Proposition 4.1.10 *For any $A \in \mathbb{K}^{m \times n}$, rank$(A^{\mathrm{T}}) = $ rank(A).*

Remark For any $A \in \mathbb{K}^{m \times n}$, denote by $R_1(A), \ldots, R_m(A) \in \mathbb{K}^{1 \times n}$ the rows of A and introduce

$$R_A := \Big\{ \sum_{k=1}^{m} \lambda_k R_k(A) \mid \lambda_1, \ldots, \lambda_m \in \mathbb{K} \Big\} \subseteq \mathbb{K}^{1 \times n}.$$

Then one has $\dim(R_A) = \text{rank}(A^{\mathrm{T}})$ and it follows from Proposition 4.1.10 that

$$\dim(R_A) = \dim(C_A).$$

Example Compute the rank of the matrix $A \in \mathbb{R}^{3 \times 4}$,

$$A = \begin{pmatrix} 1 & 0 & 1 & -1 \\ 1 & 1 & 0 & 1 \\ 1 & 1 & 1 & 2 \end{pmatrix}.$$

The vector space generated by the columns of A is

$$C_A := \Big\{ \sum_{j=1}^{4} \lambda_j C_j(A) \mid \lambda_1, \ldots, \lambda_4 \in \mathbb{R} \Big\}$$

where the columns $C_j(A) \in \mathbb{R}^{3 \times 1}$, $1 \le j \le 4$, are given by

$$C_1(A) = \begin{pmatrix} 1 \\ 1 \\ 1 \end{pmatrix}, \qquad C_2(A) = \begin{pmatrix} 0 \\ 1 \\ 1 \end{pmatrix}, \qquad C_3(A) = \begin{pmatrix} 1 \\ 0 \\ 1 \end{pmatrix}, \qquad C_4(A) = \begin{pmatrix} -1 \\ 1 \\ 2 \end{pmatrix}.$$

To compute rank(A) one has to construct a basis of C_A. We want to use Theorem 4.1.8(ii). Note that C_A is a subspace of $\mathbb{R}^{3\times 1}$ and $\dim(\mathbb{R}^{3\times 1}) = 3$. Hence the columns $C_j(A)$, $1 \leq j \leq 4$, cannot be linearly independent. To choose from $C_1(A), \ldots, C_4(A)$ a maximal set of linearly independent ones, we need to study the homogeneous linear system

$$\sum_{j=1}^{4} x_j C_j(A) = 0, \quad x = \begin{pmatrix} x_1 \\ x_2 \\ x_3 \\ x_4 \end{pmatrix} \in \mathbb{R}^{4\times 1}.$$

In matrix notation the latter linear system reads $Ax = 0$. We solve for x by Gaussian elimination. The augmented coefficient matrix is given by

$$\left[\begin{array}{cccc|c} 1 & 0 & 1 & -1 & 0 \\ 1 & 1 & 0 & 1 & 0 \\ 1 & 1 & 1 & 2 & 0 \end{array}\right].$$

Step 1. $R_2 \rightsquigarrow R_2 - R_1, R_3 \rightsquigarrow R_3 - R_1$

$$\left[\begin{array}{cccc|c} 1 & 0 & 1 & 1 & 0 \\ 0 & 1 & -1 & 2 & 0 \\ 0 & 1 & 0 & 3 & 0 \end{array}\right].$$

Step 2. $R_3 \rightsquigarrow R_3 - R_2$

$$\left[\begin{array}{cccc|c} 1 & 0 & 1 & 1 & 0 \\ 0 & 1 & -1 & 2 & 0 \\ 0 & 0 & 1 & 1 & 0 \end{array}\right].$$

We then conclude that x_4 is a free variable and

$$x_3 = -x_4; \qquad x_2 = x_3 - 2x_4 \rightsquigarrow x_2 = -3x_4; \qquad x_1 = -x_3 - x_4 = 0.$$

Choosing $x_4 = 1$ we get $x_3 = -1$, $x_2 = -3$, $x_1 = 0$ and hence

$$C_4(A) = 3C_2(A) + C_3(A).$$

We conclude that $[C_1(A), C_2(A), C_3(A)]$ is a basis of C_A. Hence $\dim(C_A) = 3$ and in turn rank$(A) = 3$. Note that since $\dim(\mathbb{R}^{3\times 1}) = 3$ one has $C_A = \mathbb{R}^{3\times 1}$.

Finally we discuss the set L of solutions of a homogeneous linear system $Ax = 0$ where A is a $m \times n$ matrix in $\mathbb{K}^{m \times n}$. Note that

$$L = \{x \in \mathbb{K}^{n \times 1} \mid Ax = 0\}$$

is a $\mathbb{K}$-subspace of $\mathbb{K}^{n \times 1}$. Let us compute its dimension. To this end we have to construct a basis of L.

Recall that by using Gaussian elimination, L can be conveniently parametrized by transforming the augmented coefficient matrix $[A \parallel 0]$ into reduced echelon form (see (1.17) in Sect. 1.3). More precisely, by renumbering the variables $x_1, \ldots, x_n$, as $y_1, \ldots, y_n$, if needed, the augmented coefficient matrix $[A \parallel 0]$ of the system $Ax = 0$ can be brought into the reduced echelon form $[\widehat{A} \parallel 0]$ by the row operations $(R1)$–$(R3)$,

$$\left[\begin{array}{ccccccccc|c}
1 & 0 & \cdots & 0 & 0 & \widehat{a}_{1(k+1)} & \cdots & \widehat{a}_{1n} & & 0 \\
0 & 1 & \cdots & 0 & 0 & \widehat{a}_{2(k+1)} & \cdots & \widehat{a}_{2n} & & 0 \\
\vdots & & & & & \vdots & & \vdots & & \\
0 & \cdots & & 0 & 1 & \widehat{a}_{k(k+1)} & \cdots & \widehat{a}_{kn} & & 0 \\
0 & \cdots & & & & 0 & \cdots & 0 & & 0 \\
\vdots & & & & & & & & & \\
0 & \cdots & & & & 0 & \cdots & 0 & & 0
\end{array}\right].$$

The solutions of $\widehat{A}y = 0$ are given by (cf. (1.18) and (1.19))

$$y_1 = -\sum_{j=k+1}^{n} \widehat{a}_{1j}t_j, \ \ldots, \ y_k = -\sum_{j=k+1}^{n} \widehat{a}_{kj}t_j, \qquad y_j := t_j, \ \ k+1 \le j \le n.$$

We summarize our findings as follows.

Theorem 4.1.11 *The space of solutions of $\widehat{A}y = 0$,*

$$\widehat{L} = \{y \in \mathbb{K}^{n \times 1} \mid \widehat{A}y = 0\},$$

is a subspace of $\mathbb{K}^{n \times 1}$ with basis

$$\widehat{v}^{(1)} = \begin{pmatrix} -\widehat{a}_{1(k+1)} \\ \vdots \\ -\widehat{a}_{k(k+1)} \\ 1 \\ 0 \\ \vdots \\ 0 \end{pmatrix}, \quad \ldots, \quad \widehat{v}^{(n-k)} = \begin{pmatrix} -\widehat{a}_{1n} \\ \vdots \\ -\widehat{a}_{kn} \\ 0 \\ \vdots \\ 0 \\ 1 \end{pmatrix}.$$

By renumerating the unknowns $y_1, \ldots, y_n$, one obtains a basis $v^{(1)}, \ldots, v^{(n-k)}$ of $L = \{x \in \mathbb{K}^{n \times 1} \mid Ax = 0\}$. As a consequence, $\dim(L) = n - k$.

A corresponding result holds for an inhomogeneous system of linear equations:

Theorem 4.1.12 *Let $A \in \mathbb{K}^{m \times n}$ and $b \in \mathbb{K}^{m \times 1}$ and assume that $Ax = b$ has at least one solution, denoted by x^{part}. Then the space of solutions*

$$L = \{x \in \mathbb{K}^{n \times 1} \mid Ax = b\}$$

equals the affine subspace $x^{\text{part}} + L_{\text{hom}}$ of $\mathbb{K}^{n \times 1}$ (cf. Definition 4.1.2),

$$L = x^{\text{part}} + L_{\text{hom}}$$

where L_{hom} is the subspace of $\mathbb{K}^{n \times 1}$, given by $L_{\text{hom}} = \{x \in \mathbb{K}^{n \times 1} \mid Ax = 0\}$. The solution x^{part} of $Ax = b$ is customarily referred to as a particular solution.

Remark Note that $x^{\text{part}} + L_{\text{hom}} \subseteq L$, since for any $x \in L_{\text{hom}}$,

$$A(x^{\text{part}} + x) = A(x^{\text{part}}) + Ax = b + 0 = b.$$

Similarly, for any solution $y \in L$ one has

$$A(y - x^{\text{part}}) = Ay - Ax^{part} = b - b = 0$$

and hence $x := y - x^{\text{part}} \in L_{\text{hom}}$ or $y = x^{\text{part}} + x \in x^{\text{part}} + L_{\text{hom}}$.

Let us discuss an example.

Example We would like to compute the dimension of the space of solutions of the homogeneous system $Ax = 0$ where $A \in \mathbb{R}^{3 \times 5}$ is given by

$$A = \begin{pmatrix} 1 & 1 & 1 & 1 & 1 \\ 2 & 1 & -1 & 3 & 0 \\ 2 & 1 & 2 & 1 & 2 \end{pmatrix}.$$

Recall that $L_{\text{hom}} = \{x \in \mathbb{R}^{5 \times 1} \mid Ax = 0\}$ is a subspace of $\mathbb{R}^{5 \times 1}$, since for any $x, x' \in L_{\text{hom}}$ and any $\lambda \in \mathbb{R}$

$$A(x + x') = Ax + Ax' = 0, \qquad A(\lambda x) = \lambda Ax = 0.$$

To compute the dimension of L_{hom}, we have to find a basis of L_{hom}. The dimension of L_{hom} is then given by the number of the vectors of this basis. Use Gaussian elimination to bring the extended coefficient matrix $[A \parallel 0]$ into reduced row echelon form.

Step 1. $R_2 \rightsquigarrow R_2 - 2R_1$, $R_3 \rightsquigarrow R_3 - 2R_1$,

$$\left[\begin{array}{ccccc|c} 1 & 1 & 1 & 1 & 1 & 0 \\ 0 & -1 & -3 & 1 & -2 & 0 \\ 0 & -1 & 0 & -1 & 0 & 0 \end{array}\right].$$

Step 2. $R_3 \rightsquigarrow R_3 - R_2$,

$$\left[\begin{array}{ccccc|c} 1 & 1 & 1 & 1 & 1 & 0 \\ 0 & -1 & -3 & 1 & -2 & 0 \\ 0 & 0 & 3 & -2 & 2 & 0 \end{array}\right].$$

Step 3. $R_2 \rightsquigarrow R_2 + R_3$, $R_1 \rightsquigarrow R_1 - \frac{1}{3} R_3$,

$$\left[\begin{array}{ccccc|c} 1 & 1 & 0 & 5/3 & 1/3 & 0 \\ 0 & -1 & 0 & -1 & 0 & 0 \\ 0 & 0 & 3 & -2 & 2 & 0 \end{array}\right].$$

Step 4. $R_1 \rightsquigarrow R_1 + R_2$,

$$\left[\begin{array}{ccccc|c} 1 & 0 & 0 & 2/3 & 1/3 & 0 \\ 0 & -1 & 0 & -1 & 0 & 0 \\ 0 & 0 & 3 & -2 & 2 & 0 \end{array}\right].$$

Step 5. $R_2 \rightsquigarrow -R_2$, $R_3 \rightsquigarrow \frac{1}{3} R_3$,

$$\left[\begin{array}{ccccc|c} 1 & 0 & 0 & 2/3 & 1/3 & 0 \\ 0 & 1 & 0 & 1 & 0 & 0 \\ 0 & 0 & 1 & -2/3 & 2/3 & 0 \end{array}\right].$$

Hence x_4, x_5 are free variables and any solution of the system $Ax = 0$ has the form

$$x = \begin{pmatrix} x_1 \\ x_2 \\ x_3 \\ x_4 \\ x_5 \end{pmatrix} = \begin{pmatrix} -2/3\, x_4 - 1/3\, x_5 \\ -x_4 \\ 2/3\, x_4 - 2/3\, x_5 \\ x_4 \\ x_5 \end{pmatrix} = x_4 \begin{pmatrix} -2/3 \\ -1 \\ 2/3 \\ 1 \\ 0 \end{pmatrix} + x_5 \begin{pmatrix} -1/3 \\ 0 \\ -2/3 \\ 0 \\ 1 \end{pmatrix}.$$

It means that x is a linear combination of

$$v^{(1)} = \begin{pmatrix} -\frac{2}{3} \\ -1 \\ \frac{2}{3} \\ 1 \\ 0 \end{pmatrix}, \qquad v^{(2)} = \begin{pmatrix} -\frac{1}{3} \\ 0 \\ -\frac{2}{3} \\ 0 \\ 1 \end{pmatrix}.$$

Since x_4, x_5 are free variables, $v^{(1)}, v^{(2)}$ are linearly independent and hence form a basis of L_{hom}. One concludes that $\dim(L_{\text{hom}}) = 2$.

Problems

1. Decide which of the following subsets are linear subspaces of the corresponding $\mathbb{R}$-vector spaces.

   (i) $W = \{(x_1, x_2, x_3) \in \mathbb{R}^3 \mid 2x_1 + 3x_2 + x_3 = 0\}$.
   (ii) $V = \{(x_1, x_2, x_3, x_4) \in \mathbb{R}^4 \mid 4x_2 + 3x_3 + 2x_4 = 7\}$.
   (iii) $\mathrm{GL}_{\mathbb{R}}(3) = \{A \in \mathbb{R}^{3 \times 3} \mid A \text{ regular}\}$.
   (iv) $L = \{(x_1, x_2) \in \mathbb{R}^2 \mid x_1 x_2 = 0\}$.

2. Consider the following linear system (S),

$$\begin{cases} 3x_1 + x_2 - 3x_3 & = 4 \\ x_1 + 2x_2 + 5x_3 & = -2 \end{cases}.$$

   (i) Determine the vector space of solutions $L_{\text{hom}} \subseteq \mathbb{R}^3$ of the corresponding homogenous system

$$\begin{cases} 3x_1 + x_2 - 3x_3 & = 0 \\ x_1 + 2x_2 + 5x_3 & = 0 \end{cases}$$

   and compute its dimension.
   (ii) Determine the affine space L of solutions of (S) by finding a particular solution of (S).

3. Let $\mathcal{P}_3(\mathbb{C})$ denote the $\mathbb{C}$-vector space of polynomials of degree at most three in one complex variable,

$$\mathcal{P}_3(\mathbb{C}) = \{p(z) = a_3 z^3 + a_2 z^2 + a_1 z + a_0 \mid a_0, a_1, a_2, a_3 \in \mathbb{C}\},$$

and by $E_3(\mathbb{C})$ the subset

$$E_3(\mathbb{C}) = \{p \in \mathcal{P}_3(\mathbb{C}) \mid p(-z) = p(z), z \in \mathbb{C}\}.$$

(i) Find a basis of $\mathcal{P}_3(\mathbb{C})$ and compute $\dim \mathcal{P}_3(\mathbb{C})$.
(ii) Verify that $E_3(\mathbb{C})$ is a $\mathbb{C}$-subspace of $\mathcal{P}_3(\mathbb{C})$ and compute its dimension.

4. Consider the subset W of $\mathbb{C}^{3\times 3}$,

$$W = \{(a_{ij})_{1 \le i,j \le 3} \in \mathbb{C}^{3\times 3} \mid a_{ij} = 0, 1 \le i < j \le 3\}.$$

(i) Find a basis of $\mathbb{C}^{3\times 3}$ and compute its dimension.
(ii) Verify that W is a linear subspace of $\mathbb{C}^{3\times 3}$ and compute its dimension.

5. Let $A = \begin{pmatrix} a_1 \\ a_2 \end{pmatrix} \in \mathbb{R}^{2\times 1}$.

(i) Determine the rank of $A^T A \in \mathbb{R}^{1\times 1}$ and of $A A^T \in \mathbb{R}^{2\times 2}$ in terms of A.
(ii) Decide for which A the matrix $A^T A$ and for which A the matrix $A A^T$ is regular.

4.2   Linear Maps

In this section we introduce the notion of a linear map between vector spaces. If not stated otherwise V, W are $\mathbb{K}$-vector spaces where $\mathbb{K}$ denotes either $\mathbb{R}$ (field of real numbers) or $\mathbb{C}$ (field of complex numbers).

Definition 4.2.1 A map $f \colon V \to W$ is said to be $\mathbb{K}$-*linear* (or *linear* for short) if it is compatible with the vector space structures of V and W. It means that

(L1) $f(u + v) = f(u) + f(v)$, $u, v \in V$,
(L2) $f(\lambda u) = \lambda f(u)$, $\lambda \in \mathbb{K}, u \in V$.

From (L1) and (L2) it follows that for any $v^{(1)}, \ldots, v^{(n)} \in V, \lambda_1, \ldots, \lambda_n \in \mathbb{K}$,

$$f\Big(\sum_{j=1}^{n} \lambda_j v^{(j)}\Big) = \sum_{j=1}^{n} \lambda_j f(v^{(j)}).$$

Lemma 4.2.2 *If $f \colon V \to W$ is a linear map, then $f(0_V) = 0_W$ where 0_V denotes the zero element of V and 0_W the one of W.*

Note that Lemma 4.2.2 can be verified in a straightforward way: using that $0_V = 0_V + 0_V$, it follows that $f(0_V) = f(0_V) + f(0_V) = 2f(0_V)$ implying that

$$0_W = f(0_V) - f(0_V) = 2f(0_V) - f(0_V) = f(0_V).$$

Examples

(i) Let $V = W = \mathbb{R}$ and $a \in \mathbb{R}$. Then $f: \mathbb{R} \to \mathbb{R}, x \mapsto ax$ is a linear map. Indeed, $(L1)$, $(L2)$ are clearly satisfied.

(ii) Let $V = \mathbb{R}^{n \times 1}$, $W = \mathbb{R}^{m \times 1}$. Then for any $A = (a_{ij})_{\substack{1 \le i \le m \\ 1 \le j \le n}} \in \mathbb{R}^{m \times n}$

$$f_A: \mathbb{R}^{n \times 1} \to \mathbb{R}^{m \times 1}, \quad x \mapsto Ax = \begin{pmatrix} \sum_{j=1}^{n} a_{1j} x_j \\ \vdots \\ \sum_{j=1}^{n} a_{mj} x_j \end{pmatrix}$$

is linear since $(L1)$ and $(L2)$ are clearly satisfied. The map f_A is called the linear map associated to the matrix A.

(iii) Let $f: \mathbb{R}^2 \to \mathbb{R}^2$ be the map which maps an arbitrary point $x \in \mathbb{R}^2$ to the point obtained by reflecting it at the x_1-axis. The map f is given by $f: (x_1, x_2) \to (x_1, -x_2)$. Identifying $\mathbb{R}^2$ and $\mathbb{R}^{2 \times 1}$ one sees that f can be represented as

$$f(x_1, x_2) = A \begin{pmatrix} x_1 \\ x_2 \end{pmatrix}, \quad A = \begin{pmatrix} 1 & 0 \\ 0 & -1 \end{pmatrix},$$

hence by Item 4.2, f is linear.

(iv) Let $R(\varphi)$ denote the map $\mathbb{R}^2 \to \mathbb{R}^2$, defined by rotating a given vector $x \in \mathbb{R}^2$ counterclockwise by the angle φ (modulo 2π). Note that

$$(1, 0) \mapsto (\cos\varphi, \sin\varphi), \qquad (0, 1) \mapsto (-\sin\varphi, \cos\varphi).$$

Identifying again $\mathbb{R}^2$ and $\mathbb{R}^{2 \times 1}$ one can verify in a straightforward way that

$$f(x_1, x_2) = A \begin{pmatrix} x_1 \\ x_2 \end{pmatrix}, \quad A := \begin{pmatrix} \cos\varphi & -\sin\varphi \\ \sin\varphi & \cos\varphi \end{pmatrix},$$

hence by Item 4.2, f is linear.

(v) Let $f: \mathbb{R} \to \mathbb{R}, x \mapsto x^2$. Then f is not linear since $f(2) = 2^2 = 4$, but $f(1) + f(1) = 2$, implying that $f(1 + 1) \neq 2f(1)$, which violates $(L1)$.

(vi) Let $f: \mathbb{R} \to \mathbb{R}, x \mapsto 1 + 2x$. Then f is not linear since $f(0) = 1$, $f(1) = 3$, and hence $f(1 + 0) \neq f(1) + f(0)$, violating $(L1)$.

Definition 4.2.3 The *nullspace* Null(f) and the *range* Range(f) of a linear map $f: V \to W$ are defined as

$$\text{Null}(f) := \{x \in V \mid f(x) = 0\}, \qquad \text{Range}(f) := \{f(x) \mid x \in V\}.$$

Alternatively, Null(f) is also referred as the *kernel* of f and denoted by ker(f). Note that Null(f) is a subspace of V and Range(f) one of W.

Theorem 4.2.4 *For any linear map $f : \mathbb{K}^n \to \mathbb{K}^m$, there exists a matrix $A \in \mathbb{K}^{m \times n}$ so that $f(x) = Ax$, for any $x \in \mathbb{K}^n$. (Here we identify $\mathbb{K}^{n \times 1}$ with $\mathbb{K}^n$.) The matrix A is referred to as the* matrix representation *of f with respect to the standard bases* $[e_{\mathbb{K}^n}] = [e_{\mathbb{K}^n}^{(1)}, \dots, e_{\mathbb{K}^n}^{(n)}]$ *of* $\mathbb{K}^n$ *and* $[e_{\mathbb{K}^m}] = [e_{\mathbb{K}^m}^{(1)}, \dots, e_{\mathbb{K}^m}^{(m)}]$ *of* $\mathbb{K}^m$.

The following notation for A turns out to be very convenient:

$$A = f_{[e_{\mathbb{K}^n}] \to [e_{\mathbb{K}^m}]}.$$

Remark In the sequel, we will often drop the subscript $\mathbb{K}^n$ in $e_{\mathbb{K}^n}^{(j)}$ and simply write $e^{(j)}$.

To get more familiar with the notion of a linear map, let us determine the matrix $A \in \mathbb{K}^{m \times n}$ in Theorem 4.2.4. Given $x = (x_1, \dots, x_n) \in \mathbb{K}^n$, we write $x = \sum_{j=1}^n x_j e_{\mathbb{K}^n}^{(j)}$ and get

$$f(x) = f\Big(\sum_{j=1}^n x_j e_{\mathbb{K}^n}^{(j)}\Big) = \sum_{j=1}^n x_j f(e_{\mathbb{K}^n}^{(j)}).$$

With $f(e_{\mathbb{K}^n}^{(j)}) = (a_{1j}, \dots, a_{mj}) \in \mathbb{K}^m$, one has

$$f(x) = \begin{pmatrix} \sum_{j=1}^n a_{1j} x_j \\ \vdots \\ \sum_{j=1}^n a_{mj} x_j \end{pmatrix} = Ax$$

where A is the $m \times n$ matrix with columns given by

$$C_1(A) := \begin{pmatrix} a_{11} \\ \vdots \\ a_{m1} \end{pmatrix}, \quad \dots, \quad C_n(A) := \begin{pmatrix} a_{1n} \\ \vdots \\ a_{mn} \end{pmatrix}.$$

More generally, assume that V is a $\mathbb{K}$-vector space of dimension n with basis $[v] = [v^{(1)}, \dots, v^{(n)}]$, W a $\mathbb{K}$-vector space of dimension m with basis $[w] = [w^{(1)}, \dots, w^{(m)}]$, and $f : V \to W$ a linear map. Then the image $f(v^{(j)})$ of the vector $v^{(j)}$ can be written in a unique way as a linear combination of the vectors of the basis $[w]$, namely

$$f(v^{(j)}) = \sum_{i=1}^m a_{ij} w^{(i)}.$$

Definition 4.2.5 The matrix

$$A = (a_{ij})_{\substack{1 \le i \le m \\ 1 \le j \le n}}$$

is called the *matrix representation* of f with respect to the basis $[v]$ of V and $[w]$ of W and is denoted by $f_{[v] \to [w]}$. The coefficients of the jth column $C_j(f_{[v] \to [w]})$ of $f_{[v] \to [w]}$ are given by the coefficients of $f(v^{(j)})$ with respect to the basis $[w]$. (The notation $f_{[v] \to [w]}$ coincides with the one introduced in (2.7) in Sect. 2.2 in the special case where $V = \mathbb{R}^n$, $W = V$, and f is given by the identity map.)

Theorem 4.2.6 *Assume that* $f : V \to W$ *is a linear map,* $x = \sum_{j=1}^{n} x_j v^{(j)}$, *and* $f(x) = \sum_{i=1}^{m} y_i w^{(i)}$. *Then* $y = (y_1, \dots, y_m)$ *is related to* $x = (x_1, \dots, x_n)$ *by*

$$y = f_{[v] \to [w]} x.$$

In words: the coordinates of $f(x)$ *with respect to the basis* $[w]$ *can be computed from the coordinates of* x *with respect to the basis* $[v]$ *with the help of the* $m \times n$ *matrix* $f_{[v] \to [w]}$, *i.e.,* $y = f_{[v] \to [w]} x$.

Let us verify the statement of Theorem 4.2.6: writing $f(v^{(j)})$ as a linear combination of the vectors in the basis $[w]$, $f(v^{(j)}) = \sum_{i=1}^{m} a_{ij} w^{(i)}$, one has

$$f\left(\sum_{j=1}^{n} x_j v^{(j)}\right) = \sum_{j=1}^{n} x_j f(v^{(j)}) = \sum_{j=1}^{n} x_j \sum_{i=1}^{m} a_{ij} w^{(i)} = \sum_{i=1}^{m} \left(\sum_{j=1}^{n} a_{ij} x_j\right) w^{(i)}.$$

So $\sum_{j=1}^{n} a_{ij} x_j$ is the ith coordinate of the vector $f\left(\sum_{j=1}^{n} x_j v^{(j)}\right)$ with respect to the basis $[w]$.

Linear maps between two vector spaces can be added and multiplied by a scalar, resulting again in linear maps.

Proposition 4.2.7 *Let V and W be $\mathbb{K}$-vector spaces. Then the following holds:*

(i) If $f, g : V \to W$ are linear maps and $\lambda \in \mathbb{K}$, then so are

$$f + g : V \to W, x \mapsto f(x) + g(x), \qquad \lambda f : V \to W, x \mapsto \lambda f(x).$$

(ii) If $f : V \to W$ is a bijective linear map, then its inverse $f^{-1} : W \to V$ is also linear.

The matrix representation of the linear maps of Proposition 4.2.7 can be computed as follows.

Theorem 4.2.8 *Let V and W be $\mathbb{K}$-vector spaces and let $[v] = [v^{(1)}, \dots, v^{(n)}]$ be a basis of V and $[w] = [w^{(1)}, \dots, w^{(m)}]$ be one of W. Then the following holds:*

(i) If $f, g: V \to W$ are linear maps and $\lambda \in \mathbb{K}$, then

$$(f + g)_{[v] \to [w]} = f_{[v] \to [w]} + g_{[v] \to [w]}, \qquad (\lambda f)_{[v] \to [w]} = \lambda f_{[v] \to [w]}.$$

(ii) If $f: V \to W$ is a bijective linear map, then $\dim(V) = \dim(W)$ and $f_{[v] \to [w]} \in \mathbb{K}^{n \times n}$ is a regular. The matrix representation of the inverse $f^{-1}: W \to V$ of f satisfies

$$(f^{-1})_{[w] \to [v]} = (f_{[v] \to [w]})^{-1} \tag{4.1}$$

An important property of linear maps is that the composition of such maps is again linear, a fact which can be verified in a straightforward way.

Proposition 4.2.9 *Assume that V, W, U are $\mathbb{K}$-vector spaces and $f: V \to W$, $g: W \to U$ are linear maps. Then the composition*

$$g \circ f: V \to U, x \mapsto g(f(x)),$$

is a linear map.

We now discuss the relation between the composition of linear maps and matrix multiplication, already mentioned in Sect. 2.1. Assume that V is a $\mathbb{K}$-vector space of dimension n with basis $[v] = [v^{(1)}, \dots, v^{(n)}]$, W a $\mathbb{K}$-vector space of dimension m with basis $[w] = [w^{(1)}, \dots, w^{(m)}]$ and U a $\mathbb{K}$-vector space of dimension k with basis $[u] = [u^{(1)}, \dots, u^{(k)}]$. Furthermore, assume that

$$f: V \to W, \qquad g: W \to U,$$

are linear maps with matrix representations $A = f_{[v] \to [w]}$ and $B = g_{[w] \to [u]}$. The following theorem says how to compute the matrix representation $(g \circ f)_{[v] \to [u]}$ of the linear map

$$g \circ f: V \xrightarrow{f} W \xrightarrow{g} U.$$

Theorem 4.2.10 *Under the above assumptions,*

$$(g \circ f)_{[v] \to [u]} = g_{[w] \to [u]} \cdot f_{[v] \to [w]}.$$

Remark The formula (4.1) for the matrix representation of the inverse $f^{-1}: W \to V$ follows from Theorem 4.2.10,

$$\mathrm{Id}_{n \times n} = \mathrm{Id}_{[v] \to [v]} = (f^{-1} \circ f)_{[v] \to [v]} = (f^{-1})_{[w] \to [v]} \cdot f_{[v] \to [w]}.$$

To get more familiar with the composition of linear maps, let us verify the statement of Theorem 4.2.10. With $x = \sum_{j=1}^{n} x_j v^{(j)}$ and $f(v^{(j)}) = \sum_{i=1}^{m} a_{ij} w^{(i)}$ one gets

$$f(x) = \sum_{i=1}^{m} \left(\sum_{j=1}^{n} a_{ij} x_j \right) w^{(i)}.$$

Writing $g(w^{(i)}) = \sum_{\ell=1}^{k} b_{\ell i} u^{(\ell)}$, one then concludes that

$$g(f(x)) = \sum_{i=1}^{m} \left(\sum_{j=1}^{n} a_{ij} x_j \right) g(w^{(i)}) = \sum_{i=1}^{m} \left(\sum_{j=1}^{n} a_{ij} x_j \right) \sum_{\ell=1}^{k} b_{\ell i} u^{(\ell)}$$

$$= \sum_{\ell=1}^{k} \left(\sum_{j=1}^{n} \left(\sum_{i=1}^{m} b_{\ell i} a_{ij} \right) x_j \right) u^{(\ell)}.$$

But

$$\sum_{i=1}^{m} b_{\ell i} a_{ij} = (BA)_{\ell j}, \qquad A = f_{[v] \to [w]}, \quad B = g_{[w] \to [u]},$$

and hence

$$g(f(x)) = \sum_{\ell=1}^{k} z_\ell u^{(\ell)}, \qquad z_\ell = \sum_{j=1}^{n} (BA)_{\ell j} x_j = (BAx)_\ell, \quad 1 \le \ell \le k.$$

Let us now consider the special case where $V = W$. Assume that V is a $\mathbb{K}$-vector space of dimension n. In this case one often chooses the same basis $[v] = [v^{(1)}, \ldots, v^{(n)}]$ in the domain and the target of a linear map $f : V \to V$ and $f_{[v] \to [v]}$ is said to be the *matrix representation* of f with respect to $[v]$. The $n \times n$ matrix $A := f_{[v] \to [v]}$ has columns $C_j(A)$, $1 \le j \le n$, whose coefficients are the coordinates of $f(v^{(j)})$ with respect to the basis $[v]$,

$$f(v^{(j)}) = \sum_{i=1}^{n} a_{ij} v^{(i)}.$$

Examples

(i) Let us consider the rotation $R(\varphi) \colon \mathbb{R}^2 \to \mathbb{R}^2$ by the angle φ in counterclockwise direction, introduced earlier, and denote by $[e] = [e^{(1)}, e^{(2)}]$ the standard basis of $\mathbb{R}^2$. The two columns $C_1(A)$, $C_2(A)$ of $A := R(\varphi)_{[e] \to [e]}$ are then computed as follows (we identify $\mathbb{R}^2$ with $R^{2\times 1}$):

$$R(\varphi)e^{(1)} = \begin{pmatrix} \cos\varphi \\ \sin\varphi \end{pmatrix} = \cos\varphi e^{(1)} + \sin\varphi e^{(2)},$$

implying that $C_1(A) = \begin{pmatrix} \cos\varphi \\ \sin\varphi \end{pmatrix}$. Similarly

$$R(\varphi)e^{(2)} = \begin{pmatrix} -\sin\varphi \\ \cos\varphi \end{pmatrix} = -\sin\varphi e^{(1)} + \cos\varphi e^{(2)},$$

yielding $C_2(A) = \begin{pmatrix} -\sin\varphi \\ \cos\varphi \end{pmatrix}$. Altogether one obtains

$$R(\varphi)_{[e]\to[e]} = \begin{pmatrix} \cos\varphi & -\sin\varphi \\ \sin\varphi & \cos\varphi \end{pmatrix}.$$

(ii) Let $w^{(1)} = (1,0)$, $w^{(2)} = (1,1)$. To compute $R(\varphi)_{[w]\to[w]}$ we proceed as follows: write

$$R(\varphi)w^{(1)} = \begin{pmatrix} \cos\varphi \\ \sin\varphi \end{pmatrix} = a_{11}w^{(1)} + a_{21}w^{(2)}$$

and determine a_{11}, a_{21} by solving the linear system

$$\begin{cases} a_{11} + a_{21} = \cos\varphi \\ \qquad\quad a_{21} = \sin\varphi \end{cases}.$$

Hence $a_{21} = \sin\varphi$ and $a_{11} = \cos\varphi - \sin\varphi$. Similarly, to find the coordinates a_{21}, a_{22} of

$$R(\varphi)w^{(2)} = \begin{pmatrix} \cos\varphi - \sin\varphi \\ \sin\varphi + \cos\varphi \end{pmatrix}$$

with respect to the basis $[w]$ we need to solve the system

$$\begin{cases} a_{12} + a_{22} = \cos\varphi - \sin\varphi \\ \qquad\quad a_{22} = \sin\varphi + \cos\varphi \end{cases},$$

whose solution is $a_{22} = \sin\varphi + \cos\varphi$ and $a_{12} = -2\sin\varphi$. Altogether, one then obtains

$$R(\varphi)_{[w]\to[w]} = \begin{pmatrix} a_{11} & a_{12} \\ a_{21} & a_{22} \end{pmatrix} = \begin{pmatrix} \cos\varphi - \sin\varphi & -2\sin\varphi \\ \sin\varphi & \sin\varphi + \cos\varphi \end{pmatrix}.$$

With the introduced notations we can also express the matrix representation of the identity map with respect to bases $[v]$ and $[w]$ of a given vector space. To be more precise, assume that V is a $\mathbb{K}$-vector space of dimension n and that

$$[v] = [v^{(1)}, \ldots, v^{(n)}], \qquad [w] = [w^{(1)}, \ldots, w^{(n)}],$$

are two bases of V. The matrix representation of the identity map $\mathrm{Id}\colon V \to V$ with respect to the bases $[v]$ and $[w]$ is then given by

$$\mathrm{Id}_{[v]\to[w]}.$$

The meaning of this matrix is the following one: given any vector b in V, write b as a linear combination with respect to the two bases $[v]$ and $[w]$, $b = \sum_{j=1}^{n} x_j v^{(j)}$ and $b = \sum_{j=1}^{n} y_j w^{(j)}$. The coordinate vectors $x = (x_j)_{1\le j\le n}$ and $y = (y_j)_{1\le j\le n}$ are then related by the formula $y = \mathrm{Id}_{[v]\to[w]}\, x$ and the matrix $\mathrm{Id}_{[v]\to[w]}$ is referred to as the *matrix of the change of the basis* $[v]$ to the basis $[w]$. We remark that the matrix $\mathrm{Id}_{[v]\to[w]}$ was already introduced in (2.7) in Sect. 2.2.

The jth column of $\mathrm{Id}_{[v]\to[w]}$ is given by the coordinates of $v^{(j)}$ with respect to the basis $[w]$, $v^{(j)} = \sum_{i=1}^{n} s_{ij} w^{(i)}$. Note that $\mathrm{Id}_{[v]\to[v]} = \mathrm{Id}_{n\times n}$ where $\mathrm{Id}_{n\times n}$ is the standard $n \times n$ identity matrix in $\mathbb{K}^{n\times n}$ and $\mathrm{Id} \circ \mathrm{Id} = \mathrm{Id}$. Hence by the previous remark, we get

$$\mathrm{Id}_{[v]\to[w]} \cdot \mathrm{Id}_{[w]\to[v]} = \mathrm{Id}_{[v]\to[v]} = \mathrm{Id}_{n\times n},$$

yielding $\mathrm{Id}_{[w]\to[v]} = (\mathrm{Id}_{[v]\to[w]})^{-1}$.

Theorem 4.2.11 *Assume that V is a $\mathbb{K}$-vector space of dimension n, $f\colon V \to V$ is a linear map, and $[v]$, $[w]$ are bases of V. Then*

$$f_{[w]\to[w]} = \mathrm{Id}_{[v]\to[w]} \cdot f_{[v]\to[v]} \cdot \mathrm{Id}_{[w]\to[v]} = (\mathrm{Id}_{[w]\to[v]})^{-1} \cdot f_{[v]\to[v]} \cdot \mathrm{Id}_{[w]\to[v]}.$$

Example Assume that $V = W = \mathbb{R}^2$ and let $[v] = [v^{(1)}, v^{(2)}]$ be the basis of $\mathbb{R}^2$ with $v^{(1)} = (1, 1)$, $v^{(2)} = (1, -1)$. As usual, $[e] = [e^{(1)}, e^{(2)}]$ denotes the standard basis of $\mathbb{R}^2$.

(i) Compute $\mathrm{Id}_{[v]\to[e]}$.
   Write $v^{(1)} = 1 \cdot e^{(1)} + 1 \cdot e^{(2)}$ and $v^{(2)} = 1 \cdot e^{(1)} + (-1) \cdot e^{(2)}$. Hence

$$\mathrm{Id}_{[v]\to[e]} = \left(v^{(1)}\ v^{(2)}\right) = \begin{pmatrix} 1 & 1 \\ 1 & -1 \end{pmatrix}.$$

(ii) Compute $\mathrm{Id}_{[e]\to[v]}$.
   It is to compute $(\mathrm{Id}_{[v]\to[e]})^{-1}$, which is given by $\begin{pmatrix} 1/2 & 1/2 \\ 1/2 & -1/2 \end{pmatrix}$.

(iii) Consider the counterclockwise rotation $R(\varphi)$ in $\mathbb{R}^2$ by the angle φ.

   (iii1) Compute $R(\varphi)_{[e]\to[e]}$.
      We have

$$R(\varphi)e^{(1)} = \cos\varphi \cdot e^{(1)} + \sin\varphi \cdot e^{(2)}, \quad R(\varphi)e^{(2)} = -\sin\varphi \cdot e^{(1)} + \cos\varphi \cdot e^{(2)},$$

      yielding

$$R(\varphi)_{[e]\to[e]} = \begin{pmatrix} \cos\varphi & -\sin\varphi \\ \sin\varphi & \cos\varphi \end{pmatrix}.$$

   (iii2) Compute $R(\varphi)_{[v]\to[e]}$.
      We have

$$R(\varphi)_{[v]\to[e]} = R(\varphi)_{[e]\to[e]} \cdot \mathrm{Id}_{[v]\to[e]}.$$

      Combining 4.2 and (iii1) we get

$$R(\varphi)_{[v]\to[e]} = \begin{pmatrix} \cos\varphi & -\sin\varphi \\ \sin\varphi & \cos\varphi \end{pmatrix} \begin{pmatrix} 1 & 1 \\ 1 & -1 \end{pmatrix}$$

$$= \begin{pmatrix} \cos\varphi - \sin\varphi & \cos\varphi + \sin\varphi \\ \sin\varphi + \cos\varphi & \sin\varphi - \cos\varphi \end{pmatrix}.$$

   (iii3) Compute $R(\varphi)_{[v]\to[v]}$.
      We have

$$R(\varphi)_{[v]\to[v]} = \mathrm{Id}_{[e]\to[v]} \cdot R(\varphi)_{[e]\to[e]} \cdot \mathrm{Id}_{[v]\to[e]}$$

$$= \begin{pmatrix} 1/2 & 1/2 \\ 1/2 & -1/2 \end{pmatrix} \begin{pmatrix} \cos\varphi - \sin\varphi & \cos\varphi + \sin\varphi \\ \sin\varphi + \cos\varphi & \sin\varphi - \cos\varphi \end{pmatrix}$$

$$= \begin{pmatrix} \cos\varphi & \sin\varphi \\ -\sin\varphi & \cos\varphi \end{pmatrix}.$$

We point out that one has to distinguish between linear maps and their matrix representations. Linear maps are independent of a choice of bases and a matrix can be the representation of many different linear maps, depending on the choice of bases made. In the case where f is a linear map of a vector space V into itself, Theorem 4.2.11 says how the matrix representations with respect to different bases of V are related to each other, motivating the following

Definition 4.2.12 Two matrices $A, B \in \mathbb{K}^{n \times n}$ are said to be *similar* if there exists a matrix $S \in \mathrm{GL}_{\mathbb{K}}(n)$ so that $B = S^{-1}AS$ (and hence $A = SBS^{-1}$).

Note that the matrix S can be interpreted as $\mathrm{Id}_{[v] \to [e]}$, with $[v] = [v^{(1)}, \ldots, v^{(n)}]$ given by the columns of S, implying that B is the matrix representation of the linear map $f_A \colon \mathbb{K}^n \to \mathbb{K}^n$, $x \mapsto Ax$ with respect to the basis $[v]$.

Examples

(i) For $A = \lambda \, \mathrm{Id}_{n \times n}$, $\lambda \in \mathbb{R}$, the set of all matrices similar to A is determined as follows: for any $S \in \mathrm{GL}_{\mathbb{R}}(n)$,

$$S^{-1}AS = S^{-1}\lambda \, \mathrm{Id}_{n \times n} \, S = \lambda S^{-1}S = \lambda \, \mathrm{Id}_{n \times n} = A.$$

In words: the matrix representation of $f_{\lambda \, \mathrm{Id}_{n \times n}} \colon \mathbb{R}^n \to \mathbb{R}^n$, $x \mapsto \lambda x$, with respect to any basis of $\mathbb{R}^n$ is the matrix $\lambda \, \mathrm{Id}_{n \times n}$.

(ii) Let A and B be the 2×2 matrices

$$A = \begin{pmatrix} 1 & 0 \\ 0 & 2 \end{pmatrix}, \qquad B = \begin{pmatrix} 3 & 2 \\ -1 & 0 \end{pmatrix}.$$

It turns out that the regular matrix $S = \begin{pmatrix} 1 & 2 \\ -1 & -1 \end{pmatrix}$ satisfies $B = S^{-1}AS$ and $A = SBS^{-1}$, implying that A and B are similar. We will learn in Chap. 5 how to find out whether two given matrices of the same dimension are similar and if so how to obtain a matrix S with the above properties.

Definition 4.2.13 Let V be a $\mathbb{K}$-vector space of dimension n and $f \colon V \to V$ a linear map. Then f is said to be *diagonalizable* if there exists a basis $[v] = [v^{(1)}, \ldots, v^{(n)}]$ of V so that the matrix representation $f_{[v] \to [v]}$ of f with respect to the basis $[v]$ is diagonal.

Note that in the Example 4.2 above, the linear map

$$f_B \colon \mathbb{R}^2 \to \mathbb{R}^2, \qquad x \mapsto Bx,$$

is diagonalizable. Indeed let $[v] = [v^{(1)}, v^{(2)}]$ be the basis of $\mathbb{R}^2$ with $v^{(1)}, v^{(2)}$ given by the columns of

$$S^{-1} = \begin{pmatrix} -1 & -2 \\ 1 & 1 \end{pmatrix}.$$

Then $S^{-1} = \mathrm{Id}_{[v] \to [e]}$ and since $B = (f_B)_{[e] \to [e]}$ it follows that

$$(f_B)_{[v] \to [v]} = \mathrm{Id}_{[e] \to [v]} \, B \, \mathrm{Id}_{[v] \to [e]} = SBS^{-1} = A.$$

In Sect. 5.1, we will describe a whole class of linear maps $f : \mathbb{C}^n \to \mathbb{C}^n$ which are diagonalizable and in Sect. 5.2 a corresponding class of linear maps from $\mathbb{R}^n$ to $\mathbb{R}^n$.

Problems

1. Verify that the following maps $f : \mathbb{R}^n \to \mathbb{R}^m$ are linear and determine their matrix representations $f_{[e] \to [e]}$ with respect to the standard bases.

   (i) $f : \mathbb{R}^4 \to \mathbb{R}^2$, $(x_1, x_2, x_3, x_4) \mapsto (x_1 + 3x_2, -x_1 + 4x_2)$.
   (ii) Let $f : \mathbb{R}^2 \to \mathbb{R}^2$ be the map acting on a vector $x = (x_1, x_2) \in \mathbb{R}^2$ as follows: first x is scaled by the factor 5 and then it is reflected at the x_2-axis.

2. Let $\mathcal{P}_7$ be the $\mathbb{R}$-vector space of polynomials of degree at most 7 with real coefficients. Denote by p' the first and by p'' the second derivative of $p \in \mathcal{P}_7$ with respect to x. Show that the following maps map $\mathcal{P}_7$ into itself and decide whether they are linear.

   (i) $T(p)(x) := 2p''(x) + 2p'(x) + 5p(x)$.
   (ii) $S(p)(x) := x^2 p''(x) + p(x)$.

3. Consider the bases in $\mathbb{R}^2$, $[e] = [e^{(1)}, e^{(2)}]$, $[v] = [e^{(2)}, e^{(1)}]$, and

   $$[w] = [w^{(1)}, w^{(2)}], \qquad w^{(1)} = (1, 1), \qquad w^{(2)} = (1, -1),$$

   and let $f : \mathbb{R}^2 \to \mathbb{R}^2$ be the linear map with matrix representation

   $$f_{[e] \to [e]} = \begin{pmatrix} 1 & 2 \\ 3 & 4 \end{pmatrix}.$$

   (i) Determine $\mathrm{Id}_{[v] \to [e]}$ and $\mathrm{Id}_{[e] \to [v]}$.
   (ii) Determine $f_{[v] \to [e]}$ and $f_{[v] \to [v]}$.
   (iii) Determine $f_{[w] \to [w]}$.

4. Consider the basis $[v] = [v^{(1)}, v^{(2)}, v^{(3)}]$ of $\mathbb{R}^3$, defined as

$$v^{(1)} = (1, 0, -1), \qquad v^{(2)} = (1, 2, 1), \qquad v^{(3)} = (-1, 1, 1),$$

and the basis $[w] = [w^{(1)}, w^{(2)}]$ of $\mathbb{R}^2$, given by

$$w^{(1)} = (1, -1), \qquad w^{(2)} = (2, -1).$$

Determine the matrix representations $T_{[v] \to [w]}$ of the following linear maps.

(i) $T : \mathbb{R}^3 \to \mathbb{R}^2$, $(x_1, x_2, x_3) \mapsto (2x_3, x_1)$.
(ii) $T : \mathbb{R}^3 \to \mathbb{R}^2$, $(x_1, x_2, x_3) \mapsto (x_1 - x_2, x_1 + x_3)$.

5. Decide whether the following assertions are true or false and justify your answers.

(i) There exists a linear map $T : \mathbb{R}^3 \to \mathbb{R}^7$ so that $\{T(x) \mid x \in \mathbb{R}^3\} = \mathbb{R}^7$.
(ii) For any linear map $f : \mathbb{R}^n \to \mathbb{R}^n$, f is bijective if and only if $\det(f_{[e] \to [e]}) \neq 0$.

4.3 Inner Products

Often in applications we are given a vector space with additional geometric structures, allowing e.g. to measure the length of a vector or the angle between two nonzero vectors. In this section, we introduce such an additional geometric structure, called *inner product*. Linear maps between vector spaces with inner products, which leave the latter invariant, will also be discussed. It turns out that that inner products on $\mathbb{R}$- and $\mathbb{C}$-vector spaces have slightly different properties and hence we discuss them separately.

Inner Products on $\mathbb{R}$-Vector Spaces

Throughout this paragraph, V denotes a $\mathbb{R}$-vector space.

Definition 4.3.1 A map $\langle \cdot, \cdot \rangle : V \times V \to \mathbb{R}$ is said to be an *inner product* or *scalar product* on V if the following conditions hold:

$(IP1)$ For any v, w in V,

$$\langle v, w \rangle = \langle w, v \rangle.$$

$(IP2)$ For any v, w, u in V and any λ in $\mathbb{R}$,

$$\langle v + w, u \rangle = \langle v, u \rangle + \langle w, u \rangle, \qquad \langle \lambda v, w \rangle = \lambda \langle v, w \rangle.$$

$(IP3)$ For any v in V,

$$\langle v, v \rangle \geq 0, \qquad \langle v, v \rangle = 0 \text{ if and only if } v = 0.$$

In words: $\langle \cdot, \cdot \rangle$ is a symmetric, positive definite, real valued, bilinear map. In case $\dim(V) < \infty$, the pair $\left(V, \langle \cdot, \cdot \rangle\right)$ is said to be a finite dimensional $\mathbb{R}$-*Hilbert space* or *Hilbert space* for short.

Definition 4.3.2 On a vector space V with an inner product $\langle \cdot, \cdot \rangle$ one can define the notion of the length of a vector,

$$\|v\| := \langle v, v \rangle^{\frac{1}{2}}.$$

The nonnegative number $\|v\|$ is referred to as the *norm* of v, associated to $\langle \cdot, \cdot \rangle$. By $(IP3)$ it follows that for any $v \in V$,

$$\|v\| = 0 \text{ if and only if } v = 0$$

and by $(IP2)$,

$$\|\lambda v\| = |\lambda| \|v\|, \quad \lambda \in \mathbb{R}, \quad v \in V.$$

In the remaining part of this paragraph, we will always assume that the $\mathbb{R}$-vector space V is equipped with an inner product $\langle \cdot, \cdot \rangle$. The following fundamental inequality holds.

Lemma 4.3.3 (Cauchy-Schwarz Inequality)

$$|\langle v, w \rangle| \leq \|v\| \|w\|, \quad v, w \in V.$$

Remark The Cauchy-Schwarz inequality can be verified as follows: obviously, it holds if $v = 0$ or $w = 0$. In the case $v \neq 0$, $w \neq 0$, let us consider the vector $v + tw$ where $t \in \mathbb{R}$ is arbitrary. Then

$$0 \leq \|v + tw\|^2 = \langle v + tw, v + tw \rangle = \langle v, v \rangle + 2t \langle v, w \rangle + t^2 \langle w, w \rangle.$$

Let us find the minimum of $\|v + tw\|^2$ when viewed as a function of t. To this end we determine the critical values of t, i.e., the numbers t satisfying

$$\frac{d}{dt} \left(\langle v, v \rangle + 2t \langle v, w \rangle + t^2 \langle w, w \rangle \right) = 0.$$

Actually, there is unique such t, given by

$$t_0 = -\frac{\langle v, w \rangle}{\|w\|^2},$$

and we conclude that $0 \leq \langle v, v \rangle + 2t_0 \langle v, w \rangle + t_0^2 \langle w, w \rangle$ yielding the Cauchy-Schwarz inequality $0 \leq \|v\|^2 \|w\|^2 - \langle v, w \rangle^2$ or, written differently,

$$|\langle v, w \rangle| \leq \|v\|^2 \|w\|^2.$$

Note that for $v, w \in V$ with $\|v\| = \|w\| = 1$ the Cauchy-Schwarz inequality yields

$$|\langle v, w \rangle| \leq 1 \qquad \text{or} \qquad -1 \leq \langle v, w \rangle \leq 1.$$

Hence there exists a unique angle $0 \leq \varphi \leq \pi$ with

$$\langle v, w \rangle = \cos \varphi.$$

Definition 4.3.4 For any vectors $v, w \in V \setminus \{0\}$, the uniquely determined real number $0 \leq \varphi \leq \pi$ with

$$\cos \varphi = \langle \frac{v}{\|v\|}, \frac{w}{\|w\|} \rangle$$

is said to be the *non-oriented angle* between v and w. (The non-oriented angle does not allow to determine whether $\frac{v}{\|v\|}$ needs to be rotated clockwise or counterclockwise to be mapped to $\frac{w}{\|w\|}$.)

As a consequence, one has for any $v, w \in V \setminus \{0\}$,

$$\langle v, w \rangle = \|v\| \|w\| \cos \varphi.$$

Definition 4.3.5 The vectors $v, w \in V$ are said to be *orthogonal* to each other if $\langle v, w \rangle = 0$.

Note that if $v = 0$ or $w = 0$, one always has $\langle v, w \rangle = 0$, whereas for $v, w \in V \setminus \{0\}$

$$0 = \langle v, w \rangle = \|v\| \|w\| \cos \varphi$$

implies $\varphi = \pi/2$. As an easy application of the notion of an inner product one obtains the following well known result:

Theorem 4.3.6 (Pythagorean Theorem) *For any $v, w \in V$, which are orthogonal to each other, one has*

$$\|v + w\|^2 = \|v\|^2 + \|w\|^2.$$

This identity follows immediately from the assumption $\langle v, w \rangle = 0$ and

$$\|v + w\|^2 = \langle v + w, v + w \rangle = \langle w, w \rangle + 2\langle v, w \rangle + \langle v, v \rangle.$$

Theorem 4.3.7 (Law of Cosines) *For any $v, w \in V \setminus \{0\}$ one has*

$$\|v - w\|^2 = \|v\|^2 + \|w\|^2 - 2\|v\|\|w\| \cos \varphi$$

where φ is the non-oriented angle between v and w.

The above theorems are indeed generalizations of the corresponding well known results in the Euclidean plane $\mathbb{R}^2$. This is explained with the following

Example Let $V = \mathbb{R}^2$ and define

$$\langle \cdot, \cdot \rangle_{Euclid} : \mathbb{R}^2 \times \mathbb{R}^2 \to \mathbb{R}, \quad (x, y) \mapsto x_1 y_1 + x_2 y_2$$

where $x = (x_1, x_2)$, $y = (y_1, y_2)$. One easily verifies that $\langle \cdot, \cdot \rangle_{Euclid}$ is an inner product for $\mathbb{R}^2$. The corresponding norm $\|x\|$ is given by

$$\|x\| := \langle x, x \rangle^{\frac{1}{2}} = (x_1^2 + x_2^2)^{\frac{1}{2}}$$

and the angle φ between vectors $x, y \in \mathbb{R}^2$ of norm 1 can be computed as follows. Using polar coordinates one has

$$x = (\cos \theta, \sin \theta), \quad y = (\cos \psi, \sin \psi), \quad 0 \leq \theta, \psi < 2\pi.$$

Then the angle $0 \leq \varphi \leq \pi$ defined by

$$\langle x, y \rangle = \|x\|\|y\| \cos \varphi = \cos \varphi$$

is determined as follows: by the addition formula for the cosine,

$$\cos \varphi = \langle x, y \rangle = \cos \theta \cos \psi + \sin \theta \sin \psi = \cos(\psi - \theta),$$

yielding $\psi - \theta = \varphi \pmod{2\pi}$ or $\psi - \theta = -\varphi \pmod{2\pi}$. (Note that $\cos x$ is an even and 2π-periodic function and that $\cos : [0, \pi] \to [-1, 1]$ is bijective. On the other hand, $\psi - \theta$ need not to be an element in the interval $[0, \pi]$.) For notational convenience, we will simply write $\langle \cdot, \cdot \rangle$ for $\langle \cdot, \cdot \rangle_{Euclid}$ in the sequel. As an application of the inner product, let us compute the area F of a triangle ABC in $\mathbb{R}^2$ where

$A = (0, 0)$ is the origin,

$$B = x = (x_1, x_2) \neq (0, 0), \qquad C = y = (y_1, y_2) \neq (0, 0),$$

and the oriented angle ψ between x and y is assumed to satisfy $0 < \psi < \pi$. Recall that

$$F = \frac{1}{2} \text{ length of } \overline{AB} \cdot \text{height}$$

where the height is given by $\|y\| \sin \psi$ and the length of $\overline{AB}$ is $\|x\|$. Since $0 < \sin \psi \leq 1$ one then has

$$F = \frac{1}{2} \|x\| \|y\| \sin \psi = \frac{1}{2} \|x\| \|y\| \sqrt{1 - \cos^2 \psi}, \qquad \cos \psi = \frac{\langle x, y \rangle}{\|x\| \|y\|}.$$

Thus

$$F = \frac{1}{2} \sqrt{\|x\|^2 \|y\|^2 - \langle x, y \rangle^2}.$$

By an elementary computation,

$$\|x\|^2 \|y\|^2 - \langle x, y \rangle^2 = (x_1 y_2 - x_2 y_1)^2$$

or

$$\sqrt{\|x\|^2 \|y\|^2 - \langle x, y \rangle^2} = |x_1 y_2 - x_2 y_1| = \left| \det \begin{pmatrix} x_1 & y_1 \\ x_2 & y_2 \end{pmatrix} \right|.$$

The absolute value of the determinant can thus be interpreted as the area of the parallelogram spanned by x and y.

Examples

(i) Assume that $V = \mathbb{R}^n$. Then the map

$$\mathbb{R}^n \times \mathbb{R}^n \to \mathbb{R}, (x, y) \mapsto \sum_{i=1}^{n} x_i y_i$$

is an inner product on $\mathbb{R}^n$, referred to as *Euclidean inner product*. The corresponding norm of a vector $x \in \mathbb{R}^n$ is given by

$$\|x\| = \left(\sum_{i=1}^{n} x_i^2 \right)^{\frac{1}{2}}$$

and the non-oriented angle $0 \leq \varphi \leq \pi$ between two vectors $x, y \in \mathbb{R}^n \setminus \{0\}$ is defined by

$$\cos \varphi = \frac{1}{\|x\| \|y\|} \sum_{i=1}^{n} x_i y_i.$$

(ii) There exist many inner products on $\mathbb{R}^n$. Customarily we denote them by the same symbol. One verifies in a straightforward way that the following map defines an inner product on $\mathbb{R}^2$,

$$\mathbb{R}^2 \times \mathbb{R}^2 \to \mathbb{R}, \quad (x, y) \mapsto \langle x, y \rangle := 2x_1 y_1 + 5x_2 y_2,$$

and so does the map

$$\mathbb{R}^2 \times \mathbb{R}^2 \to \mathbb{R}, \quad (x, y) \mapsto \langle x, y \rangle := 3x_1 y_1 + x_1 y_2 + x_2 y_1 + 3x_2 y_2.$$

We will see in Sect. 5.3 how inner products such as the two examples above can be found.

An important property of inner products is the so called triangle inequality.

Lemma 4.3.8 (Triangle Inequality)   *For any $v, w \in V$*

$$\|v + w\| \leq \|v\| + \|w\|.$$

We remark that the triangle inequality follows in a straightforward way from the Cauchy-Schwarz inequality,

$$\|v + w\|^2 = \|v\|^2 + \|w\|^2 + 2\langle v, w \rangle \leq \|v\|^2 + \|w\|^2 + 2\|v\| \|w\| = (\|v\| + \|w\|)^2.$$

Isometries and Orthogonal Matrices

Assume that V is a $\mathbb{R}$-vector space equipped with an inner product $\langle \cdot, \cdot \rangle$.

Definition 4.3.9   A linear map $f : V \to V$ is said to be *isometric* (or an *isometry*) with respect to $\langle \cdot, \cdot \rangle$ if

$$\langle f(v), f(w) \rangle = \langle v, w \rangle, \quad v, w \in V.$$

In words: f preserves the length of vectors and the non-oriented angle between nonzero vectors.

Definition 4.3.10 A matrix $A \in \mathbb{R}^{n \times n}$ is said to be *orthogonal* if

$$\langle Ax, Ay \rangle = \langle x, y \rangle, \quad x, y \in \mathbb{R}^n$$

where $\langle \cdot, \cdot \rangle$ denotes the Euclidean inner product in $\mathbb{R}^n$,

$$\langle x, y \rangle := \sum_{j=1}^{n} x_j y_j, \quad x, y \in \mathbb{R}^n.$$

Lemma 4.3.11 *For any $A \in \mathbb{R}^{n \times n}$, A is orthogonal if and only if $A^{\mathrm{T}} A = \mathrm{Id}_{n \times n}$.*

Remark Lemma 4.3.11 can be verified in a straightforward way. Indeed, if A is orthogonal one has for any vectors $e^{(i)}$, $e^{(j)}$ of the standard basis $[e] = [e^{(1)}, \ldots, e^{(n)}]$,

$$\langle Ae^{(i)}, Ae^{(j)} \rangle = 0, \quad i \neq j, \qquad \langle Ae^{(i)}, Ae^{(i)} \rangle = 1.$$

Since for any $1 \leq j \leq n$, the jth column $C_j(A)$ of A equals $Ae^{(j)}$, one has

$$\langle C_i(A), C_j(A) \rangle = \langle Ae^{(i)}, Ae^{(j)} \rangle = \sum_{k=1}^{n} a_{ki} a_{kj} = (A^{\mathrm{T}} A)_{ij}$$

and thus indeed $A^{\mathrm{T}} A = \mathrm{Id}_{n \times n}$. Conversely, if $A^{\mathrm{T}} A = \mathrm{Id}_{n \times n}$, then for any $x, y \subset \mathbb{R}^n$

$$\langle x, y \rangle = \langle x, A^{\mathrm{T}} Ay \rangle = \langle Ax, Ay \rangle.$$

Note that Lemma 4.3.11 implies that any orthogonal $n \times n$ matrix A is invertible since

$$1 = \det(\mathrm{Id}_{n \times n}) = \det(A^{\mathrm{T}} A) = \det(A^{\mathrm{T}}) \det(A) = (\det(A))^2.$$

The following theorem states various properties of orthogonal matrices, which can be verified in a straightforward way.

Theorem 4.3.12 *For any $A \in \mathbb{R}^{n \times n}$, the following holds:*

(i) If A is orthogonal, then A is regular and $A^{-1} = A^{\mathrm{T}}$.
(ii) If A is orthogonal so is A^{-1} (and hence A^{T} by Item 4.3.12).
(iii) $\mathrm{Id}_{n \times n}$ is orthogonal.
(iv) If $A, B \in \mathbb{R}^{n \times n}$ are orthogonal, so is AB.
(v) If A is orthogonal, then $\det(A) \in \{-1, 1\}$.

The set of orthogonal $n \times n$ matrices is customarily denoted by $O(n)$ and by $SO(n)$ the subset of orthogonal matrices whose determinant equals one,

$$O(n) := \{A \in \mathbb{R}^{n \times n} \mid A \text{ is orthogonal}\}, \quad SO(n) := \{A \in O(n) \mid \det(A) = 1\}.$$

Remark From Theorem 4.3.12 it follows that $O(n)$ and $SO(n)$ are *groups*.

Definition 4.3.13 Let V be a $\mathbb{R}$-vector space of dimension n, equipped with an inner product $\langle \cdot, \cdot \rangle$. A basis $[v] = [v^{(1)}, \ldots, v^{(n)}]$ of V is said to be *orthonormal* if

$$\langle v^{(i)}, v^{(j)} \rangle = \delta_{ij}, \quad 1 \leq i, j \leq n$$

where δ_{ij} is the *Kronecker delta*, defined as

$$\delta_{ij} = 1, \quad i \neq j, \quad \delta_{ii} = 1.$$

In words: the vectors $v^{(i)}$, $1 \leq i \leq n$, have length 1 and are pairwise orthogonal.

Examples

(i) Let $V = \mathbb{R}^n$ and let $\langle \cdot, \cdot \rangle$ be the Euclidean inner product. Then the standard basis $[e] = [e^{(1)}, \ldots, e^{(n)}]$ is an orthonormal basis.
(ii) Let $V = \mathbb{R}^2$ and let $\langle \cdot, \cdot \rangle$ be the Euclidean inner product. Define

$$v^{(1)} = (\cos \varphi, \sin \varphi), \quad v^{(2)} = (-\sin \varphi, \cos \varphi)$$

where $0 \leq \varphi < 2\pi$. Then $[v] = [v^{(1)}, v^{(2)}]$ is an orthonormal basis of $\mathbb{R}^2$.
(iii) Assume that $V = \mathbb{R}^n$ and $\langle \cdot, \cdot \rangle$ is the Euclidean inner product of $\mathbb{R}^n$. The columns $C_1(A), \ldots, C_n(A)$ of an arbitrary orthogonal $n \times n$ matrix A form an orthonormal basis of $\mathbb{R}^n$. (To verify this statement use Lemma 4.3.11.)

The following assertions are frequently used in applications. They can be verified in a straightforward way.

Proposition 4.3.14 *Assume that $\left(V, \langle \cdot, \cdot \rangle\right)$ is a $\mathbb{R}$-Hilbert space of dimension n. Then the following holds:*

(i) *Let $v^{(1)}, \ldots, v^{(n)}$ be an orthonormal basis of V. If $f : V \to V$ is an isometry, then $f(v^{(1)}), \ldots, f(v^{(n)})$ is also an orthonormal basis of V.*
(ii) *Assume that $[v] = [v^{(1)}, \ldots, v^{(n)}]$ is an orthonormal basis of a V. Then for any $w \in V$,*

$$w = \sum_{j=1}^{n} \langle w, v^{(j)} \rangle v^{(j)},$$

i.e., the coordinates $x_1, \ldots, x_n$ of w with respect to the orthonormal basis $[v]$ are given by $x_j = \langle w, v^{(j)} \rangle$, $1 \leq j \leq n$.

There is a close connection between orthogonal $n \times n$ matrices and isometries of $\mathbb{R}$-Hilbert spaces of dimension n

Theorem 4.3.15 *Assume that* $V \equiv (V, \langle \cdot, \cdot \rangle)$ *is a* $\mathbb{R}$-*Hilbert space of dimension* n, $[v]$ *an orthonormal basis of* V *and* $f : V \to V$ *an isometry. Then* $f_{[v] \to [v]}$ *is an orthogonal matrix.*

Finally we introduce the notion of orthogonal complement.

Definition 4.3.16 Assume that $M \subseteq V$ is a subset of the $\mathbb{R}$-Hilbert space $(V, \langle \cdot, \cdot \rangle)$. Then

$$M^{\perp} := \{ w \in V \mid \langle w, v \rangle = 0, v \in M \}$$

is referred to as the *orthogonal complement* of M.

Theorem 4.3.17 *Assume that* $(V, \langle \cdot, \cdot \rangle)$ *is a* $\mathbb{R}$-*Hilbert space of dimension* n *and* $M \subseteq V$. *Then the following holds:*

 (i) $M^{\perp}$ *is a subspace of* V. *In particular,* $\{0\}^{\perp} = V$ *and* $V^{\perp} = \{0\}$.
 (ii) *Assume that* $M \subseteq V$ *is a subspace of* V *and* $v^{(1)}, \ldots, v^{(m)}$ *is an orthonormal basis of* M. *Then there are vectors* $v^{(m+1)}, \ldots, v^{(n)} \in V$ *so that* $[v^{(1)}, \ldots, v^{(n)}]$ *is an orthonormal basis of* V *and* $[v^{(m+1)}, \ldots, v^{(n)}]$ *is an orthonormal basis of* $M^{\perp}$.
 (iii) *If* $M \subseteq V$ *is a subspace of* V, *then* $\dim(V) = \dim(M) + \dim(M^{\perp})$.

Inner Products on $\mathbb{C}$-Vector Spaces

The inner product on a $\mathbb{C}$-vector space V is defined as follows.

Definition 4.3.18 The map $\langle \cdot, \cdot \rangle : V \times V \to \mathbb{C}$ is said to be an *inner product* if the following is satisfied

$(IP1)_c$ $\langle v, w \rangle = \overline{\langle w, v \rangle}$ $v, w \in V$.
$(IP2)_c$ $\langle \alpha v + \beta w, u \rangle = \alpha \langle v, u \rangle + \beta \langle w, u \rangle$ $v, w, u \in V, \alpha, \beta \in \mathbb{C}$.
$(IP3)_c$ For any $v \in V$,

$$\langle v, v \rangle \geq 0, \qquad \langle v, v \rangle = 0 \text{ if and only if } v = 0.$$

Remark

 (i) Note that $(IP1)_c$ and $(IP2)_c$ imply that $\langle \cdot, \cdot \rangle$ is antilinear in the second argument, i.e.,

$$\langle u, \alpha v + \beta w \rangle = \overline{\alpha} \langle u, v \rangle + \overline{\beta} \langle u, w \rangle, \quad v, w, u \in V, \quad \alpha, \beta \in \mathbb{C}.$$

(ii) Alternatively, instead of $(IP2)_c$, one could require that $\langle \cdot, \cdot \rangle$ is $\mathbb{C}$-linear with respect to the second argument,

$$\langle u, \alpha v + \beta w \rangle = \alpha \langle u, v \rangle + \beta \langle u, w \rangle, \quad v, w, u \in V, \quad \alpha, \beta \in \mathbb{C}.$$

In that case, $\langle \cdot, \cdot \rangle$ is antilinear in the first argument.

Again one can define the *norm* of a vector $v \in V$, induced by $\langle \cdot, \cdot \rangle$,

$$\|v\| := \langle v, v \rangle^{\frac{1}{2}}.$$

It has the following properties: for any $v, w \in V$,

$$\|\alpha v\| = |\alpha| \|v\|, \quad \alpha \in \mathbb{C}, \qquad \|v + w\| \leq \|v\| + \|w\|,$$

and

$$\|v\| \geq 0, \qquad \|v\| = 0 \text{ if and only if } v = 0.$$

Furthermore, the Cauchy-Schwarz inequality continues to hold,

$$|\langle v, w \rangle| \leq \|v\| \|w\|, \quad v, w \in V.$$

However note that in this case, $\langle v, w \rangle \in \mathbb{C}$ and hence we cannot define an angle.

Example For any $v = (v_1, \ldots, v_n)$, $w = (w_1 \ldots, w_n) \in \mathbb{C}^n$, let

$$\langle v, w \rangle := \sum_{j=1}^{n} v_j \overline{w}_j.$$

It is straightforward to verify that $(IP1)_c$ to $(IP3)_c$ are satisfied. This inner product is referred to as the *Euclidean inner product* of $\mathbb{C}^n$.

In case V is a $\mathbb{C}$-vector space of dimension n, equipped with an inner product $\langle \cdot, \cdot \rangle$, $\left(V, \langle \cdot, \cdot \rangle \right)$ is referred to as a n-dimensional $\mathbb{C}$-*Hilbert space*. A basis $[v] = [v^{(1)}, \ldots, v^{(n)}]$ of the $\mathbb{C}$-vector space V of dimension n with inner product $\langle \cdot, \cdot \rangle$ is said to be an *orthonormal basis* if

$$\langle v^{(i)}, v^{(j)} \rangle = \delta_{ij}, \quad 1 \leq i, j \leq n$$

where δ_{ij} denotes the Kronecker delta.

Definition 4.3.19

(i) A matrix $A \in \mathbb{C}^{n \times n}$ is said to be *unitary* if

$$\langle Av, Aw \rangle = \langle v, w \rangle, \quad v, w \in \mathbb{C}^n.$$

Here $\langle \cdot, \cdot \rangle$ denotes the Euclidean inner product on $\mathbb{C}^n$.

(ii) A linear map $f : V \to V$ on a n-dimensional $\mathbb{C}$-Hilbert space $V \equiv \left(V, \langle \cdot, \cdot \rangle \right)$ is said to be an *isometry* if

$$\langle f(v), f(w) \rangle = \langle v, w \rangle, \quad v, w \in V.$$

Note that $A \in \mathbb{C}^{n \times n}$ is unitary if and only if

$$f_A : \mathbb{C}^n \to \mathbb{C}^n, \quad v \mapsto Av$$

is an isometry. Furthermore, similarly as in the case of orthogonal matrices (cf. Lemma 4.3.11), one has that for any $A \in \mathbb{C}^{n \times n}$,

$$A \text{ is unitary if and only if } \overline{A}^{\mathrm{T}} A = \mathrm{Id}_{n \times n}.$$

Here $\overline{A}^{\mathrm{T}}$ is the conjugate transpose of A,

$$(\overline{A}^{\mathrm{T}})_{ij} = \overline{a}_{ji}, \quad 1 \leq i, j \leq n \quad \text{where} \quad A = (a_{ij})_{1 \leq i, j \leq n}.$$

Theorem 4.3.20 *Assume that* $[v] = [v^{(1)}, \ldots, v^{(n)}]$ *is an orthonormal basis of the* $\mathbb{C}$-*Hilbert space* $\left(V, \langle \cdot, \cdot \rangle \right)$ *and* $f : V \to V$ *a linear map. Then* f *is an isometry if and only if* $f_{[v] \to [v]}$ *is unitary.*

Vector Product in $\mathbb{R}^3$

The Euclidean space $\mathbb{R}^3$ is special in the sense that a multiplication of vectors with good properties can be defined, which turns out to be useful.

Definition 4.3.21 The *vector product* in $\mathbb{R}^3$, also referred to as cross product), is the map

$$\mathbb{R}^3 \times \mathbb{R}^3 \to \mathbb{R}^3, \quad (v, w) \mapsto v \times w := (v_2 w_3 - v_3 w_2, -v_1 w_3 + v_3 w_1, v_1 w_2 - v_2 w_1).$$

Example For $v = (0, 1, 0)$, $w = (1, 0, 0)$, one has $v \times w = (0, 0, -1)$.

Theorem 4.3.22 *The vector product in $\mathbb{R}^3$, with $\mathbb{R}^3$ being equipped with the Euclidean inner product $\langle \cdot, \cdot \rangle$, has the following properties:*

(i) $v \times w = -w \times v, \quad v, w \in \mathbb{R}^3.$
(ii) *For any $v, w, u \in \mathbb{R}^3, \alpha, \beta \in \mathbb{R}$,*

$$u \times (\alpha v + \beta w) = \alpha(u \times v) + \beta(u \times w),$$

$$(\alpha v + \beta w) \times u = \alpha(v \times u) + \beta(w \times u).$$

(iii) $\langle v \times w, v \rangle = 0, \langle v \times w, w \rangle = 0, v, w \in \mathbb{R}^3.$
(iv) *If $v \times w = 0$, then v and w are linearly dependent. In more detail,*

$$v = 0 \quad or \quad w = 0 \quad or \quad v = \alpha w, \alpha \in \mathbb{R}.$$

(v) *Let $0 \leq \varphi \leq \pi$ be the angle given by $\langle v, w \rangle = \|v\| \|w\| \cos \varphi$. Then*

$$\|v \times w\| = \|v\| \|w\| \sin \varphi.$$

(vi) *The vectors $v, w, v \times w$ are positively oriented, meaning that $\det(A) \geq 0$ where A is the 3×3 matrix whose columns are given by $v, w, v \times w$ (right thumb rule).*

Corollary 4.3.23 *Assume that $v^{(1)}, v^{(2)}$ are vectors in $\mathbb{R}^3$ with $\|v^{(1)}\| = \|v^{(2)}\| = 1$ and $\langle v^{(1)}, v^{(2)} \rangle = 0$. Then $[v^{(1)}, v^{(2)}, v^{(1)} \times v^{(2)}]$ is an orthonormal basis of $\mathbb{R}^3$.*

Remark Corollary 4.3.23 can be verified in a straightforward way. Since $\langle v^{(1)}, v^{(2)} \rangle = 0$ one has $\langle v^{(1)}, v^{(2)} \rangle = \cos \varphi$ with $\varphi = \pi/2$, implying that

$$\|v^{(1)} \times v^{(2)}\| = \|v^{(1)}\| \|v^{(2)}\| \sin \varphi = 1.$$

By Theorem 4.3.22(iii), it then follows that $[v^{(1)}, v^{(2)}, v^{(1)} \times v^{(2)}]$ is an orthonormal basis.

Example Let $v^{(1)} = \frac{1}{\sqrt{6}} (2, 1, -1)$. Then $\|v^{(1)}\| = 1$. We choose a vector $v^{(2)}$ of norm one, which is orthogonal to $v^{(1)}$, $v^{(2)} = \frac{1}{\sqrt{5}} (1, -2, 0)$. Then $[v^{(1)}, v^{(2)}, v^{(3)}]$, $v^{(3)} = v^{(1)} \times v^{(2)}$ is an orthonormal basis of $\mathbb{R}^3$. One computes $v^{(3)} = \frac{1}{\sqrt{30}} (-2, -1, -5)$.

Problems

1. Let $B \colon \mathbb{R}^2 \times \mathbb{R}^2 \to \mathbb{R}$ be given by

$$B(x, y) = 3x_1 y_1 + x_1 y_2 + x_2 y_1 + 2x_2 y_2$$

where $x = (x_1, x_2)$, $y = (y_1, y_2) \in \mathbb{R}^2$.

  (i) Verify that B is an inner product on $\mathbb{R}^2$.

 (ii) Verify that the vectors $a = (\frac{1}{\sqrt{3}}, 0)$ and $b = (0, \frac{1}{\sqrt{2}})$ have length 1 with respect to the inner product B.

(iii) Compute the cosine of the (unoriented) angle between the vectors a and b of (ii).

2. Let $\langle \cdot, \cdot \rangle$ be the Euclidean inner product of $\mathbb{R}^3$ and $v^{(1)}, v^{(2)}, v^{(3)}$ be the following vectors in $\mathbb{R}^3$, $v^{(1)} = \left(-\frac{1}{\sqrt{6}}, -\frac{2}{\sqrt{6}}, \frac{1}{\sqrt{6}} \right)$, $v^{(2)} = \left(\frac{1}{\sqrt{2}}, 0, \frac{1}{\sqrt{2}} \right)$, $v^{(3)} = \left(-\frac{1}{\sqrt{3}}, \frac{1}{\sqrt{3}}, \frac{1}{\sqrt{3}} \right)$.

  (i) Verify that $[v] = [v^{(1)}, v^{(2)}, v^{(3)}]$ is an orthonormal basis of $\mathbb{R}^3$.

 (ii) Compute $\mathrm{Id}_{[v] \to [e]}$, verify that it is an orthonormal 3×3 matrix, and then compute $\mathrm{Id}_{[e] \to [v]}$. Here $[e] = [e^{(1)}, e^{(2)}, e^{(3)}]$ denotes the standard basis of $\mathbb{R}^3$.

(iii) Represent the vectors $a = (1, 2, 1)$ and $b = (1, 0, 1)$ as linear combinations of $v^{(1)}, v^{(2)}$, and $v^{(3)}$.

3. Let $\langle \cdot, \cdot \rangle$ be the Euclidean inner product on $\mathbb{R}^2$ and $v^{(1)} = \left(\frac{1}{\sqrt{2}}, -\frac{1}{\sqrt{2}} \right)$.

  (i) Determine all possible vectors $v^{(2)} \in \mathbb{R}^2$ so that $[v^{(1)}, v^{(2)}]$ is an orthonormal basis of $\mathbb{R}^2$.

 (ii) Determine the matrix representation $R(\varphi)_{[v] \to [e]}$ of the linear map

$$R(\varphi) \colon \mathbb{R}^2 \to \mathbb{R}^2, \quad (x_1, x_2) \mapsto (\cos \varphi \cdot x_1 - \sin \varphi \cdot x_2, \sin \varphi \cdot x_1 + \cos \varphi \cdot x_2)$$

with respect to an orthonormal basis $[v] = [v^{(1)}, v^{(2)}]$ of $\mathbb{R}^2$, found in Item 3.

4. (i) Let $T \colon \mathbb{R}^3 \to \mathbb{R}^3$ be the rotation by the angle $\frac{\pi}{3}$ in clockwise direction in the $x_1 x_2$-plane with rotation axis $\{0\} \times \{0\} \times \mathbb{R}$. Determine $T_{[e] \to [e]}$ where $[e] = [e^{(1)}, e^{(2)}, e^{(3)}]$ is the standard basis of $\mathbb{R}^3$.

 (ii) Let $S \colon \mathbb{R}^3 \to \mathbb{R}^3$ be the rotation by the angle $\frac{\pi}{3}$ in counterclockwise direction in the $x_2 x_3$-plane with rotation axis $\mathbb{R} \times \{0\} \times \{0\}$. Determine $S_{[e] \to [e]}$ and verify that it is an orthogonal 3×3 matrix.

(iii) Compute $(S \circ T)_{[e] \to [e]}$ and $(T \circ S)_{[e] \to [e]}$.

5. Decide whether the following assertions are true or false and justify your answers.

  (i) There exists an orthogonal 2×2 matrix A with $\det(A) = -1$.

 (ii) Let $v^{(1)}, v^{(2)}, v^{(3)}, v^{(4)}$ be vectors in $\mathbb{R}^4$ so that

$$\langle v^{(i)}, v^{(j)} \rangle = \delta_{ij}$$

for any $1 \leq i, j \leq 4$ where $\langle \cdot, \cdot \rangle$ denotes the Euclidean inner product in $\mathbb{R}^4$ and δ_{ij} is the *Kronecker delta*, defined as

$$
\delta_{ij} = \begin{cases} 0 & \text{if } i \neq j, \\ 1 & \text{if } i = j. \end{cases}
$$

Then $[v^{(1)}, v^{(2)}, v^{(3)}, v^{(4)}]$ is a basis of $\mathbb{R}^4$.

Chapter 5
Eigenvalues and Eigenvectors

In this chapter we introduce the important notions of eigenvalue and eigenvector of a linear map $f : V \to V$ on a vector space V of finite dimension. Since the case where V is a $\mathbb{C}$-vector space is somewhat simpler, we first treat this case.

5.1 Eigenvalues and Eigenvectors of $\mathbb{C}$-Linear Maps

Assume that V is a $\mathbb{C}$-vector space and f a $\mathbb{C}$-linear map. We restrict ourselves to the case where V is of finite dimension. If not stated otherwise, we assume that $\dim(V) = n \in \mathbb{N}$.

Definition 5.1.1 A complex number $\lambda \in \mathbb{C}$ is said to be an *eigenvalue* of f if there exists $w \in V \setminus \{0\}$ such that

$$f(w) = \lambda w.$$

The vector w is called an *eigenvector* of f for the eigenvalue λ. Geometrically, it means that in the direction w, f is a dilation by the complex number λ.

A complex number $\lambda \in \mathbb{C}$ is said to be an *eigenvalue of the matrix $A \in \mathbb{C}^{n \times n}$*, if λ is an eigenvalue of the linear map $f_A : \mathbb{C}^n \to \mathbb{C}^n$. An eigenvector $w \in \mathbb{C}^n \setminus \{0\}$ of f_A for λ is also referred to as an *eigenvector* of A for the eigenvalue λ.

Example Let $V = \mathbb{C}^n$ and let $f : \mathbb{C}^n \to \mathbb{C}^n$ be the linear map

$$f(v) = \mathrm{diag}(\lambda_1, \ldots, \lambda_n)v$$

where $\lambda_1, \ldots, \lambda_n$ are given complex numbers. Then $f(e^{(j)}) = \lambda_j e^{(j)}$ for any $1 \le j \le n$ where $e^{(1)}, \ldots, e^{(n)}$ is the standard basis of $\mathbb{C}^n$. Since $e^{(j)} \ne 0$, $e^{(j)}$ is an eigenvector of f for the eigenvalue λ_j.

© The Author(s), under exclusive license to Springer Nature Switzerland AG 2023
M. Benz, T. Kappeler, *Linear Algebra for the Sciences*, La Matematica
per il 3+2 151, https://doi.org/10.1007/978-3-031-27220-2_5

How to compute the eigenvalues of the linear map $f : V \to V$ when V is of dimension n? Choose a basis $[v] = [v^{(1)}, \ldots, v^{(n)}]$ of V and consider the matrix representation $f_{[v] \to [v]} \in \mathbb{C}^{n \times n}$ of f with respect to $[v]$. Assume that $w \in V \setminus \{0\}$ is an eigenvector of f for the eigenvalue λ. Then w can be uniquely represented as a linear combination of $v^{(1)}, \ldots, v^{(n)}$,

$$w = \sum_{j=1}^{n} x_j v^{(j)}.$$

Then $x := (x_1, \ldots, x_n) \in \mathbb{C}^n \setminus \{0\}$ and $f(w) = \lambda w$ implies that

$$f_{[v] \to [v]} x = \lambda x \qquad \text{or} \qquad \left(f_{[v] \to [v]} - \lambda \, \mathrm{Id}_{n \times n} \right) x = 0.$$

It means that $f_{[v] \to [v]} - \lambda \, \mathrm{Id}_{n \times n}$ is not regular and hence

$$\det(f_{[v] \to [v]} - \lambda \, \mathrm{Id}_{n \times n}) = 0.$$

We have therefore obtained the following

Theorem 5.1.2 *Assume that V is a $\mathbb{C}$-vector space of dimension n with basis $[v] = [v^{(1)}, \ldots, v^{(n)}]$ and $f : V \to V$ is $\mathbb{C}$-linear. Then $\lambda \in \mathbb{C}$ is an eigenvalue of f if and only if*

$$\det(f_{[v] \to [v]} - \lambda \, \mathrm{Id}_{n \times n}) = 0.$$

Let us now investigate the function

$$\mathbb{C} \to \mathbb{C}, \quad z \mapsto \det(f_{[v] \to [v]} - z \, \mathrm{Id}_{n \times n}).$$

Consider first the case where

$$f_{[v] \to [v]} = (a_{ij})_{1 \le i, j \le n}$$

is a diagonal matrix, i.e., $a_{ij} = 0$ for $i \neq j$. Then

$$f_{[v] \to [v]} - z \, \mathrm{Id}_{n \times n} = \mathrm{diag}(a_{11} - z, \cdots, a_{nn} - z)$$

and hence $\det(f_{[v] \to [v]} - z \, \mathrm{Id}_{n \times n})$ is given by

$$(a_{11} - z) \cdots (a_{nn} - z) = (-1)^n z^n + (-1)^{n-1}(a_{11} + \cdots + a_{nn}) z^{n-1} + \ldots + \det(f_{[v] \to [v]}),$$

which is a polynomial of degree n in z. Actually this holds in general. As an illustration, consider the case when $n = 2$,

$$\det(f_{[v]\to[v]} - z\,\mathrm{Id}_{2\times2}) = \det\begin{pmatrix} a_{11} - z & a_{12} \\ a_{21} & a_{22} - z \end{pmatrix} = z^2 - (a_{11} + a_{22})z + \det(A).$$

More generally, one has the following

Theorem 5.1.3 *Assume that V is a $\mathbb{C}$-vector space of dimension n with basis $[v] = [v^{(1)}, \ldots, v^{(n)}]$ and $f : V \to V$ is $\mathbb{C}$-linear. Then the following holds:*

(i) $\chi(z) := \det(f_{[v]\to[v]} - z\,\mathrm{Id}_{n\times n})$ *is a polynomial in z of degree n of the form*

$$\chi(z) = p_n z^n + \cdots + p_1 z + p_0 \quad \text{with} \quad p_n := (-1)^n,\ p_0 := \det(f_{[v]\to[v]}),$$

and $p_j \in \mathbb{C}, 0 \le j < n$.
(ii) Every root of χ is an eigenvalue of f.

Remark The polynomial $\chi(z)$ is independent of the choice of the basis $[v]$. This can be verified as follows: Assume that $[w] = [w^{(1)}, \ldots, w^{(n)}]$ is another basis of V. Since

$$f_{[w]\to[w]} = \mathrm{Id}_{[v]\to[w]}\, f_{[v]\to[v]}\, \mathrm{Id}_{[w]\to[v]}, \qquad \mathrm{Id}_{[v]\to[w]} = (\mathrm{Id}_{[w]\to[v]})^{-1},$$

it follows that $\chi(z) := \det(f_{[w]\to[w]} - z\,\mathrm{Id}_{n\times n})$ can be computed as

$$\begin{aligned}
\chi(z) &= \det((\mathrm{Id}_{[w]\to[v]})^{-1} f_{[v]\to[v]}\, \mathrm{Id}_{[w]\to[v]} - z(\mathrm{Id}_{[w]\to[v]})^{-1}\, \mathrm{Id}_{[w]\to[v]}) \\
&= \det((\mathrm{Id}_{[w]\to[v]})^{-1}(f_{[v]\to[v]} - z\,\mathrm{Id}_{n\times n})\, \mathrm{Id}_{[w]\to[v]}) \\
&= \det((\mathrm{Id}_{[w]\to[v]})^{-1})\det(f_{[v]\to[v]} - z\,\mathrm{Id}_{n\times n})\det(\mathrm{Id}_{[w]\to[v]}) \\
&= \det(f_{[v]\to[v]} - z\,\mathrm{Id}_{n\times n})
\end{aligned}$$

where we have used that the determinant is multiplicative and hence in particular

$$\det((\mathrm{Id}_{[w]\to[v]})^{-1}) = \big(\det(\mathrm{Id}_{[w]\to[v]})\big)^{-1}.$$

Since $\chi(z)$ depends only on f we denote it also by $\chi_f(z)$. It is referred to as the *characteristic polynomial* of f.

How many eigenvalues does f have? Or equivalently, how many roots does χ_f have? Since χ_f is a polynomial of degree n, Theorem 3.2.3 applies and we conclude that it has n complex roots, when counted with their multiplicities. The multiplicity of a root of χ_f is also referred to as the *algebraic multiplicity* of the corresponding eigenvalue of f.

We summarize our discussion as follows:

Theorem 5.1.4 *Assume that V is a $\mathbb{C}$-vector space of dimension n and $f : V \to V$ is linear. Then f has precisely n eigenvalues, when counted with their algebraic multiplicities.*

Definition 5.1.5 Assume that V is a $\mathbb{C}$-vector space of dimension n and $f : V \to V$ is linear. The collection of all the eigenvalues of f, listed with their algebraic multiplicities, is called the *spectrum* of f and denoted by $\text{spec}(f)$. For any eigenvalue λ of f, we denote by m_λ its algebraic multiplicity. The eigenvalue λ of f is said to be *simple* if $m_\lambda = 1$. The map f is said to have a *simple spectrum* if each eigenvalue of f is simple.

The spectrum of a matrix $A \in \mathbb{C}^{n \times n}$ is defined to be the spectrum of f_A and denoted by $\text{spec}(A)$. An eigenvalue λ of $A \in \mathbb{C}^{n \times n}$ is called simple if $m_\lambda = 1$. The spectrum of A is called simple if each eigenvalue of A is simple.

Examples

(i) Let $A = \text{diag}(4, 4, 4, 7, 3, 3) \in \mathbb{C}^{6 \times 6}$. Then $\chi_A(z) = (4 - z)^3(7 - z)(3 - z)^2$ and the spectrum of f is given by

$$\lambda_1 = \lambda_2 = \lambda_3 = 4, \qquad \lambda_4 = 7, \qquad \lambda_5 = \lambda_6 = 3.$$

One has $m_{\lambda_1} = 3$, $m_{\lambda_4} = 1$, and $m_{\lambda_5} = 2$. In particular, $\text{spec}(A)$ is not simple. Note that $m_{\lambda_1} + m_{\lambda_4} + m_{\lambda_5} = 6 = \dim(\mathbb{C}^6)$.

(ii) Assume that V is a $\mathbb{C}$-vector space of dimension n and f is the identity map, $f = \text{Id}$. Choose an arbitrary basis $[v]$ of V. Then $f_{[v] \to [v]} = \text{Id}_{n \times n}$ and hence

$$\det(f_{[v] \to [v]} - z\,\text{Id}_{n \times n}) = (1 - z)^n.$$

Hence Id has the eigenvalues $\lambda_j = 1$, $1 \le j \le n$, and $m_{\lambda_1} = n$.

Definition 5.1.6 Given any eigenvalue λ of a linear map $f : V \to V$, the set

$$E_\lambda \equiv E_\lambda(f) := \{v \in V \mid f(v) = \lambda v\}$$

is said to be the *eigenspace* of f for the eigenvalue λ. Note that

$$E_\lambda = \{0\} \cup \{v \in V \mid v \text{ eigenvector of } f \text{ for the eigenvalue } \lambda\}.$$

Lemma 5.1.7 *For any eigenvalue λ of f, $E_\lambda(f)$ is a $\mathbb{C}$-subspace of V.*

Remark Note that $E_\lambda(f)$ can be interpreted as the nullspace of the linear map $f - \lambda\,\text{Id}$, which implies that it is $\mathbb{C}$-subspace of V (cf. Definition 4.2.3).

Definition 5.1.8 Let V be a vector space of dimension n. For an eigenvalue λ of a linear map $f : V \to V$, $\dim(E_\lambda(f))$ is referred to as the *geometric multiplicity* of λ.

Theorem 5.1.9 *Assume that V is a $\mathbb{C}$-vector space of dimension n and $f: V \to V$ a linear map. Then the following holds:*

(i) For any eigenvalue $\lambda \in \mathbb{C}$ of f,

$$\dim(E_\lambda(f)) \leq m_\lambda, \quad m_\lambda \leq n$$

where m_λ denotes the algebraic multiplicity of λ.

(ii) If $\dim(E_\lambda(f)) = m_\lambda$ for any eigenvalue λ of f, then V admits a basis consisting of eigenvectors of f.

Definition 5.1.10 A $\mathbb{C}$-linear map $f: V \to V$ on a $\mathbb{C}$-vector space of dimension n is said to be *diagonalizable* if there exists a basis $[v] = [v^{(1)}, \ldots, v^{(n)}]$ of eigenvectors of f. In such a case

$$f_{[v] \to [v]} = \operatorname{diag}(\lambda_1, \ldots, \lambda_n)$$

where for any $1 \leq i \leq n$, $\lambda_i \in \mathbb{C}$ is the eigenvalue of f, corresponding to the eigenvector $v^{(i)}$.

Examples Let $V = \mathbb{C}^2$. We consider examples of linear maps $f: \mathbb{C}^2 \to \mathbb{C}^2$ of the form f_A with $A \in \mathbb{C}^{2 \times 2}$. Recall that $(f_A)_{[e] \to [e]} = A$. Hence

$$\chi_{f_A}(z) = \det(A - \lambda \operatorname{Id}_{2 \times 2}).$$

We refer to $\chi_{f_A}(z)$ also as the characteristic polynomial of A and for notational convenience denote it by $\chi_A(z)$. For each of the 2×2 matrices A in Item (i)–Item (iii) below, we determine its spectrum and the corresponding eigenspaces. Recall that the eigenvalues of A are defined to be the eigenvalues of f_A and the eigenvectors of A the ones of f_A. To make our computations fairly simple, the matrices A considered have real coefficients.

(i)

$$A = \begin{pmatrix} 2 & -1 \\ 1 & 2 \end{pmatrix}.$$

Step 1. Computation of the spectrum of A.

The characteristic polynomial of A is given by

$$\chi_A(z) = \det \begin{pmatrix} 2 - z & -1 \\ 1 & 2 - z \end{pmatrix} = (2 - z)(2 - z) + 1 = z^2 - 4z + 5.$$

The roots are $z_\pm = 2 \pm \frac{1}{2}\sqrt{16 - 20} = 2 \pm i$. Hence the eigenvalues of A and their algebraic multiplicities are given by

$$\lambda_1 = 2 + i, \quad m_{\lambda_1} = 1, \qquad \lambda_2 = 2 - i, \quad m_{\lambda_2} = 1.$$

In particular, the spectrum of A is simple.

Step 2. Computation of the eigenspaces of λ_1 and λ_2.

$E_{\lambda_1}(A) \equiv E_{\lambda_1}(f_A)$: an eigenvector $v = (v_1, v_2) \in \mathbb{C}^2$ of A for the eigenvalue λ_1 solves the following linear system

$$\begin{pmatrix} 2 - \lambda_1 & -1 \\ 1 & 2 - \lambda_1 \end{pmatrix} \begin{pmatrix} v_1 \\ v_2 \end{pmatrix} = \begin{pmatrix} (2 - \lambda_1)v_1 - v_2 \\ v_1 + (2 - \lambda_1)v_2 \end{pmatrix} = \begin{pmatrix} 0 \\ 0 \end{pmatrix}.$$

To find a nontrivial solution of the latter linear system, consider the first equation

$$(2 - \lambda_1)v_1 - v_2 = 0.$$

Since $2 - \lambda_1 = -i$, we may choose $v_1 = 1$ and get $v_2 = -i$. One verifies that the second equation is then verified as well,

$$v_1 + (2 - \lambda_1)v_2 = 1 + (-i)(-i) = 1 - 1 = 0.$$

Hence $v^{(1)} = (1, -i)$ is an eigenvector for λ_1. Since $m_{\lambda_1} = 1$ one has

$$E_{\lambda_1}(A) = \{\alpha v^{(1)} \mid \alpha \in \mathbb{C}\}, \qquad \dim(E_{\lambda_1}(A)) = 1.$$

$E_{\lambda_2}(A) \equiv E_{\lambda_2}(f_A)$: an eigenvector $v = (v_1, v_2) \in \mathbb{C}^2$ of A for the eigenvalue λ_2 solves the following linear system:

$$\begin{pmatrix} 2 - \lambda_2 & -1 \\ 1 & 2 - \lambda_2 \end{pmatrix} \begin{pmatrix} v_1 \\ v_2 \end{pmatrix} = \begin{pmatrix} (2 - \lambda_2)v_1 - v_2 \\ v_1 + (2 - \lambda_2)v_2 \end{pmatrix} = \begin{pmatrix} 0 \\ 0 \end{pmatrix}.$$

To find a nontrivial solution of the latter system, consider the first equation $(2 - \lambda_2)v_1 - v_2 = 0$. Since $2 - \lambda_2 = i$, we may again choose $v_1 = 1$ and get $v_2 = i$. One verifies that the second equation is then verified as well,

$$v_1 + (2 - \lambda_2)v_2 = 1 + i \cdot i = 0.$$

Hence $v^{(2)} = (1, i)$ is an eigenvector for the eigenvalue λ_2. Since $m_{\lambda_2} = 1$ one has

$$E_{\lambda_2}(A) = \{\alpha v^{(2)} \mid \alpha \in \mathbb{C}\}, \qquad \dim(E_{\lambda_2}(A)) = 1.$$

The vectors $v^{(1)}$, $v^{(2)}$ form a basis $[v]$ of $\mathbb{C}^2$. (Since they are eigenvectors of f_A corresponding to distinct eigenvalues, they are linearly independent.) One has

$$(f_A)_{[v] \to [v]} = \begin{pmatrix} \lambda_1 & 0 \\ 0 & \lambda_2 \end{pmatrix}.$$

Since $A = (f_A)_{[e] \to [e]}$ it follows that A is diagonalizable,

$$A = \text{Id}_{[v] \to [e]} (f_A)_{[v] \to [v]} \, \text{Id}_{[e] \to [v]}$$

where $\text{Id}_{[v] \to [e]} = \begin{pmatrix} v^{(1)} & v^{(2)} \end{pmatrix}$ is the 2×2 matrix with first column $v^{(1)}$ and second column $v^{(2)}$, and $\text{Id}_{[e] \to [v]} = (\text{Id}_{[v] \to [e]})^{-1}$.

(ii)

$$A = \begin{pmatrix} 1 & 1 \\ 0 & 2 \end{pmatrix}.$$

Step 1. Computation of the spectrum of A.

The characteristic polynomial of A is given by

$$\chi_A(z) = \det \begin{pmatrix} 1 - z & 1 \\ 0 & 2 - z \end{pmatrix} = (1 - z)(2 - z).$$

Hence

$$\lambda_1 = 1, \quad m_{\lambda_1} = 1, \quad \lambda_2 = 2, \quad m_{\lambda_2} = 1.$$

In particular, the spectrum of A is simple.

Step 2. Computation of the eigenspaces of λ_1 and λ_2.

$E_{\lambda_1}(A) \equiv E_{\lambda_1}(f_A)$: an eigenvector $v = (v_1, v_2) \in \mathbb{C}^2$ of A for the eigenvalue λ_1 solves the following linear system

$$\begin{pmatrix} 1 - 1 & 1 \\ 0 & 2 - 1 \end{pmatrix} \begin{pmatrix} v_1 \\ v_2 \end{pmatrix} = \begin{pmatrix} 0 \cdot v_1 + v_2 \\ 0 \cdot v_1 + v_2 \end{pmatrix} = \begin{pmatrix} 0 \\ 0 \end{pmatrix}.$$

To find a nontrivial solution of the latter linear system, consider the first equation $0 \cdot v_1 + v_2 = 0$. It has the solution $v_2 = 0$, $v_1 = 1$. Then the second equation $0 \cdot v_1 + v_2 = 0$ is also satisfied and hence $v^{(1)} = (1, 0)$ is an eigenvector for the eigenvalue λ_1. Since $m_{\lambda_1} = 1$ one has

$$E_{\lambda_1}(A) = \{\alpha(1, 0) \mid \alpha \in \mathbb{C}\}, \qquad \dim(E_{\lambda_1}(A)) = 1.$$

$E_{\lambda_2}(A) \equiv E_{\lambda_2}(f_A)$: an eigenvector $v = (v_1, v_2) \in \mathbb{C}^2$ of A for the eigenvalue λ_2 solves the following linear system

$$\begin{pmatrix} 1-2 & 1 \\ 0 & 2-2 \end{pmatrix} \begin{pmatrix} v_1 \\ v_2 \end{pmatrix} = \begin{pmatrix} -v_1 + v_2 \\ 0 \cdot v_1 + 0 \cdot v_2 \end{pmatrix} = \begin{pmatrix} 0 \\ 0 \end{pmatrix}.$$

To find a nontrivial solution of the latter linear system, it suffices to consider the first equation $-v_1 + v_2 = 0$. It has the solution $v_1 = 1$, $v_2 = 1$. The second equation $0 \cdot v_1 + 0 \cdot v_2$ is trivially satisfied and hence $v^{(2)} = (1, 1)$ is an eigenvector for the eigenvalue λ_2. Since $m_{\lambda_2} = 1$ one has

$$E_{\lambda_2}(A) = \{\alpha(1, 1) \mid \alpha \in \mathbb{C}\}, \qquad \dim(E_{\lambda_2}(A)) = 1.$$

We conclude that $[v] = [v^{(1)}, v^{(2)}]$ is a basis of $\mathbb{C}^2$ and

$$(f_A)_{[v] \to [v]} = \begin{pmatrix} 1 & 0 \\ 0 & 2 \end{pmatrix}.$$

Since $A = (f_A)_{[e] \to [e]}$ it follows that A is diagonalizable,

$$A = \mathrm{Id}_{[v] \to [e]} (f_A)_{[v] \to [v]} \, \mathrm{Id}_{[e] \to [v]}$$

where $\mathrm{Id}_{[v] \to [e]} = \begin{pmatrix} v^{(1)} & v^{(2)} \end{pmatrix}$ and $\mathrm{Id}_{[e] \to [v]} = (\mathrm{Id}_{[v] \to [e]})^{-1}$.

(iii)

$$A = \begin{pmatrix} 1 & 1 \\ 0 & 1 \end{pmatrix}.$$

Step 1. Computation of the spectrum of A.
The characteristic polynomial of A is given by

$$\chi_A(z) = \det(A - z \, \mathrm{Id}_{2 \times 2}) = \det \begin{pmatrix} 1-z & 1 \\ 0 & 1-z \end{pmatrix} = (1-z)^2.$$

Hence

$$\lambda_1 = 1, \qquad \lambda_2 = \lambda_1, \qquad m_{\lambda_1} = 2.$$

The spectrum of A is therefore not simple.
Step 2. Computation of the eigenspace of λ_1.
$E_{\lambda_1}(A) \equiv E_{\lambda_1}(f_A)$: an eigenvector $v = (v_1, v_2) \in \mathbb{C}^2$ of A for the eigenvalue λ_1 solves the following linear system

$$\begin{pmatrix} 0 & 1 \\ 0 & 0 \end{pmatrix} \begin{pmatrix} v_1 \\ v_2 \end{pmatrix} = \begin{pmatrix} 0 \cdot v_1 + v_2 \\ 0 \cdot v_1 + 0 \cdot v_2 \end{pmatrix} = \begin{pmatrix} 0 \\ 0 \end{pmatrix}.$$

The solutions of the latter system are of the form $(v_1, 0)$ with $v_1 \in \mathbb{C}$ arbitrary. Hence

$$E_{\lambda_1}(A) = \{\alpha(1, 0) \mid \alpha \in \mathbb{C}\}$$

is of dimension 1,

$$\dim(E_{\lambda_1}(A)) = 1 < 2 = m_{\lambda_1}.$$

As a consequence, there does not exist a basis of $\mathbb{C}^2$, consisting of eigenvectors of A. The matrix A *cannot* be diagonalized.

For special types of matrices, the eigenvalues are quite easy to compute. Let us discuss the following ones.

Upper Triangular Matrices A matrix $A \in \mathbb{C}^{n \times n}$ is said to be *upper triangular* if it is of the form

$$A = (a_{ij})_{1 \le i, j \le n}, \qquad a_{ij} = 0, \quad i > j.$$

Then the coefficients $a_{11}, \dots, a_{nn}$ of the diagonal of A are the eigenvalues of f_A, counted with their algebraic multiplicities. Indeed

$$\det(A - z \operatorname{Id}_{n \times n}) = (a_{11} - z) \cdots (a_{nn} - z).$$

The spectrum of an upper triangular matrix A can therefore be directly read from the diagonal of A.

Lower Triangular Matrices A matrix $A \in \mathbb{C}^{n \times n}$ is said to be *lower triangular* if it is of the form

$$A = (a_{ij})_{1 \le i, j \le n}, \qquad a_{ij} = 0, \quad i < j.$$

As in the case of upper triangular matrices, the spectrum of a lower triangular matrix $A \in \mathbb{C}^{n \times n}$ can be read of from the diagonal of A.

Invertible Matrices Assume that $A \in \mathbb{C}^{n \times n}$ is *invertible*. Then the eigenvalues $\lambda_1, \dots, \lambda_n$ of A (listed with their algebraic multiplicities) are in $\mathbb{C} \setminus \{0\}$ and $\lambda_1^{-1}, \dots, \lambda_n^{-1}$ are the eigenvalues of A^{-1}, again listed with their algebraic multiplicities. If $v^{(j)}$ is an eigenvector of A for λ_j, then $A^{-1} v^{(j)} = \lambda_j^{-1} v^{(j)}$, i.e., $v^{(j)}$ is also an eigenvector of A^{-1}, corresponding to λ_j^{-1}.

Transpose of a Matrix Given a matrix $A \in \mathbb{C}^{n \times n}$, the eigenvalues of its transpose A^T coincide with the eigenvalues of A. Indeed

$$\det(A^T - z \operatorname{Id}_{n \times n}) = \det([A - z \operatorname{Id}_{n \times n}]^T) = \det(A - z \operatorname{Id}_{n \times n}),$$

i.e., A and A^T have the same characteristic polynomial and hence the same spectrum.

Theorem 5.1.11 *Let $A \in \mathbb{C}^{n \times n}$. Then $f_A \colon \mathbb{C}^n \to \mathbb{C}^n$, $x \mapsto Ax$ has a basis of eigenvectors if and only if there is a regular matrix $S \in \mathbb{C}^{n \times n}$ such that $S^{-1} A S$ is a diagonal matrix. The elements of the diagonal of $S^{-1} A S$ are the eigenvalues of A.*

Remark If $[v] = [v^{(1)}, \ldots, v^{(n)}]$ is a basis of $\mathbb{C}^n$ consisting of eigenvectors of $f_A \colon \mathbb{C}^n \to \mathbb{C}^n$, then $S = \operatorname{Id}_{[v] \to [e]}$ is regular and $S^{-1} A S$ is a diagonal matrix. Conversely, let $S \in \mathbb{C}^{n \times n}$ be regular and $S^{-1} A S$ a diagonal matrix $B = \operatorname{diag}(\lambda_1, \ldots, \lambda_n)$. Then the columns $C_1(S), \ldots, C_n(S)$ of S form a basis of $\mathbb{C}^n$. Since $AS = SB$ it follows that for any $1 \leq j \leq n$

$$\lambda_j C_j(S) = \lambda_j S e^{(j)} = S(\lambda_j e^{(j)}) = SB(e^{(j)}) = AS(e^{(j)}) = AC_j(S).$$

Hence $C_j(S)$ is an eigenvector of A for the eigenvalue λ_j.

Theorem 5.1.12 *Assume that $f \colon V \to V$ is a $\mathbb{C}$-linear map and V a $\mathbb{C}$-vector space of dimension n. If f has simple spectrum, i.e., has n distinct eigenvalues $\lambda_1, \ldots, \lambda_n \in \mathbb{C}$, then f is diagonalizable.*

Example Computation of the spectrum and the eigenspaces of

$$A = \begin{pmatrix} 2 & 1 & 0 \\ 0 & 1 & -1 \\ 0 & 2 & 4 \end{pmatrix} \in \mathbb{C}^{3 \times 3}.$$

Step 1. Computation of the eigenvalues of A.
    The characteristic polynomial of A is given by

$$\det(A - z \operatorname{Id}_{3 \times 3}) = (2 - z)\big((1 - z)(4 - z) + 2\big).$$

Then $\lambda_1 = 2$ and λ_2, λ_3 are the roots of

$$(1 - z)(4 - z) + 2 = z^2 - 5z + 6 = (z - 2)(z - 3),$$

i.e., $\lambda_2 = 2$ and $\lambda_3 = 3$. Note that

$$m_{\lambda_1} = 2, \qquad m_{\lambda_3} = 1.$$

In particular, the spectrum of A is not simple.

Step 2. Computation of the eigenspaces of λ_1 and λ_3.

$E_{\lambda_3}(A) \equiv E_{\lambda_3}(f_A)$: an eigenvector $v = (v_1, v_2, v_3) \in \mathbb{C}^3 \setminus \{0\}$ of A for the eigenvalue λ_3 solves the following linear system

$$\begin{pmatrix} -1 & 1 & 0 \\ 0 & -2 & -1 \\ 0 & 2 & 1 \end{pmatrix} \begin{pmatrix} v_1 \\ v_2 \\ v_3 \end{pmatrix} = \begin{pmatrix} 0 \\ 0 \\ 0 \end{pmatrix}.$$

Gaussian elimination ($R_3 \rightsquigarrow R_2 + R_3$) leads to

$$\left[\begin{array}{ccc|c} -1 & 1 & 0 & 0 \\ 0 & -2 & -1 & 0 \\ 0 & 0 & 0 & 0 \end{array} \right].$$

Thus the solutions are given by the vectors $\alpha(-1, -1, 2), \alpha \in \mathbb{C}$,

$$E_{\lambda_3}(A) = \{\alpha(-1, -1, 2) \mid \alpha \in \mathbb{C}\}.$$

$E_{\lambda_1}(A) \equiv E_{\lambda_1}(f_A)$: an eigenvector $v = (v_1, v_2, v_3) \in \mathbb{C}^3 \setminus \{0\}$ of A for the eigenvalue λ_1 solves the following linear system

$$\begin{pmatrix} 0 & 1 & 0 \\ 0 & -1 & -1 \\ 0 & 2 & 2 \end{pmatrix} \begin{pmatrix} v_1 \\ v_2 \\ v_3 \end{pmatrix} = \begin{pmatrix} 0 \\ 0 \\ 0 \end{pmatrix}. \tag{5.1}$$

By Gaussian elimination ($R_3 \rightsquigarrow R_3 + 2R_2, R_2 \rightsquigarrow R_2 + R_1$)

$$\left[\begin{array}{ccc|c} 0 & 1 & 0 & 0 \\ 0 & 0 & -1 & 0 \\ 0 & 0 & 0 & 0 \end{array} \right].$$

Hence any solution of (5.1) is of the form $v = \alpha(1, 0, 0), \alpha \in \mathbb{C}$, and

$$E_{\lambda_1}(A) = \{\alpha(1, 0, 0) \mid \alpha \in \mathbb{C}\}.$$

Since

$$\dim(E_{\lambda_1}(A)) < m_{\lambda_1} = 2,$$

the matrix A cannot be diagonalized.

To finish this section, let us discuss two important classes of matrices. Recall that by Definition 4.3.19, a matrix $A = (a_{ij})_{1 \leq i, j \leq n} \in \mathbb{C}^{n \times n}$ is said to be *unitary* if $\overline{A}^{\mathrm{T}} A = \mathrm{Id}_{n \times n}$ where $\overline{A}^{\mathrm{T}}$ is the transpose of $\overline{A}$ and the matrix $\overline{A}$ is given by $(\overline{a}_{ij})_{1 \le i, j \le n}$. Note that A is unitary if and only if

$$\langle Av, Aw \rangle = \langle v, w \rangle, \quad v, w \in \mathbb{C}^n$$

where $\langle v, w \rangle = \sum_{j=1}^n v_j \overline{w}_j$ denotes the Euclidean inner product on $\mathbb{C}^n$.

Theorem 5.1.13 *For any unitary matrix $A \in \mathbb{C}^{n \times n}$, the following holds:*

(i) *Every eigenvalue λ of A satisfies $|\lambda| = 1$.*
(ii) *If $\lambda, \mu \in \mathbb{C}$ are distinct eigenvalues of A, $\lambda \ne \mu$, then for any eigenvectors v, w of A for λ, respectively μ, v and w are orthogonal, i.e., $\langle v, w \rangle = 0$.*
(iii) *There exists an orthonormal basis of eigenvectors of A. In particular, A is diagonalizable.*

Remark Let us briefly discuss Items (i) and (ii). Concerning Item (i), assume that $\lambda \in \mathbb{C}$ is an eigenvalue of A and $v \in \mathbb{C}^n \setminus \{0\}$ an eigenvector of A for λ. Then

$$\langle v, v \rangle = \langle Av, Av \rangle = \langle \lambda v, \lambda v \rangle = \lambda \overline{\lambda} \langle v, v \rangle.$$

Since $v \ne 0$ and hence $\langle v, v \rangle \ne 0$ it follows that $|\lambda|^2 = 1$. Regarding (ii), assume that λ, μ are distinct eigenvalues of A, $\lambda \ne \mu$, and $v, w \in \mathbb{C}^n \setminus \{0\}$ are eigenvectors of A for λ, respectively μ. Then

$$\lambda \langle v, w \rangle = \langle \lambda v, w \rangle = \langle Av, w \rangle = \langle v, \overline{A}^{\mathrm{T}} w \rangle.$$

Since $\overline{A}^{\mathrm{T}} = A^{-1}$ and $A^{-1} w = \mu^{-1} w$ with $\mu^{-1} = \overline{\mu}$ it then follows that

$$\lambda \langle v, w \rangle = \langle v, \overline{A}^{\mathrm{T}} w \rangle = \langle v, \overline{\mu} w \rangle = \mu \langle v, w \rangle$$

or $(\lambda - \mu) \langle v, w \rangle = 0$. By assumption $\lambda - \mu \ne 0$ and hence $\langle v, w \rangle = 0$.

Example Computation of the spectrum and the eigenspaces of

$$A = \begin{pmatrix} \cos \varphi & -\sin \varphi \\ \sin \varphi & \cos \varphi \end{pmatrix}, \quad \varphi \notin \{n\pi \mid n \in \mathbb{N}\}.$$

Since under the latter assumption $\sin \varphi \ne 0$, A is not a diagonal matrix.

Step 0. Clearly, $A = \overline{A}$ and one verifies in a straighforward way that the matrix A is unitary,

$$A \overline{A}^{\mathrm{T}} = A A^{\mathrm{T}} = \mathrm{Id}_{2 \times 2}.$$

Step 1. Computation of the spectrum of A.
   The characteristic polynomial of A is given by

$$\det\begin{pmatrix} \cos\varphi - z & -\sin\varphi \\ \sin\varphi & \cos\varphi - z \end{pmatrix} = (\cos\varphi - z)^2 + \sin^2\varphi = z^2 - 2z\cos\varphi + 1.$$

The eigenvalues of A and their multiplicities are given by

$$\begin{aligned} \lambda_1 &= \cos\varphi + i\sin\varphi = e^{i\varphi}, & m_{\lambda_1} &= 1, \\ \lambda_2 &= \cos\varphi - i\sin\varphi = e^{-i\varphi}, & m_{\lambda_2} &= 1. \end{aligned}$$

In particular, the spectrum of A is simple.

Step 2. Computation of the eigenspaces of λ_1 and λ_2.

$E_{\lambda_2}(A) \equiv E_{\lambda_2}(f_A)$: an eigenvector $v = (v_1, v_2) \in \mathbb{C}^2 \setminus \{(0,0)\}$ of A for the eigenvalue λ_2 solves the following linear system

$$\begin{pmatrix} i\sin\varphi & -\sin\varphi \\ \sin\varphi & i\sin\varphi \end{pmatrix}\begin{pmatrix} v_1 \\ v_2 \end{pmatrix} = \begin{pmatrix} i\sin\varphi \cdot v_1 - \sin\varphi \cdot v_2 \\ \sin\varphi \cdot v_1 + i\sin\varphi \cdot v_2 \end{pmatrix} = \begin{pmatrix} 0 \\ 0 \end{pmatrix}.$$

To find a nontrivial solution of the latter linear system, consider the first equation $i\sin\varphi \cdot v_1 - \sin\varphi \cdot v_2 = 0$. Since by assumption $\sin\varphi \neq 0$, we can divide by $\sin\varphi$ and choose $v_1 = 1$, $v_2 = i$. The second equation $\sin\varphi \cdot v_1 + i\sin\varphi \cdot v_2 = 0$ is then also satisfied and hence

$$E_{\lambda_2}(A) = \{\alpha(1, i) \mid \alpha \in \mathbb{C}\}.$$

$E_{\lambda_1}(A) \equiv E_{\lambda_1}(f_A)$: an eigenvector $v = (v_1, v_2) \in \mathbb{C}^2 \setminus \{0\}$ of A for the eigenvalue λ_1 solves the following linear system

$$\begin{pmatrix} -i\sin\varphi & -\sin\varphi \\ \sin\varphi & -i\sin\varphi \end{pmatrix}\begin{pmatrix} v_1 \\ v_2 \end{pmatrix} = \begin{pmatrix} -i\sin\varphi \cdot v_1 - \sin\varphi \cdot v_2 \\ \sin\varphi \cdot v_1 - i\sin\varphi \cdot v_2 \end{pmatrix} = \begin{pmatrix} 0 \\ 0 \end{pmatrix}.$$

To find a nontrivial solution of the latter linear system, consider the first equation $-i\sin\varphi \cdot v_1 - \sin\varphi \cdot v_2 = 0$. Since by assumption $\sin\varphi \neq 0$, we can divide by $\sin\varphi$ and choose $v_1 = 1$, $v_2 = -i$. The second equation $\sin\varphi \cdot v_1 - i\sin\varphi \cdot v_2 = 0$ is then also satisfied and hence

$$E_{\lambda_1}(A) = \{\alpha(1, -i) \mid \alpha \in \mathbb{C}\}.$$

Then $v^{(1)} = (1, -i)$, $v^{(2)} = (1, i)$ form a basis $[v]$ of eigenvectors of $\mathbb{C}^2$ with

$$(f_A)_{[v] \to [v]} = \operatorname{diag}(e^{i\varphi}, e^{-i\varphi})$$

and $A = \operatorname{Id}_{[v] \to [e]} \operatorname{diag}(e^{i\varphi}, e^{-i\varphi})(\operatorname{Id}_{[v] \to [e]})^{-1}$ can be computed as

$$A = \begin{pmatrix} 1 & 1 \\ -i & i \end{pmatrix}\begin{pmatrix} e^{i\varphi} & 0 \\ 0 & e^{-i\varphi} \end{pmatrix}\frac{1}{2i}\begin{pmatrix} i & -1 \\ i & 1 \end{pmatrix}.$$

The second class of matrices we want to discuss are the Hermitian matrices.

Definition 5.1.14 A matrix $A \in \mathbb{C}^{n \times n}$ is said to be *Hermitian* if $\overline{A}^T = A$ or, equivalently,

$$\langle Av, w \rangle = \langle v, Aw \rangle, \quad v, w \in \mathbb{C}^n$$

where $\langle \cdot, \cdot \rangle$ denotes the Euclidean inner product on $\mathbb{C}^n$.

Theorem 5.1.15 *For any Hermitian matrix $A \in \mathbb{C}^{n \times n}$, the following holds:*

(i) Every eigenvalue of A is real.
(ii) Any two eigenvectors $v, w \in \mathbb{C}^n \setminus \{0\}$ of A for distinct eigenvalues λ, μ of A, $\lambda \neq \mu$, are orthogonal, i.e., $\langle v, w \rangle = 0$.
(iii) There exists an orthonormal basis of $\mathbb{C}^n$ consisting of eigenvectors of A. In particular, A is diagonalizable.

Remark Let us briefly discuss Items (i) and (ii). Concerning (i), let λ be an eigenvalue of A and $v \in \mathbb{C}^n \setminus \{0\}$ a corresponding eigenvector, $Av = \lambda v$. Since

$$\lambda \langle v, v \rangle = \langle \lambda v, v \rangle = \langle Av, v \rangle = \langle v, Av \rangle = \langle v, \lambda v \rangle = \overline{\lambda} \langle v, v \rangle,$$

it then follows that $\lambda = \overline{\lambda}$. Regarding Item (ii), assume that λ, μ are distinct eigenvalues of A, $\lambda \neq \mu$, and $v, w \in \mathbb{C}^n \setminus \{0\}$ eigenvectors for λ, respectively μ. Then

$$\lambda \langle v, w \rangle = \langle \lambda v, w \rangle = \langle Av, w \rangle = \langle v, Aw \rangle = \langle v, \mu w \rangle = \overline{\mu} \langle v, w \rangle.$$

Since μ is real and by assumption $\lambda \neq \mu$, it then follows that $\langle v, w \rangle = 0$.

Problems

1. Let $A = \begin{pmatrix} -1 + 2\,\mathrm{i} & 1 + \mathrm{i} \\ 2 + 2\,\mathrm{i} & 2 - \mathrm{i} \end{pmatrix} \in \mathbb{C}^{2 \times 2}$.

   (i) Compute the eigenvalues of A.
   (ii) Compute the eigenspaces of the eigenvalues of A.
   (iii) Find a regular 2×2 matrix $S \in \mathbb{C}^{2 \times 2}$ so that $S^{-1}AS$ is diagonal.

2. Let $A = \begin{pmatrix} 1 & -1 \\ 2 & -1 \end{pmatrix}$, viewed as an element in $\mathbb{C}^{2 \times 2}$.

   (i) Compute the eigenvalues of A.
   (ii) Compute the eigenspaces (in $\mathbb{C}^2$) of the eigenvalues of A.
   (iii) Find a regular 2×2 matrix $S \in \mathbb{C}^{2 \times 2}$ so that $S^{-1}AS$ is diagonal.

3. Let $A = \begin{pmatrix} 1/2 + i/2 & 1/2 + i/2 \\ -1/2 - i/2 & 1/2 + i/2 \end{pmatrix} \in \mathbb{C}^{2 \times 2}$.

   (i) Verify that A is unitary.
   (ii) Compute the spectrum of A.
   (iii) Find an orthonormal basis of $\mathbb{C}^2$, consisting of eigenvectors of A.

4. Let $A = \begin{pmatrix} 2 & i & 1 \\ -i & 2 & -i \\ 1 & i & 2 \end{pmatrix} \in \mathbb{C}^{3 \times 3}$.

   (i) Verify that A is Hermitian.
   (ii) Compute the spectrum of A.
   (iii) Find a unitary 3×3 matrix $S \in \mathbb{C}^{3 \times 3}$ so that $S^{-1} A S$ is diagonal.

5. Decide whether the following assertions are true or false and justify your answers.

   (i) Assume that $A, B \in \mathbb{C}^{2 \times 2}$ and that $\lambda \in \mathbb{C}$ is an eigenvalue of A and $\mu \in \mathbb{C}$ is an eigenvalue of B. Then $\lambda + \mu$ is an eigenvalue of $A + B$.
   (ii) For any $A \in \mathbb{C}^{2 \times 2}$, A and A^{T} have the same eigenspaces.

5.2 Eigenvalues and Eigenvectors of $\mathbb{R}$-Linear Maps

The aim of this section is to discuss issues in spectral theory, arising when one considers $\mathbb{R}$-linear maps on a $\mathbb{R}$-vector space V. To simplify the exposition we limit ourselves to the case where $V = \mathbb{R}^n$. Note that any $\mathbb{R}$-linear map $f : \mathbb{R}^n \to \mathbb{R}^n$ is of the form $f_A : \mathbb{R}^n \to \mathbb{R}^n$, $x \to Ax$ where $A = f_{[e] \to [e]} \in \mathbb{R}^{n \times n}$ and that we can extend the $\mathbb{R}$-linear map f_A to a $\mathbb{C}$-linear map on $\mathbb{C}^n$, $f_A : \mathbb{C}^n \to \mathbb{C}^n$, $x \mapsto Ax$ and then apply the results obtained in Sect. 5.1. It turns out that square matrices with real coefficients have special features and new questions arise which we now would like to discuss in some detail. For $A \in \mathbb{R}^{n \times n}$, consider the characteristic polynomial

$$\chi_A(z) = \det(A - z \, \mathrm{Id}_{n \times n}).$$

It is not difficult to see that the polynomial $\chi_A(z)$ has real coefficients. However, note that in general this does not imply that the roots of χ_A are real numbers.

Lemma 5.2.1 *Assume that $A \in \mathbb{R}^{n \times n}$ and $v \in \mathbb{C}^n \setminus \{0\}$ is an eigenvector of A for the eigenvalue $\lambda \in \mathbb{C}$. Then $\bar{\lambda}$ is an eigenvalue of A as well and $\bar{v}$ is a corresponding eigenvector.*

Indeed, since $\overline{A} = A$ and $\overline{Av} = \overline{A}\overline{v}$, one has $A\bar{v} = \overline{Av} = \overline{\lambda v} = \bar{\lambda}\bar{v}$.

Remark Actually one can show that for any eigenvalue $\lambda \in \mathbb{C}$ of $A \in \mathbb{R}^{n \times n}$,

$$m_\lambda = m_{\overline{\lambda}}, \qquad \dim(E_\lambda(A)) = \dim(E_{\overline{\lambda}}(A)),$$

i.e., λ and $\overline{\lambda}$ have the same algebraic and geometric multiplicities. For notational convenience, we write $E_\lambda(A)$ for $E_\lambda(f_A)$.

Lemma 5.2.1 can be applied as follows: assume that $\lambda \in \mathbb{C} \setminus \mathbb{R}$ is an eigenvalue of A with eigenvector $v = (v_j)_{1 \leq j \leq n} \in \mathbb{C}^n$. Then

$$Av = \lambda v, \qquad A\overline{v} = \overline{\lambda}\overline{v}.$$

Note that $v + \overline{v} = \big(2\operatorname{Re}(v_j)\big)_{1 \leq j \leq n} \in \mathbb{R}^n$ and $\mathrm{i}(\overline{v} - v) = \big(2\operatorname{Im}(v_j)\big)_{1 \leq j \leq n} \in \mathbb{R}^n$. Hence with $\lambda_1 = \operatorname{Re}(\lambda)$, $\lambda_2 = \operatorname{Im}(\lambda)$, one has

$$A(v + \overline{v}) = (\lambda_1 + \mathrm{i}\lambda_2)v + (\lambda_1 - \mathrm{i}\lambda_2)\overline{v} = \lambda_1(v + \overline{v}) - \mathrm{i}\lambda_2(\overline{v} - v)$$

and similarly,

$$A\big(\mathrm{i}(\overline{v}-v)\big) = \mathrm{i}\,A\overline{v} - \mathrm{i}\,Av = \mathrm{i}(\lambda_1 - \mathrm{i}\lambda_2)\overline{v} - \mathrm{i}(\lambda_1 + \mathrm{i}\lambda_2)v = \lambda_1\,\mathrm{i}(\overline{v}-v) + \lambda_2(v + \overline{v}).$$

In the case where $v + \overline{v}$ and $\mathrm{i}(\overline{v} - v)$ are linearly independent vectors in $\mathbb{R}^n$, we consider the two dimensional subspace $W := \{\alpha_1 v^{(1)} + \alpha_2 v^{(2)} \mid \alpha_1, \alpha_2 \in \mathbb{R}\}$ of $\mathbb{R}^n$, generated by

$$v^{(1)} := v + \overline{v}, \qquad v^{(2)} := \mathrm{i}(\overline{v} - v).$$

Then $Aw \in W$ for any $w \in W$ and

$$(f_A|W)_{[v^{(1)}, v^{(2)}] \to [v^{(1)}, v^{(2)}]} = \begin{pmatrix} \operatorname{Re}(\lambda) & \operatorname{Im}(\lambda) \\ -\operatorname{Im}(\lambda) & \operatorname{Re}(\lambda) \end{pmatrix}.$$

We have the following

Lemma 5.2.2 *Assume that $A \in \mathbb{R}^{n \times n}$ and $v \in \mathbb{C}^n$ is an eigenvector of A with eigenvalue $\lambda \in \mathbb{C} \setminus \mathbb{R}$. Then the following holds:*

(i) $v^{(1)} = v + \overline{v}$ *and* $v^{(2)} = \mathrm{i}(\overline{v} - v)$ *are linearly independent vectors in* $\mathbb{R}^n$.
(ii) *The two dimensional subspace W, spanned by $v^{(1)}$ and $v^{(2)}$, is invariant under f_A, i.e., for any $\alpha_1, \alpha_2 \in \mathbb{R}$, $A(\alpha_1 v^{(1)} + \alpha_2 v^{(2)}) \in W$.*
(iii) *The matrix representation of the restriction $f_A|_W : W \to W$ with respect to the basis $[v^{(1)}, v^{(2)}]$ is given by*

$$(f_A|W)_{[v^{(1)}, v^{(2)}] \to [v^{(1)}, v^{(2)}]} = \begin{pmatrix} \operatorname{Re}(\lambda) & \operatorname{Im}(\lambda) \\ -\operatorname{Im}(\lambda) & \operatorname{Re}(\lambda) \end{pmatrix} = |\lambda| \begin{pmatrix} \cos\theta & \sin\theta \\ -\sin\theta & \cos\theta \end{pmatrix}$$

where $|\lambda|e^{\mathrm{i}\theta}$ is the polar representation of the complex number λ. Since $\lambda \notin \mathbb{R}$ one has $\sin\theta \neq 0$.

Remark The above computations imply that for any real eigenvalue λ of $A \in \mathbb{R}^{n \times n}$, there exists an eigenvector $v \in \mathbb{R}^n$, $Av = \lambda v$. Indeed, assume that $v \in \mathbb{C}^n \setminus \{0\}$ is an eigenvector for the eigenvalue λ of A, $Av = \lambda v$. Then $A\bar{v} = \lambda \bar{v}$ and hence

$$A(v + \bar{v}) = \lambda(v + \bar{v}), \qquad A\big(\mathrm{i}(\bar{v} - v)\big) = \lambda\, \mathrm{i}(\bar{v} - v).$$

Since $0 \neq 2v = (v + \bar{v}) + \mathrm{i}\big(\mathrm{i}(\bar{v} - v)\big)$ and $v \neq 0$, either $v + \bar{v} \neq 0$ or $\mathrm{i}(\bar{v} - v) \neq 0$.

To finish this section, we consider two special classes of matrices in $\mathbb{R}^{n \times n}$.

Orthogonal Matrices   A matrix $A \in \mathbb{R}^{n \times n}$ is said to be *orthogonal* if $A^{\mathrm{T}} A = \mathrm{Id}_{n \times n}$. Equivalently, it means that

$$\langle Ax, Ay \rangle = \langle x, y \rangle, \quad x, y \in \mathbb{R}^n$$

where $\langle \cdot, \cdot \rangle$ denotes the Euclidean scalar product on $\mathbb{R}^n$, $\langle x, y \rangle = \sum_{k=1}^{n} x_k y_k$.

Theorem 5.2.3 *For any orthogonal matrix $A \in \mathbb{R}^{n \times n}$, the following holds:*

(i) *Any eigenvalue λ of A satisfies $|\lambda| = 1$.*
(ii) *If $v, w \in \mathbb{C}^n$ are eigenvectors for distinct eigenvalues λ, μ of A, then $\sum_{k=1}^{n} v_k \bar{w}_k = 0$.*
(iii) *$\mathbb{C}^n$ admits a basis of eigenvectors of A.*

Remark Since any orthogonal matrix is unitary, Theorem 5.1.13 applies.

Example The 2×2 matrix $R(\varphi) = \begin{pmatrix} \cos\varphi & \sin\varphi \\ -\sin\varphi & \cos\varphi \end{pmatrix}$, describing the rotation in $\mathbb{R}^2$ by the angle φ, is orthogonal. Its eigenvalues are $\lambda_1 = e^{\mathrm{i}\varphi}$, $\lambda_2 = e^{-\mathrm{i}\varphi}$.

Symmetric Matrices   A matrix $A \in \mathbb{R}^{n \times n}$ is said to be *symmetric* if $A^{\mathrm{T}} = A$. Equivalently, it means that

$$\langle Ax, y \rangle = \langle x, Ay \rangle, \quad x, y \in \mathbb{R}^n.$$

Theorem 5.2.4 *For any symmetric matrix $A \in \mathbb{R}^{n \times n}$, the following holds:*

(i) *Every eigenvalue of A is real and hence admits an eigenvector x in $\mathbb{R}^n$, $Ax = \lambda x$.*
(ii) *If λ, μ are eigenvalues of A with $\lambda \neq \mu$, then $\langle x, y \rangle = 0$ for any eigenvector $x \in \mathbb{R}^n$ [$y \in \mathbb{R}^n$] of A for the eigenvalue λ [μ].*
(iii) *$\mathbb{R}^n$ admits an orthonormal basis of eigenvectors of A.*

Remark To see that Item (i) of Theorem 5.2.4 holds, note that since a symmetric matrix $A \in \mathbb{R}^{n \times n}$ is Hermitian, it follows by Theorem 5.1.15 that every eigenvalue of A is real. By the Remark after Lemma 5.2.2, there exists $x \in \mathbb{R}^n \setminus \{0\}$ so that $Ax = \lambda x$. Concerning Item (ii), one argues as in the last Remark of the previous section.

In the case $A \in \mathbb{R}^{n \times n}$ is symmetric, one can define the geometric multiplicity of an eigenvalue λ in the following alternative way: consider

$$E_{\lambda, \mathbb{R}^n}(A) := \{x \in \mathbb{R}^n \mid Ax = \lambda x\} = E_\lambda(A) \cap \mathbb{R}^n.$$

It is straightforward to verify that $E_{\lambda, \mathbb{R}^n}(A)$ is a subspace of $\mathbb{R}^n$. By Theorem 5.2.4 we know that it is a non trivial subspace of $\mathbb{R}^n$. It can be proved that

$$\dim_{\mathbb{R}}(E_{\lambda, \mathbb{R}^n}(A)) = \dim_{\mathbb{C}}(E_\lambda(A))$$

and hence the geometric multiplicity of λ is given by $\dim_{\mathbb{R}}(E_{\lambda, \mathbb{R}^n}(A))$.

Theorem 5.2.5 *Any symmetric matrix $A \in \mathbb{R}^{n \times n}$ can be diagonalized in the following sense: there exists an orthonormal basis $[v] = [v^{(1)}, \ldots, v^{(n)}]$ of $\mathbb{R}^n$, consisting of eigenvectors of A, $Av^{(j)} = \lambda_j v^{(j)}$, $1 \leq j \leq n$, so that*

$$(f_A)_{[v] \to [v]} = S^{\mathsf{T}} A S$$

where S is the orthogonal matrix in $\mathbb{R}^{n \times n}$, given by $\mathrm{Id}_{[v] \to [e]}$. Hence the jth column of S is given by $v^{(j)}$. Furthermore, for any eigenvalue λ of A, the algebraic multiplicity m_λ coincides with the geometric multiplicity $\dim_{\mathbb{R}}(E_{\lambda, \mathbb{R}^n}(A))$.

Problems

1. Let $A = \begin{pmatrix} 1 & 4 \\ 2 & 3 \end{pmatrix} \in \mathbb{R}^{2 \times 2}$.

    (i) Compute the spectrum of A.
    (ii) Find eigenvectors $v^{(1)}, v^{(2)} \in \mathbb{R}^2$ of A, which form a basis of $\mathbb{R}^2$.
    (iii) Find a regular 2×2 matrix $S \in \mathbb{R}^{2 \times 2}$ so that $S^{-1}AS$ is diagonal.

2. Let $A = \begin{pmatrix} 2 & 1 & 0 \\ 0 & 1 & -1 \\ 0 & 2 & -1 \end{pmatrix}$, viewed as an element in $\mathbb{C}^{3 \times 3}$.

    (i) Compute the spectrum of A.
    (ii) For each eigenvalue of A, compute the eigenspace.
    (iii) Find a regular matrix $S \in \mathbb{C}^{3 \times 3}$ so that $S^{-1}AS$ is diagonal.

3. Determine for each of the following symmetric matrices

    (i) $A = \begin{pmatrix} 2 & 1 \\ 1 & 2 \end{pmatrix}$ (ii) $B = \begin{pmatrix} 2 & 1 & 1 \\ 1 & 3 & -2 \\ 1 & -2 & 3 \end{pmatrix}$

    its spectrum and find an orthogonal matrix, which diagonalizes it.

4. (i) Assume that A is a $n \times n$ matrix with real coefficients, $A \in \mathbb{R}^{n \times n}$, satisfying $A^2 = A$. Verify that each eigenvalue of A is either 0 or 1.
   (ii) Let $A \in \mathbb{R}^{n \times n}$. For any $a \in \mathbb{R}$, compute the eigenvalues and the eigenspaces of $A + a \operatorname{Id}_{n \times n}$ in terms of the eigenvalues and the eigenspaces of A.
5. Decide whether the following assertions are true or false and justify your answers.

   (i) Any matrix $A \in \mathbb{R}^{3 \times 3}$ has at least one real eigenvalue.
   (ii) There exists a symmetric matrix $A \in \mathbb{R}^{5 \times 5}$, which admits an eigenvalue, whose geometric multiplicity is 1, but its algebraic multiplicity is 2.

5.3 Quadratic Forms on $\mathbb{R}^n$

In this section we introduce the notion of quadratic forms on $\mathbb{R}^n$ and discuss applications to geometry.

Definition 5.3.1 We say that a function $Q \colon \mathbb{R}^n \to \mathbb{R}$ is a *quadratic form* on $\mathbb{R}^n$ if for any $x = (x_1, \ldots, x_n) \in \mathbb{R}^n$,

$$Q(x) = \sum_{1 \le i, j \le n} a_{ij} x_i x_j, \quad a_{ij} \in \mathbb{R}.$$

It means that Q is a polynomial homogeneous of degree 2 in the variables $x_1, \ldots, x_n$ with real coefficients a_{ij}, $1 \le i, j \le n$.

Since

$$Q(x) = \sum_{1 \le i, j \le n} \frac{a_{ij} + a_{ji}}{2} x_i x_j = \sum_{1 \le i, j \le n} \frac{1}{2} (A + A^{\mathrm{T}})_{ij} x_i x_j,$$

we can assume without loss of generality that A is symmetric, i.e., $A = A^{\mathrm{T}}$. We say that $Q = Q_A$ is the quadratic form associated to the symmetric matrix A. One can represent the quadratic form Q_A with the help of the Euclidean inner product on $\mathbb{R}^n$,

$$Q_A(x) = \langle Ax, x \rangle, \quad x \in \mathbb{R}^n.$$

Example Assume that $Q(x) = 3x_1^2 + 5x_1 x_2 + 8x_2^2$. Then $Q(x) = Q_A(x) = \langle Ax, x \rangle$ for any $x \in \mathbb{R}^2$ where

$$A = \begin{pmatrix} 3 & 5/2 \\ 5/2 & 8 \end{pmatrix} \in \mathbb{R}^{2 \times 2}.$$

Quadratic forms can be classified as follows:

Definition 5.3.2 A quadratic form $Q: \mathbb{R}^n \to \mathbb{R}$ is said to be *positive definite* (*positive semi-definite*) if $Q(x) > 0$ ($Q(x) \geq 0$) for any $x \in \mathbb{R}^n \setminus \{0\}$. It is said to be *negative definite* (*negative semi-definite*) if $Q(x) < 0$ ($Q(x) \leq 0$) for any $x \in \mathbb{R}^n \setminus \{0\}$. Finally, Q is said to be *indefinite*, if there exists $x, y \in \mathbb{R}^n$ such that $Q(x) > 0$ and $Q(y) < 0$.

Note that for a positive definite quadratic form Q, $x = 0$ is a global strict minimum of Q. If Q is indefinite, then $x = 0$ is a saddle point. To decide the type of a given quadratic form Q_A, it is useful to introduce the following classification of symmetric $n \times n$ matrices.

Definition 5.3.3 A symmetric $n \times n$ matrix $A \in \mathbb{R}^{n \times n}$ is said to be *positive definite* (*positive semi-definite*) if any eigenvalue λ of A satisfies $\lambda > 0$ ($\lambda \geq 0$). Similarly A is said to be *negative definite* (*negative semi-definite*) if any eigenvalue of A satisfies $\lambda < 0$ ($\lambda \leq 0$). Finally, A is said to be *indefinite* if it possesses eigenvalues λ, μ satisfying $\lambda < 0 < \mu$.

It turns out that the classifications of quadratic forms and symmetric matrices are closely related. For any quadratic form Q_A with $A \in \mathbb{R}^{n \times n}$ symmetric, it follows from Theorem 5.2.5 that there exists an orthogonal matrix $S \in \mathbb{R}^{n \times n}$ and real numbers $\lambda_1, \ldots, \lambda_n$ such that $A = SBS^T$ where $B = \text{diag}(\lambda_1, \ldots, \lambda_n)$. Here, $\lambda_1, \ldots, \lambda_n$ are the eigenvalues of A and the jth column of S is a normalized eigenvector of A for λ_j. As a consequence

$$Q_A(x) = \langle Ax, x \rangle = \langle SBS^T x, x \rangle = \langle BS^T x, S^T x \rangle.$$

With $y = S^T x$, $x \in \mathbb{R}^n$ one then can write

$$Q_A(x) = \sum_{j=1}^{n} \lambda_j y_j^2, \quad y = (y_1, \ldots, y_n) = S^T x.$$

It yields the following relationship between the classification of quadratic forms Q_A and the one of symmetric matrices A.

 (i) Q_A is positive definite (positive semi-definite) if and only if A is positive definite (positive semi-definite).
 (ii) Q_A is negative definite (negative semi-definite) if and only if A is negative definite (negative semi-definite).
 (iii) Q_A is indefinite if and only if A is indefinite.

To decide the type of a given symmetric $n \times n$ matrix A, it is not always necessary to compute the spectrum of A. The following result characterizes positive definite symmetric matrices.

Theorem 5.3.4 *Assume that $A \in \mathbb{R}^{n \times n}$ is symmetric and denote by $A^{(k)}$ the $k \times k$ matrix $A^{(k)} = (a_{ij})_{1 \leq i, j \leq k}$. Then A is positive definite if and only if $\det(A^{(k)}) > 0$ for any $1 \leq k \leq n$.*

Example Consider the symmetric 3×3 matrix

$$A = \begin{pmatrix} 1 & 1 & 0 \\ 1 & 2 & -1 \\ 0 & -1 & 2 \end{pmatrix}.$$

Then $A^{(1)} = 1$, $A^{(2)} = \begin{pmatrix} 1 & 1 \\ 1 & 2 \end{pmatrix}$ and $A^{(3)} = A$. One computes $\det A^{(1)} = 1 > 0$, $\det A^{(2)} = 1 > 0$, $\det A^{(3)} = 1 > 0$.

We finish this section with an application to geometry. Consider a function $f : \mathbb{R}^2 \to \mathbb{R}$ of the form

$$f(x_1, x_2) = ax_1^2 + bx_1x_2 + cx_2^2 + dx_1 + ex_2 + k \tag{5.2}$$

with real coefficients a, b, c, d, e, and k. It means that f is a polynomial of degree 2 in the variables x_1, x_2. The coefficients of the polynomial f are assumed to be real. The quadratic form

$$Q_A(x) = \langle x, Ax \rangle, \qquad A = \begin{pmatrix} a & b/2 \\ b/2 & c \end{pmatrix} \in \mathbb{R}^{2 \times 2}$$

is referred to as the *quadratic form associated to f*.

Definition 5.3.5 A conic section is a subset of $\mathbb{R}^2$ of the form

$$K_f = \{(x_1, x_2) \in \mathbb{R}^2 \mid f(x_1, x_2) = 0\}$$

where f is a polynomial of the form (5.2).

Examples

(i) For $f(x_1, x_2) = \frac{1}{4}x_1^2 + x_2^2 - 1$,

$$K_f = \{(x_1, x_2) \in \mathbb{R}^2 \mid \frac{1}{4}x_1^2 + x_2^2 = 1\},$$

i.e., K_f is an ellipse centered at $(0, 0)$ with half axes of length 2 and 1.

(ii) For $f(x_1, x_2) = x_1^2 - x_2$,

$$K_f = \{(x_1, x_2) \in \mathbb{R}^2 \mid x_2 = x_1^2\}$$

is a parabola with vertex $(0, 0)$.

(iii) For $f(x_1, x_2) = x_1x_2 - 1$,

$$K_f = \{(x_1, x_2) \in \mathbb{R}^2 \mid x_2 = 1/x_1\}$$

is a hyperbola with two branches.

(iv) For $f(x_1, x_2) = (x_1 + x_2)^2 + 1$, one has $K_f = \emptyset$.

(v) For $f(x_1, x_2) = (x_1 + x_2)^2 - 1$,

$$K_f = \{(x_1, x_2) \in \mathbb{R}^2 \mid x_1 + x_2 = 1 \text{ or } x_1 + x_2 = -1\},$$

i.e., K_f is the union of two straight lines.

(vi) For $f(x_1, x_2) = (x_1 + x_2)^2$,

$$K_f = \{(x_1, x_2) \in \mathbb{R}^2 \mid x_1 + x_2 = 0\},$$

i.e., K_f is one straight line.

(vii) For $f(x_1, x_2) = x_1^2 + x_2^2$, one has $K_f = \{(0, 0)\}$.

Remark A conic section K_f is said to be *degenerate* if K_f is empty or a point set or a straight line or a union of two straight lines. Otherwise it is called *nondegenerate*.

Theorem 5.3.6 *By means of translations and/or rotations in $\mathbb{R}^2$, any nondegenerate conic section can be brought into one of the following forms, referred to as canonical forms:*

(i) *Ellipse with center* $(0, 0)$,

$$\frac{x_1^2}{a_1^2} + \frac{x_2^2}{a_2^2} = 1, \quad a_1, a_2 \in \mathbb{R}_{>0}.$$

The associated symmetric matrix A equals $\mathrm{diag}(\frac{1}{a_1^2}, \frac{1}{a_2^2})$, *hence* $\det A > 0$.

(ii) *Hyperbola, centered at* $(0, 0)$, *with two branches,*

$$\frac{x_1^2}{a_1^2} - \frac{x_2^2}{a_2^2} = 1, \quad a_1, a_2 \in \mathbb{R}_{>0}.$$

The associated symmetric matrix A equals $\mathrm{diag}(\frac{1}{a_1^2}, -\frac{1}{a_2^2})$, *hence* $\det A < 0$.

(iii) *Parabola with vertex* $(0, 0)$,

$$x_1^2 = ax_2 \quad \text{or} \quad x_2^2 = ax_1, \quad a \in \mathbb{R} \setminus \{0\}.$$

The associated symmetric matrix A is given by $A = \mathrm{diag}(1, 0)$ *or* $A = \mathrm{diag}(0, 1)$. *Hence* $\det A = 0$.

Examples

(i) Consider the polynomial

$$f(x_1, x_2) = 9x_1^2 - 18x_1 + 4x_2^2 + 16x_2 - 11.$$

Bring K_f in canonical form, if possible.

In a first step we complete squares in the following expressions,

$$9x_1^2 - 18x_1 = 9(x_1^2 - 2x_1) = 9\left((x_1 - 1)^2 - 1\right) = 9(x_1 - 1)^2 - 9,$$

$$4x_2^2 + 16x_2 = 4(x_2^2 + 4x_2) = 4\left((x_2 + 2)^2 - 4\right) = 4(x_2 + 2)^2 - 16,$$

to get

$$f(x_1, x_2) = 9(x_1 - 1)^2 + 4(x_2 + 2)^2 - 36.$$

Then $f(x_1, x_2) = 0$ if and only if

$$\frac{9(x_1 - 1)^2}{36} + \frac{4(x_2 + 2)^2}{36} = 1$$

or

$$\frac{(x_1 - 1)^2}{2^2} + \frac{(x_2 + 2)^2}{3^2} = 1.$$

Hence by the translation $x \mapsto y := x - (1, -2)$ one gets

$$\frac{y_1^2}{a_1^2} + \frac{y_2^2}{a_2^2} = 1, \quad a_1 = 2, a_2 = 3,$$

i.e., K_f is an ellipse with center $(1, -2)$.

(ii) Consider

$$f(x_1, x_2) = 3x_1^2 + 2x_1x_2 + 3x_2^2 - 8.$$

Decide whether K_f is nondegenerate and if so, bring it into canonical form. First note that

$$f(x) = \langle x, Ax \rangle - 8, \quad A = \begin{pmatrix} 3 & 1 \\ 1 & 3 \end{pmatrix}.$$

The eigenvalues of A can be computed to be $\lambda_1 = 2$, $\lambda_2 = 4$. Hence $\det A = 2 \cdot 4 = 8 > 0$. This shows that K_f is an ellipse. To bring it in canonical form, we have to diagonalize A. One verifies that

$$v^{(1)} = \frac{1}{\sqrt{2}}(1, -1), \quad v^{(2)} = \frac{1}{\sqrt{2}}(1, 1),$$

are normalized eigenvectors of A for λ_1 and λ_2. Then $S := \mathrm{Id}_{[v] \to [e]}$ is the orthogonal 2×2 matrix

$$S = \frac{1}{\sqrt{2}} \begin{pmatrix} 1 & 1 \\ -1 & 1 \end{pmatrix}$$

and $A = S \operatorname{diag}(2, 4) S^T$. Hence

$$\langle x, Ax \rangle = \langle S^T x, \begin{pmatrix} 2 & 0 \\ 0 & 4 \end{pmatrix} S^T x \rangle, \quad x \in \mathbb{R}^2.$$

Thus we have shown that $K_f = \{ x = Sy \mid \frac{y_1^2}{4} + \frac{y_2^2}{2} = 1 \}$. When, expressed in the y coordinates, K_f is an ellipse with axes of length 2 and $\sqrt{2}$.

One can investigate the zero sets of polynomials also in higher dimensions. In the case $n = 3$, one considers polynomials $f : \mathbb{R}^3 \to \mathbb{R}$ of the form

$$f(x) = \langle x, Ax \rangle + \langle b, x \rangle + c$$

where $A \in \mathbb{R}^{3 \times 3}$ is symmetric, $b \in \mathbb{R}^3$ and $c \in \mathbb{R}$. The zero set $K_f = \{ x \in \mathbb{R}^3 \mid f(x) = 0 \}$ is called a quadric surface or a quadric. Types of non degenerate quadrics in $\mathbb{R}^3$ are the following surfaces:

(i) *ellipsoid.* For any $a_1, a_2, a_3 > 0$,

$$a_1^2 x_1^2 + a_2^2 x_2^2 + a_3^2 x_3^2 = 1.$$

(ii) *hyperboloid with one sheet.* For any $a_1, a_2, a_3 > 0$,

$$- a_1^2 x_1^2 + a_2^2 x_2^2 + a_3^2 x_3^2 = 1;$$

(iii) *hyperboloid with two sheets.* For any $a_1, a_2, a_3 > 0$,

$$- a_1^2 x_1^2 - a_2^2 x_2^2 + a_3^2 x_3^2 = 1;$$

(iv) *elliptic paraboloid.* For any $a_3 \neq 0$ and $a_1, a_2 > 0$,

$$a_3 x_3 = a_1^2 x_1^2 + a_2^2 x_2^2;$$

(v) *hyperbolic paraboloid.* For any $a_3 \neq 0$ and $a_1, a_2 > 0$,

$$a_3 x_3 = -a_1^2 x_1^2 + a_2^2 x_2^2.$$

Problems

1. (i) Decide, which of the following matrices $A \in \mathbb{R}^{n \times n}$ are symmetric and which are not.

$$\text{(a)} \begin{pmatrix} 3 & 2 & 1 \\ 2 & 1 & 3 \\ 1 & 3 & 2 \end{pmatrix} \qquad \text{(b)} \begin{pmatrix} 1 & 2 \\ -2 & 1 \end{pmatrix} \qquad \text{(c)} \begin{pmatrix} 1 & 2 & 3 \\ 2 & -1 & 3 \\ 3 & 4 & 1 \end{pmatrix}$$

   (ii) Determine for the following quadratic forms Q the symmetric matrices $A \in \mathbb{R}^{3 \times 3}$ so that $Q(x) = \langle x, Ax \rangle$ for any $x = (x_1, x_2, x_3) \in \mathbb{R}^3$.
   (a) $Q(x_1, x_2, x_3) = 2x_1^2 + 3x_2^2 + x_3^2 + x_1 x_2 - 2x_1 x_3 + 3x_2 x_3$.
   (b) $Q(x_1, x_2, x_3) = 8x_1 x_2 + 10x_1 x_3 + x_1^2 - x_3^2 + 5x_2^2 + 7x_2 x_3$.

2. Find a coordinate transformation of $\mathbb{R}^2$ (translation and/or rotation) so that the conic section $K_f = \{ f(x_1, x_2) = 0 \}$ is in canonical form where

$$f(x_1, x_2) = 3x_1^2 + 8x_1 x_2 - 3x_2^2 + 28.$$

3. Verify that the conic section $K_f = \{ f(x) = 0 \}$ is a parabola where f is given by

$$f(x) = x_1^2 + 2x_1 x_2 + x_2^2 + 3x_1 + x_2 - 1, \quad x = (x_1, x_2) \in \mathbb{R}^2.$$

4. (i) Determine symmetric matrices $A, B \in \mathbb{R}^{2 \times 2}$ so that A and B have the same eigenvalues, but not the same eigenspaces.
   (ii) Assume that $A, B \in \mathbb{C}^{2 \times 2}$ have the same eigenvalues and the same eigenspaces. Decide whether in such a case $A = B$.

5. Decide whether the following assertions are true or false and justify your answers.

   (i) For any eigenvalue of a symmetric matrix $A \in \mathbb{R}^{n \times n}$, its algebraic multiplicity equals its geometric multiplicity.
   (ii) The linear map $T : \mathbb{R}^2 \to \mathbb{R}^2$, $(x_1, x_2) \mapsto (x_2, x_1)$ is a rotation.
   (iii) The linear map $R : \mathbb{R}^2 \to \mathbb{R}^2$, $(x_1, x_2) \mapsto (-x_2, x_1)$ is orthogonal.

Chapter 6
Differential Equations

The aim of this chapter is to present a brief introduction to the theory of ordinary differential equations. The main focus is on systems of linear differential equations of first order in $\mathbb{R}^n$ with constant coefficients. They can be solved by the means of linear algebra. Hence this chapter is an application of what we have learned so far to a topic in the field of analysis.

In this chapter, we assume some basic knowledge of analysis such as the notion of a continuous or the one of a differentiable function $f : \mathbb{R} \to \mathbb{R}^n$, the fundamental theorem of calculus, the exponential function $t \mapsto e^t$ or the logarithm $t \mapsto \log t$ among others.

6.1 Introduction

Mathematical models describing the dynamics of systems, considered in the sciences, are often expressed in terms of differential equations, which relate the quantities describing the essential features of the system. Such models are introduced and analyzed with the aim of predicting how these systems evolve in time. Prominent examples are models of mechanical systems, such as the motion of a particle of mass m in the space $\mathbb{R}^3$, models for radioactive decay, population models etc.

Motion of a Particle in $\mathbb{R}^3$ The motion of a particle in $\mathbb{R}^3$ can be described by a curve in $\mathbb{R}^3$,

$$y : \mathbb{R} \to \mathbb{R}^3, \quad t \mapsto y(t)$$

M. Benz, T. Kappeler, *Linear Algebra for the Sciences*, La Matematica
per il 3+2 151, https://doi.org/10.1007/978-3-031-27220-2_6

where t (independent variable) denotes time and $y(t)$ (dependent variable) the position of the particle at time t. In case the particle has mass m and a force F acts on it, Newton's law says that

$$m \cdot y''(t) = F$$

where $y'(t) = \frac{d}{dt} y(t)$ is the velocity and $y''(t) = \frac{d^2}{dt^2} y(t)$ is the acceleration of the particle at time t. Here F is the vector resulting from all the forces acting on the particle at a given time. The equation $my''(t) = F$ is a *differential equation* in case the force F only depends on t, on the position $y(t)$ at time t, on the velocity $y'(t)$ at time $t, \ldots$, but not on the values $y(s)$ or one of its derivatives for $s \neq t$.

Radioactive Decay A substance, such as radium, decays by a stochastic process. It is assumed that the probability P of the decay of an atom of the substance in an infinitesimal time interval $[t, t + \Delta t]$ is proportional to Δt, i.e., there exists a constant λ, depending on the substance considered, so that

$$P\big(\text{decay of atom in } [t, t + \Delta t]\big) = \lambda \Delta t.$$

Denote by m is the mass of a single atom of the substance considered and by $N(t)$ the number of atoms at time t. It is then expected that

$$N(t + \Delta t) - N(t) = -N(t) \cdot P\big(\text{decay of atom in } [t, t + \Delta t]\big) = -N(t)\lambda \Delta t$$

and hence that the total mass of the substance $x(t) = mN(t)$ obeys the law

$$x(t + \Delta t) - x(t) = -x(t)\lambda \Delta t.$$

Taking the limit $\Delta t \to 0$, one is led to the differential equation

$$x'(t) = -\lambda x(t).$$

More generally, if one is given three substances X, Y, Z where X decays to Y and Y decays to Z, then the total masses $x(t)$, $y(t)$, $z(t)$ of the substances X, Y, Z at time t satisfy a system of differential equations of the form

$$\begin{cases} x'(t) = -\lambda x(t) \\ y'(t) = \lambda x(t) - \mu y(t) \\ z'(t) = \mu y(t). \end{cases}$$

Population Models Denote by $N(t)$ the number of individuals of a given species at time t. We assume that $N(t)$ is very large so that it can be approximated by a real valued function $p(t)$ which is continuously differentiable in t. Often it is assumed that the growth rate at time t,

$$\frac{p'(t)}{p(t)},$$

depends on the population $p(t)$ at time t, i.e., that it is modelled by a function $r(p)$, yielding the equation

$$\frac{p'(t)}{p(t)} = r(p(t)), \quad t \in \mathbb{R}.$$

In case the function r is a constant $r \equiv \alpha > 0$, one speaks of exponential growth,

$$p'(t) = \alpha \cdot p(t), \quad t \in \mathbb{R}$$

In applications, it is often observed that the growth rate becomes negative if the population exceeds a certain threshold p_0. In such a case, one frequently chooses for r an affine function,

$$r(p) = \beta \cdot (p_0 - p), \quad \beta > 0,$$

leading to the equation

$$p'(t) = \beta \cdot \big(p_0 - p(t)\big) \cdot p(t) = \beta p_0 p(t) - \beta p(t)^2, \quad t \in \mathbb{R}.$$

The equation is referred to as the *logistic equation*.

Now assume that

$$f : \mathbb{R} \times \mathbb{R} \times \mathbb{R} \to \mathbb{R}, \quad (t, x_0, x_1) \mapsto f(t, x_0, x_1)$$

is a function which is sufficiently often differentiable for the purposes considered. We then say that

$$f\big(t, x(t), x'(t)\big) = 0, \quad t \in \mathbb{R}, \tag{6.1}$$

is an *ordinary differential equation (ODE) of first order* and that a continuously differentiable function $x : \mathbb{R} \to \mathbb{R}$ satisfying $f(t, x(t), x'(t)) = 0$ for any $t \in \mathbb{R}$ (or alternatively for any t in some open nonempty interval) is a solution of (6.1). More generally, if

$$F : \mathbb{R} \times \mathbb{R}^n \times \mathbb{R}^n \to \mathbb{R}^m$$

is a sufficiently regular vector valued map, we say that

$$F\big(t, y(t), y'(t)\big) = 0, \quad t \in \mathbb{R}, \tag{6.2}$$

is a *system of ordinary differential equations of first order* and that the continuously differentiable vector valued function

$$y: \mathbb{R} \to \mathbb{R}^n,$$

satisfying $F(t, y(t), y'(t)) = 0$ for any $t \in \mathbb{R}$, is a solution of (6.2). We say that (6.2) is in explicit form if it can be written in the form

$$y'(t) = G(t, y(t)), \quad t \in \mathbb{R}. \tag{6.3}$$

Note that in this case, $m = n$ and G is a map $G: \mathbb{R} \times \mathbb{R}^n \to \mathbb{R}^n$.

More generally, a *system of ODEs of nth order in explicit form* is an equation of the form

$$y^{(n)}(t) = G(t, y(t), y'(t), \ldots, y^{(n-1)}(t)) \tag{6.4}$$

where $y^{(j)}(t)$ denotes the jth derivative of $y: \mathbb{R} \to \mathbb{R}^n$ and

$$G: \mathbb{R} \times \mathbb{R}^n \times \cdots \times \mathbb{R}^n \to \mathbb{R}^n$$

is sufficiently regular for the purposes considered. In the sequel, we will only consider ODEs of the form (6.4). We remark that systems of order n can always be converted into systems of first order, albeit of higher dimension. See the discussion in Sect. 6.3 concerning second order ODEs. Hence, we may restrict our attention to equations of the form (6.3).

The main questions concerning the Eq. (6.3) are the *existence* and the *uniqueness* of solutions and their properties. We remark that only in very rare cases, the solution can be represented in terms of an explicit formula. Hence investigations of qualitative properties of solutions play an important role. In particular, one is interested to know whether solutions exist for all time or whether some of them blow up in finite time. Of special interest is to determine the asymptotic behaviour of solutions as $t \to +\infty$ or as t approaches the blow up time. In addition, one wants to find out if there are special solutions such as stationary solutions or periodic solutions and investigate their stability.

Associated to (6.3) is the so called initial value problem (IVP)

$$\begin{cases} y'(t) = G(t, y(t)), \quad t \in \mathbb{R}, & (6.5) \\ y(0) = y^{(0)} & (6.6) \end{cases}$$

where $y^{(0)} \in \mathbb{R}^n$ is a given vector. There are quite general results saying that under appropriate conditions of the map G, (6.5) has a unique solution at least in some time interval containing $t = 0$. An important class of equations of the form (6.3) are the so called linear ODEs. We say that (6.3) is a *linear ODE* if G is of the form

$$G(t, y) = A(t)y + f(t),$$

or written componentwise,

$$G_j(t, y) = \sum_{k=1}^{n} a_{jk}(t) y_k + f_j(t), \quad 1 \le j \le n \tag{6.7}$$

where $A: \mathbb{R} \to \mathbb{R}^{n \times n}$, $t \mapsto A(t) = \left(a_{jk}(t) \right)_{1 \le j, k \le n}$ is a matrix valued map and $f: \mathbb{R} \to \mathbb{R}^n$, $t \mapsto f(t)$ a vector valued one. Note that the notion of linear ODE concerns the (dependent) variable y and not to the (independent) variable t. The real valued functions $t \mapsto a_{jk}(t)$ are referred to as the *coefficients* of (6.7). In case $f = 0$, (6.7) is said to be a *homogeneous ODE*, otherwise a *inhomogeneous* one. We say that a linear ODE of the form (6.7) has constant coefficients if all the coefficients $a_{jk}(t)$, $1 \le j, k \le n$, are independent of t. Note that f need not to be constant in t. Linear ODEs of the form (6.3) with constant coefficients can be solved explicitly. Such ODEs will be discussed in Sect. 6.2, whereas ODEs of second order (with constant coefficients) will be studied in Sect. 6.3. We point out that the theory of ODEs is a large field within analysis and that in this chapter we only discuss a tiny, albeit important, part of it. We also mention that a field closely related to the field of ordinary differential equations is the field of *partial differential equations* (PDEs), including equations such as heat equations, transport equations, Schrödinger equations, wave equations, Maxwell's equations, and Einstein's equations. In contrast to ODEs, PDEs have more than one independent variable. In addition to the time variable, these might be space variables, or more generally, variables in a phase space. The field of partial differential equations is huge and currently a very active area of research.

6.2 Linear ODEs with Constant Coefficients of First Order

In this section we treat systems of $n \ge 1$ linear differential equations of first order with n unknowns, $y = (y_1, \ldots, y_n) \in \mathbb{R}^n$,

$$P y'(t) + Q y(t) = g(t), \quad t \in \mathbb{R} \tag{6.8}$$

where $y'(t) = \frac{d}{dt} y(t)$ denotes the derivative of y at time t, P, Q are matrices in $\mathbb{R}^{n \times n}$ with constant coefficients, and $g: \mathbb{R} \to \mathbb{R}^n$ is a continuous function. In applications, the variable t has often the meaning of time. We are looking for solutions $y: t \mapsto y(t)$ of (6.8) which are continuously differentiable. In the sequel we will always assume that P is invertible. Hence we might multiply the left and the right hand side of (6.8) by P^{-1} and are thus led to consider systems of differential equations of first order of the form

$$y'(t) = A y(t) + f(t), \quad t \in \mathbb{R}$$

where $A \in \mathbb{R}^{n \times n}$ has constant coefficients and $f \colon \mathbb{R} \to \mathbb{R}^n$ is assumed to be continuous. We first treat the case where f identically vanishes. The system

$$y'(t) = Ay(t), \quad t \in \mathbb{R}, \tag{6.9}$$

is referred to as a *homogeneous system* of linear ODEs with constant coefficients. Note that $t \mapsto y(t) = 0$ is always a solution of (6.9), referred to as the *trivial solution*. An important question is whether (6.9) has nontrivial solutions. Written componentwise, the system reads

$$
\begin{cases}
y'_1(t) = \displaystyle\sum_{j=1}^{n} a_{1j} y_j(t) \\[2mm]
\quad \vdots \\[2mm]
y'_n(t) = \displaystyle\sum_{j=1}^{n} a_{nj} y_j(t)
\end{cases}
$$

where $A = (a_{ij})_{1 \le i, j \le n}$.

Let us first consider the special case $n = 1$. Writing a for $A = (a_{11})$, Eq. (6.9) becomes $y'(t) = ay(t)$. We claim that $y(t) = ce^{at}$, $c \in \mathbb{R}$ arbitrary, is a solution. Indeed, by substituting $y(t) = ce^{at}$ into the equation $y'(t) = ay(t)$ and using that $\frac{d}{dt}(e^{at}) = ae^{at}$ one sees that this is the case.

How can one find a solution of $y'(t) = ay(t)$ without guessing? A possible way is to use the *method of separation of variables*. It consists in transforming the equation in such a way that the left hand side is an expression in y and its derivative only, whereas the right hand side does not involve y and its derivative at all. In the case at hand, we argue formally.

Divide $y'(t) = ay(t)$ by $y(t)$ to get

$$\frac{y'(t)}{y(t)} = a.$$

Since, formally, $\frac{y'(t)}{y(t)} = \frac{d}{dt}\log(y(t))$, one concludes that

$$\log(y(t)) - \log(y(0)) = \int_0^t \frac{d}{dt}\log(y(t))\, dt = \int_0^t a\, dt = at.$$

Using that $\log(y(t)) - \log(y(0)) = \log(y(t)/y(0))$, one gets $y(t) = y(0)e^{at}$ for any $t \in \mathbb{R}$. Hence for any initial value $y_0 \in \mathbb{R}$, $t \mapsto y_0 e^{at}$ is a solution of the *initial value problem*

$$
\begin{cases}
y'(t) = ay(t), \quad t \in \mathbb{R}, \\
y(0) = y_0
\end{cases}
\tag{IVP}
$$

It turns out that $y_0 e^{at}$ is the unique solution of (IVP). Indeed, assume that $z : \mathbb{R} \to \mathbb{R}$ is a continuously differentiable function such that (IVP) holds. Then consider $w(t) := e^{-at} z(t)$. By the product rule for differentiation,

$$w'(t) = -a e^{-at} z(t) + e^{-at} z'(t).$$

Since $z'(t) = az(t)$, it follows that $w'(t) = 0$ for all $t \in \mathbb{R}$. Hence $w(t)$ is constant in time. Since $w(0) = z(0) = y_0$, one has $y_0 = e^{-at} z(t)$ for any $t \in \mathbb{R}$, implying that $z(t) = y_0 e^{at}$. In summary we have seen that in the case $n = 1$, Eq. (6.9) admits a one parameter family of solutions and the initial value problem (IVP) has, for any initial value $y_0 \in \mathbb{R}$, a unique solution.

Do similar results hold in the case $n \geq 2$? First we would like to investigate if the exponential e^a can be defined when a is replaced by an arbitrary $n \times n$ matrix A. Recall that in (3.2), the exponential e^z of a complex number z is defined as a power series,

$$e^z = \sum_{n=0}^{\infty} \frac{z^n}{n!} = 1 + z + \frac{z^2}{2!} + \dots.$$

If we replace z by a $n \times n$ matrix A, then $A^2 = AA$, and inductively, for any $n \geq 1$, $A^{n+1} = AA^n$ is well defined. It can be shown that the series $\sum_{n=0}^{\infty} \frac{A^n}{n!}$ converges and defines a $n \times n$ matrix which is denoted by e^A,

$$e^A = \sum_{n=0}^{\infty} \frac{A^n}{n!}. \tag{6.10}$$

Examples

(i) For $A = \text{diag}(-1, 2)$, e^A can be computed as follows,

$$A^2 = \text{diag}((-1)^2, 2^2), \qquad A^3 = \text{diag}((-1)^3, 2^3), \qquad \dots,$$

hence

$$e^A = \sum_{n=0}^{\infty} \frac{A^n}{n!} = \sum_{n=0}^{\infty} \frac{\text{diag}((-1)^n, 2^n)}{n!}$$

$$= \begin{pmatrix} \sum_{n=0}^{\infty} \frac{(-1)^n}{n!} & 0 \\ 0 & \sum_{n=0}^{\infty} \frac{2^n}{n!} \end{pmatrix} = \begin{pmatrix} e^{-1} & 0 \\ 0 & e^2 \end{pmatrix}.$$

(ii) Let $S \in \mathbb{R}^{2 \times 2}$ be invertible and consider $B = S^{-1}AS$ where A is the diagonal matrix $A = \mathrm{diag}(-1, 2)$ of Item (i). Then B^n, $n \geq 2$, can be computed as follows:

$$B^2 = (S^{-1}AS)(S^{-1}AS) = S^{-1}A^2 S,$$

and, inductively, for any $n \geq 2$,

$$B^{n+1} = (S^{-1}AS)B^n = (S^{-1}AS)S^{-1}A^n S = S^{-1}A(SS^{-1})A^n S = S^{-1}A^{n+1}S.$$

Hence

$$e^B = \sum_{n=0}^{\infty} \frac{B^n}{n!} = \sum_{n=0}^{\infty} \frac{S^{-1}A^n S}{n!} = S^{-1}\Big(\sum_{n=0}^{\infty} \frac{A^n}{n!}\Big)S = S^{-1}e^A S.$$

By Item (i) it then follows that

$$e^B = S^{-1}\,\mathrm{diag}(e^{-1}, e^2)S.$$

Theorem 6.2.1 *For any $A \in \mathbb{R}^{n \times n}$, the map $\mathbb{R} \to \mathbb{R}^{n \times n}$, $t \mapsto e^{tA}$ is continuously differentiable and satisfies*

$$\begin{cases} \frac{\mathrm{d}}{\mathrm{d}t}e^{tA} = Ae^{tA}, & t \in \mathbb{R}, \\ e^{tA}|_{t=0} = \mathrm{Id}_{n \times n}. \end{cases}$$

Remark

(i) Let us comment on why Theorem 6.2.1 holds.
By the definition (6.10) of e^A, one has $e^{0 \cdot A} = \mathrm{Id}_{n \times n}$. Furthermore, by differentiating term by term, one gets at least formally

$$\frac{\mathrm{d}}{\mathrm{d}t}(e^{tA}) = \frac{\mathrm{d}}{\mathrm{d}t}\Big(\mathrm{Id}_{n \times n} + tA + \frac{t^2 A^2}{2!} + \dots\Big)$$

$$= A + \frac{2t A^2}{2!} + \frac{3t^2 A^3}{3!} \dots$$

$$= A\Big(\mathrm{Id}_{n \times n} + tA + \frac{t^2 A^2}{2!} + \dots\Big) = Ae^{tA}.$$

(ii) One can show that

$$e^{(t+s)A} = e^{tA}e^{sA}, \quad t, s \in \mathbb{R}.$$

Indeed for any given $s \in \mathbb{R}$, let $E(t) := e^{tA}e^{sA}$. Then $t \mapsto E(t)$ is continuously differentiable and satisfies

$$\begin{cases} E'(t) = AE(t), & t \in \mathbb{R}, \\ E(0) = e^{sA}. \end{cases} \tag{6.11}$$

On the other hand, Theorem 6.2.1 implies that

$$\frac{d}{dt}e^{(t+s)A} = Ae^{(t+s)A}, \qquad e^{(t+s)A}\big|_{t=0} = e^{sA}.$$

Furthermore, one can show that the solution of (6.11) is unique. This implies that $E(t) = e^{(t+s)A}$ for any $t \in \mathbb{R}$. Since $s \in \mathbb{R}$ is arbitrary, the claimed identity follows.

(iii) By Item (ii) applied for $s = -t$, one has

$$e^{tA}e^{-tA} = e^{(t-t)A} = \mathrm{Id}_{n \times n},$$

meaning that for any $t \in \mathbb{R}$, e^{tA} is invertible and its inverse is e^{-tA}.

Theorem 6.2.2 *For any $A \in \mathbb{R}^{n \times n}$ and $y^{(0)} \in \mathbb{R}^n$, the initial value problem*

$$\begin{cases} y'(t) = Ay(t), & t \in \mathbb{R}, \\ y(0) = y^{(0)}, \end{cases}$$

has a unique solution. It is given by $y(t) = e^{tA}y^{(0)}$. Hence the general solution of $y'(t) = Ay(t)$ is $e^{tA}v$, $v = (v_1, \ldots, v_n) \in \mathbb{R}^n$. If $t \mapsto u(t)$ and $t \mapsto v(t)$ are solutions of $y' = Ay$, so is $t \mapsto au(t) + bv(t)$ for any $a, b \in \mathbb{R}$ (superposition principle).

Remark

(i) We often write y_0 instead of $y^{(0)}$ for the initial value.

(ii) By Theorem 6.2.1, $y(t) = e^{tA}y^{(0)}$ satisfies

$$\frac{d}{dt}e^{tA}y^{(0)} = Ae^{tA}y^{(0)} = Ay(t), \quad t \in \mathbb{R}, \qquad y(0) = e^{0A}y^{(0)} = y^{(0)}.$$

To see that this is the only solution, we argue as in the case where $A \in \mathbb{R}^{n \times n}$ with $n = 1$, treated at the beginning of this section. Assume that $z: \mathbb{R} \to \mathbb{R}^n$ is another solution, i.e., $z'(t) = Az(t)$ and $z(0) = y^{(0)}$. Define $w(t) := e^{-tA}z(t)$ and note that

$$w(0) = e^{-0A}z(0) = \mathrm{Id}_{n \times n}\, y^{(0)} = y^{(0)}.$$

By the product rule for differentiation,

$$w'(t) = \frac{d}{dt}(e^{-tA})z(t) + e^{-tA}z'(t)$$

$$= -Ae^{-tA}z(t) + e^{-tA}Az(t)$$

$$= -Ae^{-tA}z(t) + Ae^{-tA}z(t) = 0$$

where we have used that $e^{-tA}A = Ae^{-tA}$ since by definition $e^{-tA} = \sum_{n=0}^{\infty} \frac{(-t)^n}{n!}A^n$. It then follows that $w(t)$ is a vector independent of t, i.e., $w(t) = w(0) = y^{(0)}$, implying that for any $t \in \mathbb{R}$,

$$y(t) = e^{tA}y^{(0)} = e^{tA}w(t) = e^{tA}e^{-tA}z(t) = z(t).$$

Examples

(i) Find the general solution of

$$\begin{cases} y_1' = 3y_1 + 4y_2 \\ y_2' = 3y_1 + 2y_2 \end{cases}$$

In matrix notation, the system reads $y'(t) = Ay(t)$ where

$$A = \begin{pmatrix} 3 & 4 \\ 3 & 2 \end{pmatrix} \in \mathbb{R}^{2\times2}.$$

The general solution is given by $y(t) = e^{tA}v$, $v = (v_1, v_2) \in \mathbb{R}^2$. To determine $y(t)$ in more explicit terms, we analyze e^{tA} as follows. The eigenvalues of A are $\lambda_1 = -1$, $\lambda_2 = 6$ with corresponding eigenvectors

$$v^{(1)} = (1, -1) \in \mathbb{R}^2, \qquad v^{(2)} = (4, 3) \in \mathbb{R}^2.$$

Note that $A^k v^{(1)} = \lambda_1^k v^{(1)}$ for any $k \geq 0$ and hence

$$e^{tA}v^{(1)} = \sum_{k=0}^{\infty} \frac{t^k}{k!}A^k v^{(1)} = \sum_{k=0}^{\infty} \frac{t^k}{k!}\lambda_1^k v^{(1)} = e^{t\lambda_1}v^{(1)}.$$

Similarly, one has $e^{tA}v^{(2)} = e^{t\lambda_2}v^{(2)}$. Since $v^{(1)}$ and $v^{(2)}$ are linearly independent, any vector $v \in \mathbb{R}^2$ can be uniquely represented as a linear combination

$$v = a_1 v^{(1)} + a_2 v^{(2)}.$$

Hence

$$e^{tA}v = e^{tA}(a_1 v^{(1)} + a_2 v^{(2)}) = a_1 e^{tA} v^{(1)} + a_2 e^{tA} v^{(2)}$$
$$= a_1 e^{t\lambda_1} v^{(1)} + a_2 e^{t\lambda_2} v^{(2)}$$
$$= (a_1 e^{-t} + 4a_2 e^{6t}, -a_1 e^{-t} + 3a_2 e^{6t}).$$

To solve the initial value problem of $y' = Ay$ with $y^{(0)} = (6, 1)$ we need to solve the linear system

$$\begin{cases} a_1 + 4a_2 = 6 \\ -a_1 + 3a_2 = 1 \end{cases}.$$

One obtains $a_1 = 2$ and $a_2 = 1$ and hence

$$y(t) = (2e^{-t} + 4e^{6t}, -2e^{-t} + 3e^{6t}).$$

The asymptotics of the solution $y(t)$ for $t \to \pm\infty$ can be described as follows:

$$y(t) \sim (2e^{-t}, -2e^{-t}) \quad t \to -\infty,$$
$$y(t) \sim (4e^{6t}, 3e^{6t}) \qquad t \to +\infty.$$

(ii) Find the general solution of

$$\begin{cases} y_1' = y_1 + y_2 \\ y_2' = -2y_1 + 3y_2 \end{cases}$$

In matrix notation, the system reads $y'(t) = Ay(t)$ where

$$A = \begin{pmatrix} 1 & 1 \\ -2 & 3 \end{pmatrix} \in \mathbb{R}^{2 \times 2}.$$

The general solution is given by $y(t) = e^{tA}v$, $v = (v_1, v_2) \in \mathbb{R}^2$. To determine the solutions $y(t)$ in more explicit terms, we analyze e^{tA} as follows. The eigenvalues of A can be computed as $\lambda_1 = 2 + i$, $\lambda_2 = \bar{\lambda}_1 = 2 - i$ with

$$v^{(1)} = (1, 1 + i) \in \mathbb{C}^2, \qquad v^{(2)} = \overline{v^{(1)}} = (1, 1 - i) \in \mathbb{C}^2,$$

being corresponding eigenvectors, which form a basis of $\mathbb{C}^2$. The general *complex* solution of $y' = Ay$ is then given by

$$y(t) = a_1 e^{\lambda_1 t} v^{(1)} + a_2 e^{\lambda_2 t} v^{(2)}, \quad a_1, a_2 \in \mathbb{C}.$$

How can we obtain the general real solution? Recall that by Euler's formula,

$$e^{\alpha+i\beta} = e^{\alpha}e^{i\beta} = e^{\alpha}(\cos\beta + i\sin\beta).$$

Hence

$$e^{\lambda_1 t} = e^{2t}(\cos t + i\sin t),$$

implying that

$$e^{\lambda_1 t}v^{(1)} = \left(e^{2t}\cos t + i e^{2t}\sin t, e^{2t}(\cos t - \sin t) + i e^{2t}(\cos t + \sin t)\right).$$

By the superposition principle, $\mathbb{C}$-linear combinations of solutions of $y'(t) = Ay(t)$ are again solutions. Since $e^{\lambda_2 t}v^{(2)} = \overline{e^{\lambda_1 t}v^{(1)}}$, one concludes that the following solutions are real,

$$\frac{1}{2}(e^{\lambda_1 t}v^{(1)} + e^{\lambda_2 t}v^{(2)}), \qquad \frac{1}{2i}(e^{\lambda_1 t}v^{(1)} - e^{\lambda_2 t}v^{(2)}).$$

They can be computed as

$$\left(e^{2t}\cos t, e^{2t}(\cos t - \sin t)\right), \qquad \left(e^{2t}\sin t, e^{2t}(\cos t + \sin t)\right).$$

Hence the general real solution is given by

$$a_1\left(e^{2t}\cos t, e^{2t}(\cos t - \sin t)\right) + a_2\left(e^{2t}\sin t, e^{2t}(\cos t + \sin t)\right), \quad a_1, a_2 \in \mathbb{R}.$$

(iii) Find the general solution of

$$\begin{cases} y_1' = y_1 + y_2 \\ y_2' = y_2. \end{cases}$$

In matrix notation, the system reads

$$y'(t) = Ay(t), \qquad A := \begin{pmatrix} 1 & 1 \\ 0 & 1 \end{pmatrix}.$$

The general solution is given by $e^{tA}v$ where $v = (v_1, v_2)$ is an arbitrary vector in $\mathbb{R}^2$. To determine the solutions in a more explicit form, we analyze e^{tA} further. Note that in this case A is not diagonalizable. To compute e^{tA} we use that for $S, T \in \mathbb{R}^{n \times n}$ with $ST = TS$, one has $e^{S+T} = e^S e^T$, an identity, which follows in a straightforward way from the definition (6.10) of the exponential function of square matrices. Since

$$A = \mathrm{Id}_{2\times2} + T, \qquad T = \begin{pmatrix} 0 & 1 \\ 0 & 0 \end{pmatrix}, \qquad \mathrm{Id}_{2\times2}\, T = T\, \mathrm{Id}_{2\times2},$$

it follows that $e^{tA} = e^{t\,\mathrm{Id}_{2\times2}} e^{tT}$. Clearly $e^{t\,\mathrm{Id}_{2\times2}} = e^t\, \mathrm{Id}_{2\times2}$ and

$$e^{tT} = \sum_{k=0}^{\infty} \frac{t^k}{k!} T^k = \mathrm{Id}_{2\times2} + tT,$$

since

$$T^2 = \begin{pmatrix} 0 & 1 \\ 0 & 0 \end{pmatrix} \begin{pmatrix} 0 & 1 \\ 0 & 0 \end{pmatrix} = \begin{pmatrix} 0 & 0 \\ 0 & 0 \end{pmatrix}.$$

Altogether

$$e^{tA} = e^t \begin{pmatrix} 1 & t \\ 0 & 1 \end{pmatrix},$$

implying that the general real solution is given by

$$y_1(t) = v_1 e^t + v_2 t e^t, \qquad y_2(t) = v_2 e^t$$

where $v_1, v_2 \in \mathbb{R}$ are arbitrary constants.

Let us now turn to the system

$$y'(t) = Ay(t) + f(t), \tag{6.12}$$

referred to as an *inhomogeneous system of linear ODEs of first order with constant coefficients*.

Theorem 6.2.3 *Assume that $A \in \mathbb{R}^{n\times n}$, $f : \mathbb{R} \to \mathbb{R}^n$ is continuous and $y_p : \mathbb{R} \to \mathbb{R}^n$ a given solution of (6.12), $y_p'(t) = Ay_p(t) + f(t)$, $t \in \mathbb{R}$. Then the following holds:*

(i) *For any solution $t \mapsto u(t)$ of the homogeneous system, $u' = Au$, $t \mapsto y_p(t) + u(t)$ is a solution of (6.12).*

(ii) *For any solution y of (6.12), $t \mapsto y(t) - y_p(t)$ is a solution of the homogeneous system $u' = Au$.*

Remark Items (i) and (ii) of Theorem 6.2.3 can be verified in a straigthforward way. Indeed, in the case of Item (i), by substituting $y_p + u$ into (6.12), one gets,

$$\frac{d}{dt}\big(y_p(t) + u(t)\big) = y_p'(t) + u'(t) = Ay_p(t) + f(t) + Au(t)$$

$$= A\big(y_p(t) + u(t)\big) + f(t)$$

and hence $t \mapsto y_p(t) + u(t)$ is a solution of (6.12). In the case of Item (ii), let $u(t) := y(t) - y_p(t)$. Then

$$u'(t) = y'(t) - y_p'(t) = Ay(t) + f(t) - Ay_p(t) - f(t)$$

$$= A\big(y(t) - y_p(t)\big) = Au(t).$$

As a consequence of Theorem 6.2.3, the general solution of (6.12) can be found as follows:

(i) Find a particular solution of (6.12).
(ii) Find the general solution of the homogeneous system $u' = Au$.

It thus remains to investigate how to find a particular solution of (6.12). For functions f of special type such as polynomials, trigonometric polynomials, or exponential functions, a particular solution can be found most efficiently by a suitable ansatz. We will discuss these cases at the end of the section.

First however, let us discuss a method, always applicable, which allows to construct the general solution of (6.12) (and hence also a particular solution) with the help of the general solution of the corresponding homogeneous system $u' = Au$. This method goes under the name of the *method of variation of the constants*. Its starting point is the ansatz

$$y(t) = e^{tA}w(t), \quad t \in \mathbb{R}.$$

In case $w(t)$ is a vector $v \in \mathbb{R}^n$ independent of t, $t \mapsto y(t)$ is a solution of the homogeneous system $u'(t) = Au(t)$. Hence the ansatz consists in replacing the constant v by a t-dependent unknown function $w(t)$, which explains the name of the method. Substituting the ansatz into the inhomogeneous equation one gets by the product rule of differentiation,

$$y'(t) = \frac{d}{dt}(e^{tA})w(t) + e^{tA}w'(t)$$

$$= Ae^{tA}w(t) + e^{tA}w'(t)$$

$$= Ay(t) + e^{tA}w'(t).$$

In case $t \mapsto y(t)$ is a solution of (6.12), it then follows that

$$Ay(t) + f(t) = Ay(t) + e^{tA}w'(t) \qquad \text{or} \qquad f(t) = e^{tA}w'(t).$$

Multiplying both sides of the latter identity by the matrix e^{-tA} one gets

$$w'(t) = e^{-tA} f(t).$$

Integrating in t, one obtains

$$w(t) - w(0) = \int_0^t w(s)\, ds = \int_0^t e^{-sA} f(s)\, ds.$$

Altogether we found

$$y(t) = e^{tA} w(t) = e^{tA} w(0) + e^{tA} \int_0^t e^{-sA} f(s)\, ds, \quad t \in \mathbb{R}.$$

Hence

$$t \mapsto y(t) = e^{tA} v + \int_0^t e^{(t-s)A} f(s)\, ds, \quad v \in \mathbb{R}^n, \tag{6.13}$$

is the general solution of (6.12). Note that $t \mapsto e^{tA} v$ is the general solution of the homogeneous system $u' = Au$, whereas $t \mapsto \int_0^t e^{(t-s)A} f(s)\, ds$ is a particular solution y_p of (6.12) with $y_p(0) = 0$.

Special Inhomogeneous Term f Assume that in the equation $y' = Ay + f$, the inhomogeneous term f is a solution of the homogeneous equation $u' = Au$, i.e., there exists $a \in \mathbb{R}$, so that $f(t) = e^{tA} a$ for any $t \in \mathbb{R}$. Substituting f into the integral $\int_0^t e^{(t-s)A} f(s)\, ds$ in (6.13) yields the particular solution $t \mapsto y_p(t)$ with $y_p(0) = 0$, given by

$$\int_0^t e^{(t-s)A} f(s)\, ds = \int_0^t e^{tA} e^{-sA} e^{sA} a\, ds = t e^{tA} a.$$

Alternatively, one verifies in a straightforward way that whenever f is a solution of the homogeneous system $u' = Au$, then $t \mapsto y_p(t) = tf(t)$ is a particular solution of $y' = Ay + f$.

Application Formula (6.13) can be used to solve the initial value problem (IVP)

$$\begin{cases} y' = Ay + f \\ y(0) = y^{(0)} \end{cases}$$

where $y^{(0)}$ is an arbitrary given vector in $\mathbb{R}^n$. The solution is given by

$$y(t) = e^{tA} y^{(0)} + \int_0^t e^{(t-s)A} f(s)\, ds, \quad t \in \mathbb{R}.$$

Example Find the general solution of

$$\begin{cases} y_1' = -y_2 \\ y_2' = y_1 + t \end{cases}.$$

In matrix notation, the system reads

$$y' = Ay + f, \qquad A = \begin{pmatrix} 0 & -1 \\ 1 & 0 \end{pmatrix}, \qquad f(t) = \begin{pmatrix} 0 \\ t \end{pmatrix}.$$

First we compute e^{tA}. The eigenvalues of A are $\lambda_1 = i$ and $\lambda_2 = -i$ and

$$v^{(1)} = \begin{pmatrix} i \\ 1 \end{pmatrix}, \qquad v^{(2)} = \begin{pmatrix} -i \\ 1 \end{pmatrix}$$

are corresponding eigenvectors. Hence

$$e^{tA} v^{(1)} = e^{it} v^{(1)}, \qquad e^{tA} v^{(2)} = e^{-it} v^{(2)}.$$

Using that $e^{\pm it} = \cos t \pm i \sin t$ (Euler's formula) and

$$\frac{v^{(1)} + v^{(2)}}{2} = \begin{pmatrix} 0 \\ 1 \end{pmatrix}, \qquad \frac{i(v^{(2)} - v^{(1)})}{2} = \begin{pmatrix} 1 \\ 0 \end{pmatrix},$$

it follows that

$$e^{tA} \begin{pmatrix} 0 \\ 1 \end{pmatrix} = \begin{pmatrix} -\sin t \\ \cos t \end{pmatrix}, \qquad e^{tA} \begin{pmatrix} 1 \\ 0 \end{pmatrix} = \begin{pmatrix} \cos t \\ \sin t \end{pmatrix},$$

implying that

$$(e^{tA})_{[e] \to [e]} = \begin{pmatrix} \cos t & -\sin t \\ \sin t & \cos t \end{pmatrix} = (\cos t)\, \mathrm{Id}_{2 \times 2} + (\sin t) A$$

where $[e] = [e^{(1)}, e^{(2)}]$ is the standard basis of $\mathbb{R}^2$. The general solution then reads

$$e^{tA} v + \int_0^t e^{(t-s)A} \begin{pmatrix} 0 \\ s \end{pmatrix} ds, \qquad t \in \mathbb{R}$$

where v is an arbitrary vector in $\mathbb{R}^2$. One has

$$e^{(t-s)A} \begin{pmatrix} 0 \\ s \end{pmatrix} = \cos(t - s) \begin{pmatrix} 0 \\ s \end{pmatrix} - \sin(t - s) \begin{pmatrix} -s \\ 0 \end{pmatrix}$$

and

$$\int_0^t -s \sin(t-s)\, ds = \left[-s \cos(t-s) - \sin(t-s) \right]_0^t = -t + \sin t,$$

$$\int_0^t s \cos(t-s)\, ds = \left[-s \sin(t-s) + \cos(t-s) \right]_0^t = 1 - \cos t.$$

Altogether we have

$$\int_0^t e^{(t-s)A} \begin{pmatrix} 0 \\ s \end{pmatrix} ds = \begin{pmatrix} -t + \sin t \\ 1 - \cos t \end{pmatrix}$$

and the general solution reads

$$v_1 \begin{pmatrix} \cos t \\ \sin t \end{pmatrix} + v_2 \begin{pmatrix} -\sin t \\ \cos t \end{pmatrix} + \begin{pmatrix} -t + \sin t \\ 1 - \cos t \end{pmatrix}.$$

As an aside, we remark that there is the following alternative way of computing the exponential e^{tA}. Note that

$$A = \begin{pmatrix} 0 & -1 \\ 1 & 0 \end{pmatrix}, \qquad A^2 = \begin{pmatrix} -1 & 0 \\ 0 & -1 \end{pmatrix}, \qquad A^3 = -A, \qquad A^4 = \mathrm{Id}_{2\times 2}, \ldots,$$

and thus

$$e^{tA} = \sum_{k=0}^{\infty} \frac{t^k}{k!} A^k = \sum_{k=0}^{\infty} \frac{t^{2k}}{(2k)!} A^{2k} + \sum_{k=0}^{\infty} \frac{t^{2k+1}}{(2k+1)!} A^{2k+1}$$

$$= \sum_{k=0}^{\infty} \frac{(-1)^k t^{2k}}{(2k)!} \mathrm{Id}_{2\times 2} + \sum_{k=0}^{\infty} \frac{(-1)^k t^{2k+1}}{(2k+1)!} A.$$

It can be shown that

$$\cos t = \sum_{k=0}^{\infty} \frac{(-1)^k t^{2k}}{(2k)!}, \qquad \sin t = \sum_{k=0}^{\infty} \frac{(-1)^k t^{2k+1}}{(2k+1)!},$$

implying that $e^{tA} = (\cos t)\,\mathrm{Id}_{2\times 2} + (\sin t) A$.

The example above shows that the computation of $\int_0^t e^{(t-s)A} f(s)\, ds$ can be quite involved. For special classes of functions $f: \mathbb{R} \to \mathbb{R}^n$ it is easier to get a particular solution by making an ansatz. We now discuss three classes of functions to illustrate this method.

Polynomials Assume that in the equation $y' = Ay + f$, each component of f is a polynomial in t of degree at most L. It means that

$$f(t) = \sum_{j=0}^{L} t^j f^{(j)}, \quad f^{(0)}, \dots, f^{(L)} \in \mathbb{R}^n.$$

We restrict ourselves to the case where the $n \times n$ matrix A is invertible and make an ansatz for a particular solution $t \mapsto y_p(t)$ of $y' = Ay + f$, assuming that each component of y_p is a polynomial in t of degree at most L, i.e., $y_p(t) = (q_1(t), \dots, q_n(t))$ where for any $1 \le k \le n$, $q_k(t)$ is a polynomial in t of degree at most L. When written in matrix notation, $y_p(t)$ is of the form

$$y_p(t) = \sum_{j=0}^{L} t^j w^{(j)}, \quad w^{(0)}, \dots, w^{(L)} \in \mathbb{R}^n.$$

Since

$$y_p'(t) = \sum_{j=1}^{L} j t^{j-1} w^{(j)} = \sum_{j=0}^{L-1} (j+1) t^j w^{(j+1)},$$

one obtains, upon substitution of y_p into the equation $y_p' = Ay_p + f$,

$$\sum_{j=0}^{L-1} (j+1) t^j w^{(j+1)} = \sum_{j=0}^{L} t^j A w^{(j)} + \sum_{j=0}^{L} t^j f^{(j)}.$$

By comparison of coefficients, one gets

$$(j+1) w^{(j+1)} = A w^{(j)} + f^{(j)}, \quad 0 \le j \le L-1, \qquad 0 = A w^{(L)} + f^{(L)}.$$

Since A is assumed to be invertible, we can solve this linear system recursively. Solving the equation $0 = A w^{(L)} + f^{(L)}$ yields $w^{(L)} = -A^{-1} f^{(L)}$, which then allows to solve the remaining equations by setting for $j = L-1, \dots, j = 0$,

$$A w^{(j)} = -f^{(j)} + (j+1) w^{(j+1)} \qquad \text{or} \qquad w^{(j)} = A^{-1}\big((j+1) w^{(j+1)} - f^{(j)}\big).$$

To illustrate how to find a particular solution y_p in the case where the components of the inhomogenous term $f(t)$ are polynomials in t, let us go back to the example treated above,

$$y'(t) = \begin{pmatrix} 0 & -1 \\ 1 & 0 \end{pmatrix} y(t) + \begin{pmatrix} 0 \\ t \end{pmatrix}. \tag{6.14}$$

Note that each component of $f(t) = (0, t)$ is a polynomial in t of degree at most 1 and that the matrix $\begin{pmatrix} 0 & -1 \\ 1 & 0 \end{pmatrix}$ is invertible. Hence we make the ansatz $y_p(t) = \big(q_1(t), q_2(t)\big)$ where the components $q_1(t)$ and $q_2(t)$ are polynomials in t of degree at most 1,

$$q_1(t) = a_0 + a_1 t, \qquad q_2(t) = b_0 + b_1 t.$$

Upon substitution of the ansatz into Eq. (6.14) and using that $y_p'(t) = (a_1, b_1)$, one gets

$$\begin{pmatrix} a_1 \\ b_1 \end{pmatrix} = \begin{pmatrix} -b_0 - b_1 t \\ a_0 + a_1 t + t \end{pmatrix}, \qquad t \in \mathbb{R}.$$

Comparing coefficients yields

$$a_1 = -b_0, \qquad 0 = -b_1, \qquad b_1 = a_0, \qquad 0 = a_1 + 1,$$

hence $a_1 = -1$, $b_1 = 0$, $a_0 = 0$, and $b_0 = 1$, yielding $y_p(t) = (-t, 1)$.

To see that in the case where the components of f are polynomials of degree at most L, the ansatz for y_p, consisting in choosing the components of y_p to be polynomials of degree at most L, not always works, consider the following scalar valued ODE

$$y'(t) = f(t), \qquad f(t) = 4 + 6t.$$

The general solution can be found by integration, $y_p(t) = c_1 + 4t + 3t^2$. Clearly, for no choice of c_1, $y_p(t)$ will be a polynomial of degree at most one. Note that in this example, when written in "matrix form" $y'(t) = Ay(t) + f(t)$, the 1×1 matrix A is zero.

Trigonometric Polynomials Assume that in the equation $y' = Ay + f$, each component of $f \colon \mathbb{R} \to \mathbb{R}^n$ is a trigonometric polynomial. It means that f can be written in the form

$$f(t) = f^{(0)} + \sum_{j=1}^{J} \cos(\xi_j t) f^{(2j-1)} + \sin(\xi_j t) f^{(2j)}, \qquad f^{(0)}, \ldots, f^{(2J)} \in \mathbb{R}^n.$$

Here $\xi_1, \ldots, \xi_J$ are in $\mathbb{R} \setminus \{0\}$. We restrict to the case where the $n \times n$ matrix A is invertible and $\pm i\xi_j$, $1 \leq j \leq J$, are not eigenvalues of A. We then make an ansatz for a particular solution $y_p(t)$ of the form

$$y_p(t) = w^{(0)} + \sum_{j=1}^{J} \cos(\xi_j t) w^{(2j-1)} + \sin(\xi_j t) w^{(2j)}, \qquad w^{(0)}, \ldots, w^{(2J)} \in \mathbb{R}^n.$$

Upon substitution of the ansatz into the equation $y' = Ay + f$, the vectors $w^{(j)}$, $0 \le j \le 2J$, can then be determined by comparison of coefficients. To illustrate the method with an example, let us consider the system

$$y'(t) = \begin{pmatrix} 0 & -1 \\ 1 & 0 \end{pmatrix} y(t) + \begin{pmatrix} \cos(2t) \\ 1 \end{pmatrix}. \tag{6.15}$$

Clearly, the 2×2 matrix $\begin{pmatrix} 0 & -1 \\ 1 & 0 \end{pmatrix}$ is invertible and its eigenvalues are i and $-i$. Following the considerations above, we make the ansatz $y_p(t) = (q_1(t), q_2(t))$ where

$$q_1(t) = a_0 + a_1 \cos(2t) + a_2 \sin(2t), \qquad q_2(t) = b_0 + b_1 \cos(2t) + b_2 \sin(2t).$$

Since

$$y_p'(t) = \begin{pmatrix} -2a_1 \sin(2t) + 2a_2 \cos(2t) \\ -2b_1 \sin(2t) + 2b_2 \cos(2t) \end{pmatrix},$$

$$Ay_p(t) = \begin{pmatrix} -b_0 - b_1 \cos(2t) - b_2 \sin(2t) \\ a_0 + a_1 \cos(2t) + a_2 \sin(2t) \end{pmatrix},$$

Eq. (6.15) leads to the following equation

$$\begin{pmatrix} -2a_1 \sin(2t) + 2a_2 \cos(2t) \\ -2b_1 \sin(2t) + 2b_2 \cos(2t) \end{pmatrix} = \begin{pmatrix} -b_0 - b_1 \cos(2t) - b_2 \sin(2t) + \cos(2t) \\ a_0 + a_1 \cos(2t) + a_2 \sin(2t) + 1 \end{pmatrix}.$$

By comparison of coefficients, one obtains the following linear system of equations,

$$0 = b_0, \quad -2a_1 = -b_2, \quad 2a_2 = -b_1+1, \quad 0 = a_0+1, \quad -2b_1 = a_2, \quad 2b_2 = a_1,$$

having the solution

$$a_0 = -1, \qquad b_0 = 0, \qquad a_1 = 0, \qquad b_2 = 0, \qquad a_2 = \frac{2}{3}, \qquad b_1 = -\frac{1}{3}.$$

Altogether, we get

$$y_p(t) = \left(-1 + \frac{2}{3} \sin(2t), -\frac{1}{3} \cos(2t) \right).$$

Exponentials   Assume that in the equation $y' = Ay + f$, each component of $f : \mathbb{R} \to \mathbb{R}^n$ is a linear combination of exponential functions, i.e., each component is an element in

$$V := \Big\{ g(t) = \sum_{j=1}^{J} a_j e^{\xi_j t} \mid a_1, \dots, a_J \in \mathbb{R} \Big\}$$

and $\xi_1, \dots, \xi_J$ are distinct real numbers. Then V is a $\mathbb{R}$-vector space of continuously differentiable functions $g : \mathbb{R} \to \mathbb{R}$ of dimension J with the property that for any $g \in V$, its derivative g' is again in V. In the case where no exponent ξ_j is an eigenvalue of A, we make the ansatz for a particular solution $y_p(t)$, by assuming that each component of $y_p(t)$ is an element in V. The particular solution $y_p(t)$ is then again computed by comparison of coefficients.

We refer to Problem 4 at the end of this section where an example of an equation of the form $y' = Ay + f$ is considered with the components of f being (linear combinations of) exponential functions.

Problems

1. Find the general solution of the following linear ODEs.

   (i) $\begin{cases} y_1'(t) = y_1(t) + y_2(t) \\ y_2'(t) = -2y_1(t) + 4y_2(t) \end{cases}$,   (ii) $\begin{cases} y_1'(t) = 2y_1(t) + 4y_2(t) \\ y_2'(t) = -y_1(t) - 3y_2(t) \end{cases}$.

2. Solve the following initial value problems.

   (i) $\begin{cases} y_1'(t) = -y_1(t) + 2y_2(t) \\ y_2'(t) = 2y_1(t) - y_2(t) \end{cases}$, $y^{(0)} = \begin{pmatrix} 2 \\ -1 \end{pmatrix}$

   (ii) $\begin{cases} y_1'(t) = 2y_1(t) - 6y_3(t) \\ y_2'(t) = y_1(t) - 3y_3(t) \\ y_3'(t) = y_2(t) - 2y_3(t) \end{cases}$, $y^{(0)} = \begin{pmatrix} 1 \\ 0 \\ -1 \end{pmatrix}$

3. Consider the linear ODE

   $$y'(t) = Ay(t), \qquad A = \begin{pmatrix} 0 & 1 & 0 \\ 0 & 0 & 1 \\ 0 & 0 & 0 \end{pmatrix} \in \mathbb{R}^{3 \times 3}.$$

   (i) Compute A^2 and A^3.
   (ii) Determine the general solution of $y'(t) = Ay(t)$.

4. Let $A = \begin{pmatrix} 0 & 2 \\ -1 & 0 \end{pmatrix} \in \mathbb{R}^{2\times 2}$. Find the solutions of the following initial value problems.

(i) $y'(t) = Ay(t) + \begin{pmatrix} e^t \\ 3 \end{pmatrix}, \qquad y^{(0)} = \begin{pmatrix} 1 \\ 0 \end{pmatrix}$

(ii) $y'(t) = Ay(t) + \begin{pmatrix} 0 \\ \cos(2t) \end{pmatrix}, \qquad y^{(0)} = \begin{pmatrix} 0 \\ 1 \end{pmatrix}$

5. Decide whether the following assertions are true or false and justify your answers.

(i) For any $A, B \in \mathbb{R}^{2\times 2}$, one has $e^{A+B} = e^A e^B$.

(ii) Let $A = \begin{pmatrix} 0 & -1 \\ 1 & 0 \end{pmatrix} \in \mathbb{R}^{2\times 2}$. Then the linear ODE $y'(t) = Ay(t) + \begin{pmatrix} t^2 \\ t \end{pmatrix}$

admits a particular solution of the form $\begin{pmatrix} a + bt + ct^2 \\ d + et \end{pmatrix}$.

6.3 Linear ODEs with Constant Coefficients of Higher Order

The main goal of this section is to discuss linear ODEs of second order with constant coefficients. They come up in many applications and therefore are particularly important. Specifically, we consider ODEs of the form

$$y''(t) = b_1 y'(t) + b_0 y(t) + f(t), \quad t \in \mathbb{R} \tag{6.16}$$

where $y'(t) = \frac{d}{dt} y(t)$, $y''(t) = \frac{d^2}{dt^2} y(t)$, $b_1, b_0 \in \mathbb{R}$ are constants, and where $f : \mathbb{R} \to \mathbb{R}$ is a continuous function. We want to discuss two methods for solving such equations.

Method 1 Convert Eq. (6.16) into a system of ODEs of first order with constant coefficients as follows: let $x(t) = (x_0(t), x_1(t)) \in \mathbb{R}^2$ be given by

$$x_0(t) := y(t), \qquad x_1(t) := y'(t).$$

Then

$$x'(t) = \begin{pmatrix} x_0'(t) \\ x_1'(t) \end{pmatrix} = \begin{pmatrix} y'(t) \\ y''(t) \end{pmatrix} = \begin{pmatrix} x_1(t) \\ b_1 y'(t) + b_0 y(t) + f(t) \end{pmatrix}$$

$$= \begin{pmatrix} x_1(t) \\ b_1 x_1(t) + b_0 x_0(t) + f(t) \end{pmatrix}$$

or, in matrix notation,

$$x'(t) = Ax(t) + \begin{pmatrix} 0 \\ f(t) \end{pmatrix}, \qquad A := \begin{pmatrix} 0 & 1 \\ b_0 & b_1 \end{pmatrix}. \tag{6.17}$$

The Eqs. (6.16) and (6.17) are equivalent in the sense that they have the same set of solutions: a solution y of (6.16) gives rise to a solution $x(t) = (y(t), y'(t))$ of (6.17) and conversely, a solution $x(t) = (x_0(t), x_1(t))$ of (6.17) yields a solution $y(t) = x_0(t)$ of (6.16). More generally, an ODE of the form

$$y^{(n)}(t) = b_{n-1}y^{(n-1)}(t) + \cdots + b_1 y'(t) + b_0, \quad t \in \mathbb{R}, \tag{6.18}$$

can be converted into a system of ODEs of first order by setting

$$x_0(t) := y(t), \qquad x_1(t) := y'(t), \qquad \dots, \qquad x_{n-1}(t) := y^{(n-1)}(t).$$

Then one gets $x_0'(t) = x_1(t), \dots, x_{n-2}'(t) = x_{n-1}(t)$, and

$$x_{n-1}'(t) = y^{(n)}(t) = b_{n-1}y^{(n-1)}(t) + \cdots + b_1 y'(t) + b_0 y + f(t).$$

In matrix notation,

$$x'(t) = Ax(t) + (0, \dots, 0, f(t)), \quad t \in \mathbb{R} \tag{6.19}$$

where

$$A := \begin{pmatrix} 0 & 1 & 0 & \dots & 0 \\ 0 & 0 & 1 & \dots & 0 \\ \vdots & \vdots & \vdots & & \vdots \\ 0 & 0 & 0 & \dots & 1 \\ b_0 & b_1 & b_2 & \dots & b_{n-1} \end{pmatrix} \in \mathbb{R}^{n \times n}. \tag{6.20}$$

Again Eqs. (6.18) and (6.19) are equivalent in the sense that they have the same set of solutions.

Example Consider

$$y''(t) = -k^2 y(t), \quad k > 0. \tag{6.21}$$

The equation is a model for the vibrations of a string without damping. Converting this equation into a 2×2 system leads to the following first order ODE,

$$x'(t) = Ax(t), \qquad x(t) = \begin{pmatrix} x_0(t) \\ x_1(t) \end{pmatrix}, \qquad A = \begin{pmatrix} 0 & 1 \\ -k^2 & 0 \end{pmatrix}.$$

Then the general solution is given by $x(t) = e^{tA}c$, $c = (c_1, c_2) \in \mathbb{R}^2$. In order to determine $y(t) = x_0(t)$ in a more explicit form, we need to analyze e^{tA} further. The eigenvalues of A are $\lambda_1 = ik$, $\lambda_2 = -ik$ and corresponding eigenvectors are

$$v^{(1)} = (1, ik), \qquad v^{(2)} = (1, -ik).$$

Note that $v^{(1)}$, $v^{(2)}$ form a basis of $\mathbb{C}^2$. The general complex solution is given by

$$a_1 e^{ikt} v^{(1)} + a_2 e^{-ikt} v^{(2)}, \quad a_1, a_2 \in \mathbb{C}, \tag{6.22}$$

and the general real solution is given by an arbitrary linear combination of the real and the imaginary part of the general complex solution of (6.22),

$$c_1 \begin{pmatrix} \cos(kt) \\ -k\sin(kt) \end{pmatrix} + c_2 \begin{pmatrix} \sin(kt) \\ k\cos(kt) \end{pmatrix}, \quad c_1, c_2 \in \mathbb{R}.$$

Method 2 With this method one finds the general solution of a homogeneous ODE of order 2, $y''(t) = b_1 y'(t) + b_0 y(t)$ or more generally, of a homogeneous ODE of order n, $y^{(n)}(t) = b_{n-1} y^{(n-1)}(t) + \cdots + b_0 y(t)$, by first finding solutions of the form $y(t) = e^{\lambda t}$ with $\lambda \in \mathbb{C}$ to be determined. Let us illustrate this method with the Eq. (6.21), $y''(t) = -k^2 y(t)$, considered above. Substituting the ansatz $y(t) = e^{\lambda t}$ into the equation $y''(t) = -k^2 y(t)$ one gets

$$\lambda^2 e^{\lambda t} = y''(t) = -k^2 y(t) = -k^2 e^{\lambda t},$$

hence

$$\lambda^2 = -k^2, \quad \text{or} \quad \lambda_1 = ik, \quad \lambda_2 = -ik.$$

The general complex solution reads

$$a_1 e^{ikt} + a_2 e^{-ikt}, \quad a_1, a_2 \in \mathbb{C},$$

whereas the general real solution is given by

$$c_1 \cos(kt) + c_2 \sin(kt), \quad c_1, c_2 \in \mathbb{R}.$$

To solve the inhomogeneous ODE of second order (cf. (6.16)),

$$y''(t) = b_1 y'(t) + b_0 y(t) + f(t), \quad t \in \mathbb{R},$$

we argue as for systems of linear ODEs of first order to conclude that the set of solutions of (6.16) is given by

$$\{y_p + u \mid u \text{ satisfies } u''(t) + au'(t) + bu(t) = 0, t \in \mathbb{R}\}$$

where y_p is a particular solution of (6.16). We illustrate this method of solving (6.16) with the following example.

Example Consider

$$y''(t) + 3y'(t) + 2y(t) = 8, \quad t \in \mathbb{R}. \tag{6.23}$$

First we consider the homogeneous ODE

$$y''(t) + 3y'(t) + 2y(t) = 0, \quad t \in \mathbb{R}. \tag{6.24}$$

By Method 2, we make the ansatz $y(t) = e^{\lambda t}$. Substituting $y(t) = e^{\lambda t}$ into (6.24) and using that $y'(t) = \lambda e^{\lambda t}$, $y''(t) = \lambda^2 e^{\lambda t}$, one is led to

$$\lambda^2 + 3\lambda + 2 = 0.$$

The roots are given by

$$\lambda_1 = -1, \qquad \lambda_2 = -2.$$

The general real solution of (6.24) is therefore

$$y(t) = c_1 e^{-t} + c_2 e^{-2t}, \quad c_1, c_2 \in \mathbb{R},$$

and in turn, the general solution of (6.23) is given by

$$\text{general solution of (6.24)} + y_p$$

where y_p is a particular solution of (6.23). As for systems of ODEs of first order, we may try to find a particular solution by making a suitable ansatz. Note that the inhomogeneous term is the constant function $f(t) = 8$, which is a polynomial of degree 0. Thus we try the ansatz $y_p = c$. Substituting the ansatz into the Eq. (6.23) one gets $0 + 3 \cdot 0 + 2c = 8$ and hence $c = 4$. Altogether we conclude that

$$c_1 e^{-t} + c_2 e^{-2t} + 4, \quad c_1, c_2 \in \mathbb{R},$$

is the general solution of (6.23).

The advantage of Method 2 over Method 1 is that the given equation has not to be converted into a system of first order and that the computations are typically shorter. However we will see that in some situations, the ansatz $y(t) = e^{\lambda t}$ has to be modified to get all solutions in this way.

As for systems of ODEs of first order, one can study the initial value problem for Eq. (6.16) or more generally for Eq. (6.18),

$$\begin{cases} y^{(n)}(t) = b_{n-1}y^{(n-1)}(t) + \cdots + b_1 y'(t) + b_0 y(t) + f(t), & t \in \mathbb{R}, \\ y(0) = a_0, \ldots, y^{(n-1)}(0) = a_{n-1} \end{cases}$$

$$\text{(IVP)}$$

where $a_0, \ldots, a_{n-1}$ are arbitrary real numbers. Note that there are n initial conditions for an equation of order n. This corresponds to a vector $a \in \mathbb{R}^n$ of initial values for the initial value problem of a $n \times n$ system of first order ODEs. As in that case, the above initial value problem has a unique solution. Let us illustrate how to find this solution with the following examples.

Examples

(i) Consider the initial value problem

$$y''(t) = -k^2 y(t), \qquad y(0) = 1, \quad y'(0) = 0$$

where we assume again that $k > 0$. We have seen that the general complex solution is given by

$$y(t) = a_1 e^{ikt} + a_2 e^{-ikt}, \quad a_1, a_2 \in \mathbb{C}.$$

We need to determine $a_1, a_2 \in \mathbb{C}$ so that $y(0) = 1$, $y'(0) = 0$, i.e.,

$$1 = y(0) = a_1 + a_2, \qquad 0 = y'(0) = ika_1 - ika_2.$$

This is a linear 2×2 system with the two unknowns a_1 and a_2. We get $a_1 = a_2$ and $a_1 = 1/2$. Hence

$$y(t) = \frac{1}{2}e^{ikt} + \frac{1}{2}e^{-ikt} = \cos(kt)$$

is the solution of the initial value problem. Note that it is automatically real valued.

(ii) Consider the initial value problem

$$y''(t) + 3y'(t) + 2y(t) = 8, \qquad y(0) = 1, \quad y'(0) = 0.$$

We have seen that the general solution of $y''(t) + 3y'(t) + 2y(t) = 8$ is given by

$$y(t) = c_1 e^{-t} + c_2 e^{-2t} + 4, \quad c_1, c_2 \in \mathbb{R}.$$

We need to determine c_1, c_2 in such a way that $y(0) = 1$, $y'(0) = 0$, i.e.,

$$1 = y(0) = c_1 + c_2 + 4, \qquad 0 = y'(0) = -c_1 - 2c_2,$$

yielding $c_1 = -6$, $c_2 = 3$ and hence

$$y(t) = -6e^{-t} + 3e^{-2t} + 4.$$

Homogeneous Equations Let us now discuss in some more detail the homogeneous equation, corresponding to (6.16). More precisely, we consider

$$y''(t) + ay'(t) + by(t) = 0, \quad a > 0, b = \omega^2, \omega > 0, \tag{6.25}$$

which is a model for the vibrations of a string with damping.

It is convenient to apply Method 2. Making the ansatz $y(t) = e^{\lambda t}$ and taking into account that $y'(t) = \lambda e^{\lambda t}$, $y''(t) = \lambda^2 e^{\lambda t}$, one is led to the equation

$$(\lambda^2 + a\lambda + b)e^{\lambda t} = 0, \quad t \in \mathbb{R}.$$

Hence $\lambda^2 + a\lambda + b = 0$ or

$$\lambda_{\pm} = -\frac{a}{2} \pm \frac{1}{2}\sqrt{a^2 - 4\omega^2}.$$

Case 1. $a^2 - 4\omega^2 > 0$, i.e., $0 < 2\omega < a$. This is the case of strong damping, leading to solutions with no oscillations. Indeed, in this case $\lambda_- < \lambda_+ < 0$ and the general real solution is given by

$$y(t) = c_+ e^{\lambda_+ t} + c_- e^{\lambda_- t}, \quad c_+, c_- \in \mathbb{R}.$$

Case 2. $a^2 - 4\omega^2 < 0$, i.e., $0 < a < 2\omega$. This is the case of weak damping, leading to solutions with oscillations. Then $\lambda_{\pm} = -\frac{a}{2} \pm i\gamma$ where $\gamma := \frac{1}{2}\sqrt{4\omega^2 - a^2}$. Then

$$e^{\lambda_{\pm} t} = e^{-\frac{a}{2}t} e^{\pm i\gamma} = e^{-\frac{a}{2}t}\big(\cos(\gamma t) + i\sin(\gamma t)\big)$$

and the general real solution is given by

$$c_1 e^{-\frac{a}{2}t}\cos(\gamma t) + c_2 e^{-\frac{a}{2}t}\sin(\gamma t), \quad c_1, c_2 \in \mathbb{R}, \quad \gamma = \frac{1}{2}\sqrt{4\omega^2 - a^2}.$$

Note that as $t \to \infty$, all solutions tend to 0, whereas for $t \to -\infty$, the nontrivial solutions are unbounded. This is in sharp contrast with the case without damping, i.e., the case where $a = 0$.

Case 3. $a^2 - 4\omega^2 = 0$, i.e., $a = 2\omega$. Then $\lambda_+ = \lambda_- = -\frac{a}{2}$. It can be verified that besides $e^{-\frac{a}{2}t}$, also $te^{-\frac{a}{2}t}$ is a solution and hence the general real solution is given by

$$c_1 e^{-\frac{a}{2}t} + c_2 t e^{-\frac{a}{2}t}, \quad c_1, c_2 \in \mathbb{R}. \tag{6.26}$$

There are no oscillations in this case, but as $t \to \infty$ the decay is weaker than in case 1. To find the general solution (6.26) without guessing, one can apply Method 1. Converting $y''(t) = -by'(t) - ay(t)$ to a 2×2 system we get $x'(t) = Ax(t)$, with

$$A = \begin{pmatrix} 0 & 1 \\ -b & -a \end{pmatrix}, \qquad b = \frac{a^2}{4}.$$

Then $\lambda_1 = \lambda_2 = -\frac{a}{2}$ are the eigenvalues of A and the geometric multiplicity of A is one. To compute e^{tA} we write A as a sum,

$$A = -\frac{a}{2}\, \mathrm{Id}_{2\times2} + B, \qquad B := \begin{pmatrix} a/2 & -1 \\ -b & -a/2 \end{pmatrix}.$$

Using that $b = (a/2)^2$ one infers that $B^2 = 0$. Since $\mathrm{Id}_{2\times2}$ and B commute, one then concludes that

$$e^{tA} = e^{-\frac{a}{2}t\, \mathrm{Id}_{2\times2}} e^{tB} = e^{-\frac{a}{2}t} e^{tB} = e^{-\frac{a}{2}t}(\mathrm{Id}_{2\times2} + tB)$$

$$= e^{-\frac{a}{2}t} \begin{pmatrix} 1 + a/2\, t & 1 \\ -bt & 1 - a/2\, t \end{pmatrix}.$$

It then follows that the general solution of (6.26) in case 3 is indeed given by (6.26).

Inhomogeneous Equations We finish this section with a discussion of how to find particular solutions of (6.16) for certain classes of functions $f : \mathbb{R} \to \mathbb{R}$ by making a suitable ansatz. For this purpose it is convenient to write (6.16) in slightly different form,

$$y''(t) + ay'(t) + by(t) = f(t), \quad t \in \mathbb{R}. \tag{6.27}$$

Polynomials In case f is a polynomial of degree L, $f(t) = f_0 + f_1 t + \ldots + f_L t^L$, we make the ansatz

$$y_p(t) = \alpha_0 + \alpha_1 t + \ldots + \alpha_{L+2} t^{L+2}.$$

Substituting the ansatz into the Eq. (6.27), one gets

$$\sum_{j=2}^{L+2} \alpha_j j(j-1)t^{j-2} + a \sum_{j=1}^{L+2} \alpha_j j t^{j-1} + b \sum_{j=0}^{L+2} \alpha_j t^j = \sum_{j=0}^{L} f_j t^j.$$

The coefficients $\alpha_0, \ldots, \alpha_{L+2}$ are then determined inductively by comparison of coefficients,

$$b\alpha_{L+2} = 0, \qquad (L+2)a\alpha_{L+2} + b\alpha_{L+1} = 0, \tag{6.28}$$

$$(j+2)(j+1)\alpha_{j+2} + (j+1)a\alpha_{j+1} + b\alpha_j = f_j, \quad 0 \le j \le L. \tag{6.29}$$

Remark

(i) To see that a particular solution of an ODE of the type (6.27) might *not* be given by a polynomial of degree L, consider

$$y''(t) = 2 + 3t^2.$$

The general solution of the latter equation can be easily found by integration,

$$y_p(t) = t^2 + \frac{1}{4}t^4 + \alpha_1 t + \alpha_0, \quad \alpha_1, \alpha_0 \in \mathbb{R}.$$

Note that in this case, $a = 0$ and $b = 0$ in (6.27) and hence the two equations in (6.28) trivially hold for any choice of α_{L+2} and α_{L+1}.

(ii) There are examples of ODEs of the type (6.27) where a particular solution is a polynomial of degree $L + 1$. Indeed, consider

$$y''(t) + 2y'(t) = t.$$

Substituting the ansatz $y_p(t) = \alpha_0 + \alpha_1 t + \alpha_2 t^2 + \alpha_3 t^3$, one gets

$$6\alpha_3 t + 2\alpha_2 + 6\alpha_3 t^2 + 4\alpha_2 t + 2\alpha_1 = t$$

and comparison of coefficients yields

$$\alpha_3 = 0, \qquad 6\alpha_3 + 4\alpha_2 = 1, \qquad 2\alpha_2 + 2\alpha_1 = 0.$$

Hence $\alpha_3 = 0, \alpha_2 = 1/4$, and $\alpha_1 = -1/4$, whereas α_0 can be chosen arbitrarily. A particular solution is hence given by the following polynomial of degree two,

$$y_p(t) = -\frac{1}{4}t + \frac{1}{4}t^2.$$

Note that in this case, $a = 2$ and $b = 0$ in (6.27). Hence the first equation in (6.28) trivially holds for any choice of α_{L+2}, whereas the second equation implies that $\alpha_{L+2} = 0$.

Solution of Homogeneous Equation Assume that f is a solution of the homogeneous equation $u''(t) + au'(t) + bu(t) = 0$. We have to consider two cases.

Case 1. $a^2 \neq 4b$. We have seen that in this case the general complex solution of $u''(t) + au'(t) + bu(t) = 0$ is given by $c_1 e^{\lambda_1 t} + c_2 e^{\lambda_2 t}$ where $c_1, c_2 \in \mathbb{C}$ and

$$\lambda_1 = -\frac{a}{2} + \frac{1}{2}\sqrt{a^2 - 4b}, \qquad \lambda_2 = -\frac{a}{2} - \frac{1}{2}\sqrt{a^2 - 4b},$$

(with $\sqrt{a^2 - 4b}$ defined as $i\sqrt{4b - a^2}$ in the case $a^2 < 4b$). By assumption, $f(t) = f_1 e^{\lambda_1 t} + f_2 e^{\lambda_2 t}$ where $f_1, f_2 \in \mathbb{C}$ are uniquely determined by

$$f(0) = f_1 + f_2, \qquad f'(0) = \lambda_1 f_1 + \lambda_2 f_2.$$

Note that f_1, f_2 might be complex numbers even if f is real valued. However, in such a case $f_2 = \overline{f_1}$ and $\lambda_2 = \overline{\lambda_1}$. Making the ansatz $y_p(t) = t(\alpha_1 e^{\lambda_1 t} + \alpha_2 e^{\lambda_2 t})$ and substituting it into (6.27), one finds by comparison of coefficients

$$\alpha_1 = \frac{f_1}{2\lambda_1 + a}, \qquad \alpha_2 = \frac{f_2}{2\lambda_2 + a}.$$

Note that $2\lambda_1 + a = \sqrt{a^2 - 4b}$ and $2\lambda_2 + a = -\sqrt{a^2 - 4b}$ do not vanish by assumption, so that α_1, α_2 are well defined. Hence

$$y_p(t) = \frac{f_1}{2\lambda_1 + a} t e^{\lambda_1 t} + \frac{f_2}{2\lambda_2 + a} t e^{\lambda_2 t}$$

is a particular real solution of (6.27).

Case 2. $a^2 = 4b$. We have seen that in this case, the general solution of $u''(t) + au'(t) + bu(t) = 0$ is given by $c_1 e^{-at/2} + c_2 t e^{-at/2}$, $c_1, c_2 \in \mathbb{R}$. By assumption, $f(t) = f_1 e^{-at/2} + f_2 t e^{-at/2}$ where f_1 and f_2 are real numbers, determined by $f(0) = f_1$, $f'(0) = (-a/2)f_1 + f_2$. Making the ansatz

$$y_p(t) = t^2(\alpha_1 + \alpha_2 t)e^{-\frac{a}{2}t}$$

and substituting it into (6.27) one finds by comparison of coefficients that $\alpha_1 = \frac{f_1}{2}, \alpha_2 = \frac{f_2}{6}$, yielding

$$y_p(t) = \left(\frac{f_1}{2} t^2 + \frac{f_2}{6} t^3\right)e^{-\frac{a}{2}t}.$$

Trigonometric Polynomials In case f is a trigonometric polynomial

$$f(t) = f_0 + \sum_{j=1}^{L} f_{2j-1} \cos(\xi_j t) + \sum_{j=1}^{L} f_{2j} \sin(\xi_j t)$$

with ξ_j real, pairwise different, and $i\xi_j \notin \{0, \lambda_1, \lambda_2\}$, we make the ansatz

$$y_p(t) = \alpha_0 + \sum_{j=1}^{L} \left(\alpha_{2j-1} \cos(\xi_j t) + \alpha_{2j} \sin(\xi_j t) \right).$$

Here λ_1, λ_2 are the zeroes of $\lambda^2 + a\lambda + b = 0$, listed with their multiplicities. Substituting the ansatz into the equation, the coefficients $\alpha_0, \ldots, \alpha_{2L}$ are then determined by comparison of coefficients.

Examples

(i) Consider

$$y''(t) + 2y'(t) + y(t) = \cos t, \quad t \in \mathbb{R}. \tag{6.30}$$

Note that $\lambda_1 = \lambda_2 = -1$ and $\xi_1 = 1$, hence $i\xi_1 \notin \{0, -1\}$. Substituting the ansatz

$$y_p(t) = \alpha_0 + \alpha_1 \cos t + \alpha_2 \sin t$$

into the equation and using that

$$y_p'(t) = -\alpha_1 \sin t + \alpha_2 \cos t, \qquad y_p''(t) = -\alpha_1 \cos t - \alpha_2 \sin t,$$

one gets

$$-\alpha_1 \cos t - \alpha_2 \sin t + 2(-\alpha_1 \sin t + \alpha_2 \cos t) + (\alpha_0 + \alpha_1 \cos t + \alpha_2 \sin t) = \cos t.$$

Hence by comparison of coefficients,

$$-\alpha_1 + 2\alpha_2 + \alpha_1 = 1, \qquad -\alpha_2 - 2\alpha_1 + \alpha_2 = 0, \qquad \alpha_0 = 0,$$

yielding $\alpha_2 = \frac{1}{2}$, $\alpha_1 = 0$, and $\alpha_0 = 0$, i.e.,

$$y_p(t) = \frac{1}{2} \sin t \tag{6.31}$$

is a particular solution of (6.30).

(ii) Consider

$$y''(t) + \omega^2 y(t) = \delta \sin(\omega t), \quad \omega > 0. \tag{6.32}$$

Since $\lambda_1 = i\omega$, $\lambda_2 = -i\omega$, and $\xi_1 = \omega$, it follows that $i\xi_1 \in \{0, \lambda_1, \lambda_2\}$. In such a case one has to modify the ansatz $\alpha_1 \cos(\omega t) + \alpha_2 \sin(\omega t)$ for a particular solution of (6.32). Note that $c_1 e^{\lambda_1 t} + c_2 e^{\lambda_2 t}$, $c_1, c_2 \in \mathbb{C}$, is the general complex solution of $u''(t) + \omega^2 u(t) = 0$. It then follows that the two solutions $\cos(\omega t)$, $\sin(\omega t)$ form a basis of the $\mathbb{R}$-vector space V of the real solutions of the homogeneous equation $u''(t) + \omega^2 u(t) = 0$,

$$V = \{c_1 \cos(\omega t) + c_2 \sin(\omega t) \mid c_1, c_2 \in \mathbb{R}\}.$$

As in the discussion above of the case where f is a solution of the homogeneous equation, we make the following ansatz for a particular solution,

$$y_p(t) = \alpha_1 t \cos(\omega t) + \alpha_2 t \sin(\omega t).$$

Substituting it into (6.32) one gets with

$$y_p'(t) = \alpha_1 \cos(\omega t) + \alpha_2 \sin(\omega t) + t\big(\alpha_1 \cos(\omega t) + \alpha_2 \sin(\omega t)\big)',$$

$$y_p''(t) = 2\big(\alpha_1 \cos(\omega t) + \alpha_2 \sin(\omega t)\big)' + t\big(\alpha_1 \cos(\omega t) + \alpha_2 \sin(\omega t)\big)'',$$

that

$$2\big(\alpha_1 \cos(\omega t) + \alpha_2 \sin(\omega t)\big)' = \delta \sin(\omega t),$$

implying that $\alpha_2 = 0$ and $-2\alpha_1 \omega = \delta$. Hence

$$y_p(t) = -\frac{\delta}{2\omega} t \cos(\omega t)$$

is a particular solution of (6.32). Note that this solution is oscillatory with unbounded amplitude. The general real solution of (6.32) is given by

$$c_1 \cos(\omega t) + c_2 \sin(\omega t) - \frac{\delta}{2\omega} t \cos(\omega t), \quad c_1, c_2 \in \mathbb{R}. \tag{6.33}$$

Remark The general solution (6.33) can be used to solve the corresponding initial value problem. To illustrate how this can be done, let us find the unique solution of (6.32), satisfying the initial conditions

$$y(0) = 0, \qquad y'(0) = 1. \tag{6.34}$$

It means that in the formula (6.33), the constants c_1, c_2 have to be determined in such a way that (6.34) holds. Therefore, we need to solve the following linear system,

$$c_1 = 0, \qquad \omega c_2 - \frac{\delta}{2\omega} = 1.$$

The solution $y(t)$ of the intial value problem is then given by

$$y(t) = \frac{1}{\omega}\left(1 + \frac{\delta}{2\omega}\right)\sin(\omega t) - \frac{\delta}{2\omega}t\cos(\omega t).$$

Exponentials   In case f is a linear combination of exponentials

$$f(t) = \sum_{j=1}^{L} f_j e^{\xi_j t}, \qquad f_1, \ldots, f_L \in \mathbb{R} \tag{6.35}$$

where $\xi_1, \ldots, \xi_L$ are in $\mathbb{R} \setminus \{\lambda_1, \lambda_2\}$ and pairwise different and λ_1, λ_2 are the zeroes of $\lambda^2 + a\lambda + b = 0$, we make the ansatz $y_p(t) = \sum_{j=1}^{L} \alpha_j e^{\xi_j t}$ for a particular solution of $y''(t) + ay'(t) + by(t) = f(t)$. Since

$$y_p'(t) = \sum_{j=1}^{L} \xi_j \alpha_j e^{\xi_j t}, \qquad y_p''(t) = \sum_{j=1}^{L} \xi_j^2 \alpha_j e^{\xi_j t},$$

one obtains, upon substitution, that

$$\sum_{j=1}^{L}(\alpha_j \xi_j^2 + a\alpha_j \xi_j + b\alpha_j)e^{\xi_j t} = \sum_{j=1}^{L} f_j e^{\xi_j t}.$$

Hence

$$\alpha_j = \frac{f_j}{\xi_j^2 + a\xi_j + b}, \qquad 1 \le j \le L,$$

and

$$y_p(t) = \sum_{j=1}^{L} \frac{f_j}{\xi_j^2 + a\xi_j + b} e^{\xi_j t}$$

is a particular solution of (6.27) with f given by (6.35). Note that $\xi_j^2 + a\xi_j + b \ne 0$ for any $1 \le j \le L$, since by assumption, $\xi_j \ne \lambda_1, \lambda_2$.

Example Consider

$$y''(t) - 5y'(t) + 4y(t) = 1 + e^t, \quad t \in \mathbb{R}.$$

Note that $\lambda^2 - 5\lambda + 4 = 0$ has the solutions $\lambda_1 = 4$, $\lambda_2 = 1$. Hence the general solution of the homogeneous equation $u''(t) - 5u'(t) + 4u(t) = 0$ is given by $c_1 e^{4t} + c_2 e^t$ with $c_1, c_2 \in \mathbb{R}$. On the other hand, $f(t) = e^{\xi_1 t} + e^{\xi_2 t}$ with $\xi_1 = 0$ and $\xi_2 = \lambda_2$. We look for a particular solution of the form

$$y_p(t) = y_p^{(1)}(t) + y_p^{(2)}(t)$$

where $y_p^{(1)}$ is a particular solution of $y''(t) - 5y'(t) + 4y(t) = 1$ and $y_p^{(2)}$ is a particular solution of $y''(t) - 5y'(t) + 4y(t) = e^t$. For $y_p^{(1)}$ we make the ansatz $y_p^{(1)}(t) = \alpha \in \mathbb{R}$. Upon substitution into the equation $y''(t) - 5y'(t) + 4y(t) = 1$ we get $y_p^{(1)}(t) = 1/4$. Since e^t is a solution of the homogeneous problem $y''(t) - 5y'(t) + 4y(t) = 0$, we make the ansatz $y_p^{(2)}(t) = \alpha t e^t$. Substituting it into $y''(t) - 5y'(t) + 4y(t) = e^t$, we obtain, with $y'(t) = \alpha e^t + \alpha t e^t$ and $y''(t) = 2\alpha e^t + \alpha t e^t$,

$$-3\alpha e^t + \alpha t (e^t - 5e^t + 4e^t) = e^t,$$

yielding $\alpha = -1/3$ and hence $y_p^{(2)}(t) = -\frac{1}{3} t e^t$. We conclude that

$$y_p(t) = \frac{1}{4} - \frac{1}{3} t e^t$$

is a solution of $y''(t) - 5y'(t) + 4y(t) = 1 + e^t$ and the general real solution is given by

$$c_1 e^{4t} + c_2 e^t + \frac{1}{4} - \frac{1}{3} t e^t, \quad c_1, c_2 \in \mathbb{R}. \tag{6.36}$$

Remark The formula (6.36) can be used to solve the initial value problem

$$\begin{cases} y''(t) - 5y'(t) + 4y(t) = 1 + e^t, & t \in \mathbb{R}, \\ y(0) = \dfrac{1}{4}, \qquad y'(0) = \dfrac{8}{3}. \end{cases}$$

Indeed, the constants $c_1, c_2 \in \mathbb{R}$ in (6.36) can be determined by solving the following linear system

$$c_1 + c_2 + \frac{1}{4} = \frac{1}{4}, \qquad 4c_1 + c_2 - \frac{1}{3} = \frac{8}{3},$$

yielding $c_1 + c_2 = 0$ and $4c_1 + c_2 = 3$. Hence $c_1 = 1$, $c_2 = -1$ and the unique solution $y(t)$ of the initial value problem is given by

$$y(t) = e^{4t} - e^t + \frac{1}{4} - \frac{1}{3} te^t.$$

Problems

1. Find the general solution of the following linear ODEs of second order.

   (i) $y''(t) + 2y'(t) + 4y(t) = 0$,
   (ii) $y''(t) + 2y'(t) - 4y(t) = t^2$.

2. Find the solutions of the following initial value problems.

   (i) $y''(t) - y'(t) - 2y(t) = e^{-\pi t}$, $y(0) = 0$, $y'(0) = 1$.
   (ii) $y''(t) + y(t) = \sin t$, $y(0) = 1$, $y'(0) = 0$.

3. Consider the following ODE:

$$\begin{cases} y_1'(t) = y_1(t) + 2y_2(t) \\ y_2'(t) = 3y_1(t) + 2y_2(t) \end{cases}.$$

   (i) Find all solutions $y(t) = (y_1(t), y_2(t))$ with the property that

$$\lim_{t \to \infty} \|y(t)\| = 0.$$

   (ii) Do there exist solutions $y(t)$ so that

$$\lim_{t \to \infty} \|y(t)\| = 0 \qquad \text{and} \qquad \lim_{t \to -\infty} \|y(t)\| = 0?$$

   Here $\|y(t)\| = \left(y_1(t)^2 + y_2(t)^2\right)^{1/2}$.

4. (i) Define for $A \in \mathbb{R}^{2 \times 2}$

$$\cos A = \sum_{k=0}^{\infty} \frac{(-1)^k}{(2k)!} A^{2k} \qquad \text{and} \qquad \sin A = \sum_{k=0}^{\infty} \frac{(-1)^k}{(2k+1)!} A^{2k+1}.$$

   Verify that $e^{iA} = \cos A + i \sin A$.

   (ii) Compute e^{tA} for $A = \begin{pmatrix} 5 & -2 \\ 2 & 5 \end{pmatrix}$.

5. Decide whether the following assertions are true or false and justify your answers.

   (i) The superposition principle holds for every ODE of the form $y''(t) + a(t)y(t) = b(t)$ where $a, b \colon \mathbb{R} \to \mathbb{R}$ are arbitrary continuous functions.
   (ii) Every solution $y(t) = \big(y_1(t), y_2(t)\big) \in \mathbb{R}^2$ of

   $$\begin{cases} y_1'(t) = 2y_1(t) + y_2(t) \\ y_2'(t) = 7y_1(t) - 3y_2(t) \end{cases}$$

   is bounded, meaning that there exists a constant $C > 0$ so that

   $$\|y(t)\|^2 = y_1(t)^2 + y_2(t)^2 \le C, \quad t \in \mathbb{R}.$$

Solutions

Solutions of Problems of Sect. 1.2

1. Determine the sets of solutions of the linear systems

   (i) $\begin{cases} x + \pi y = 1 \\ 2x + 6y = 4, \end{cases}$

   (ii) $\begin{cases} x + 2y = e \\ 2x + 3y = f. \end{cases}$

   Solutions

   (i) $L = \left\{ \left(\dfrac{2\pi - 3}{\pi - 3}, \dfrac{1}{3 - \pi} \right) \right\}$

   (ii) $L = \{(-3e + 2f, 2e - f)\}$

2. Consider the system (S) of two linear equations with two unknowns.

$$\begin{cases} 2x + y = 4 & \text{(S1)} \\ x - 4y = 2 & \text{(S2)} \end{cases}$$

   (i) Determine the sets L_1 and L_2 of solutions of (S1) and (S2), respectively and represent them geometrically as straight lines in $\mathbb{R}^2$.
   (ii) Determine the intersection $L_1 \cap L_2$ of L_1 and L_2 from their geometric representation in $\mathbb{R}^2$.

   Solutions

   (i) $L_1 = \{(x, -2x + 4) \mid x \in \mathbb{R}\}$, $L_2 = \left\{ \left(x, \tfrac{1}{4}x - \tfrac{1}{2}\right) \mid x \in \mathbb{R} \right\}$

© The Author(s), under exclusive license to Springer Nature Switzerland AG 2023
M. Benz, T. Kappeler, *Linear Algebra for the Sciences*, La Matematica
per il 3+2 151, https://doi.org/10.1007/978-3-031-27220-2

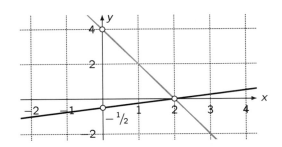

(ii) The intersection with the x-axis is at the point $(2, 0)$. Indeed, an algebraic check confirms this result.

3. Compute the determinants of the following matrices and decide for which values of the parameter $a \in \mathbb{R}$, they vanish.

(i) $A = \begin{pmatrix} 1 - a & 4 \\ 1 & 1 - a \end{pmatrix}$
(ii) $B = \begin{pmatrix} 1 - a^2 & a + a^2 \\ 1 - a & a \end{pmatrix}$

Solutions

(i) $\det(A) = (1 - a)^2 - 4$. From $(1 - a)^2 - 4 = 0$ follows that the determinant vanishes for $a = -1$ and $a = 3$.

(ii) $\det(A) = (1 - a^2)a - (a + a^2)(1 - a) = 0$, i.e. $\det(A)$ vanishes for all $a \in \mathbb{R}$.

4. Solve the system of linear equations

$$\begin{cases} 3x - y = 1 \\ 5x + 3y = 2 \end{cases}$$

by Cramer's rule.

Solutions Since $\det \begin{pmatrix} 3 & -1 \\ 5 & 3 \end{pmatrix} = 14 \neq 0$, we can use Cramer's rule. Thus

$$x = \frac{\det \begin{pmatrix} 1 & -1 \\ 2 & 3 \end{pmatrix}}{\det \begin{pmatrix} 3 & -1 \\ 5 & 3 \end{pmatrix}} = \frac{5}{14}, \qquad y = \frac{\det \begin{pmatrix} 3 & 1 \\ 5 & 2 \end{pmatrix}}{\det \begin{pmatrix} 3 & -1 \\ 5 & 3 \end{pmatrix}} = \frac{1}{14}.$$

Hence, $L = \{(\frac{5}{14}, \frac{1}{14})\}$.

5. Decide whether the following assertions are true or false and justify your answers.

(i) For any given values of the coefficients $a, b, c, d \in \mathbb{R}$, the linear system

$$\begin{cases} ax + by = 0 \\ cx + dy = 0 \end{cases}$$

has at least one solution.

(ii) There exist real numbers a, b, c, d so that the linear system of (a) has infinitely many solutions.

(iii) The system of equations

$$\begin{cases} x_1 + x_2 = 0 \\ x_1^2 + x_2^2 = 1 \end{cases}$$

is a linear system.

Solutions

(i) True. $x = 0, y = 0$ is always a solution.

(ii) True. Choose $a = 0, b = 0, c = 0, d = 0$. Then any point $(x, y) \in \mathbb{R}^2$ is a solution.

(iii) False. The second equation reads $q(x_1, x_2) = 0$ where $q(x_1, x_2) = x_1^2 + x_2^2 - 1$ is a polynomial of degree two. Hence the system is not linear.

Solutions of Problems of Sect. 1.3

1. Determine the augmented coefficient matrices of the following linear systems and transform them in row echelon form by using Gaussian elimination.

(i) $\begin{cases} x_1 + 2x_2 + x_3 = 0 \\ 2x_1 + 6x_2 + 3x_3 = 4 \\ 2x_2 + 5x_3 = -4 \end{cases}$ (ii) $\begin{cases} x_1 - 3x_2 + x_3 = 1 \\ 2x_1 + x_2 - x_3 = 2 \\ x_1 + 4x_2 - 2x_3 = 1 \end{cases}$

Solutions

(i)

$$\text{augmented coefficient matrix} \quad \begin{bmatrix} 1 & 2 & 1 & \big|\big| & 0 \\ 2 & 6 & 3 & \big|\big| & 4 \\ 0 & 2 & 5 & \big|\big| & -4 \end{bmatrix}$$

$$R_2 \leadsto R_2 - 2R_1 \quad \begin{bmatrix} 1 & 2 & 1 & \big|\big| & 0 \\ 0 & 2 & 1 & \big|\big| & 4 \\ 0 & 2 & 5 & \big|\big| & -4 \end{bmatrix}$$

$$R_3 \leadsto R_3 - R_2 \quad \begin{bmatrix} 1 & 2 & 1 & \big|\big| & 0 \\ 0 & 2 & 1 & \big|\big| & 4 \\ 0 & 0 & 4 & \big|\big| & -8 \end{bmatrix}$$

(ii)

$$\text{augmented coefficient matrix} \quad \begin{bmatrix} 1 & -3 & 1 & \Big\| & 1 \\ 2 & 1 & -1 & \Big\| & 2 \\ 1 & 4 & -2 & \Big\| & 1 \end{bmatrix}$$

$$R_2 \rightsquigarrow R_2 - 2R_1, R_3 \rightsquigarrow R_3 - R_1 \quad \begin{bmatrix} 1 & -3 & 1 & \Big\| & 1 \\ 0 & 7 & -3 & \Big\| & 0 \\ 0 & 7 & -3 & \Big\| & 0 \end{bmatrix}$$

$$R_3 \rightsquigarrow R_3 - R_1 \quad \begin{bmatrix} 1 & -3 & 1 & \Big\| & 1 \\ 0 & 7 & -3 & \Big\| & 0 \\ 0 & 0 & 0 & \Big\| & 0 \end{bmatrix}$$

2. Transform the augmented coefficient matrices of the following linear systems into reduced echelon form and find a parameter representation of the sets of its solutions.

(i) $\begin{cases} x_1 - 3x_2 + 4x_3 = 5 \\ x_2 - x_3 = 4 \\ 2x_2 + 4x_3 = 2 \end{cases}$ 　　(ii) $\begin{cases} x_1 + 3x_2 + x_3 + x_4 = 3 \\ 2x_1 - x_2 + x_3 + 2x_4 = 8 \\ x_1 - 5x_2 + x_4 = 5 \end{cases}$

Solutions

(i)

$$\text{augmented coefficient matrix} \quad \begin{bmatrix} 1 & -3 & 4 & \Big\| & 5 \\ 0 & 1 & -1 & \Big\| & 4 \\ 0 & 2 & 4 & \Big\| & 2 \end{bmatrix}$$

$$R_3 \rightsquigarrow R_3 - 2R_2 \quad \begin{bmatrix} 1 & -3 & 4 & \Big\| & 5 \\ 0 & 1 & -1 & \Big\| & 4 \\ 0 & 0 & 6 & \Big\| & -6 \end{bmatrix}$$

$$R_3 \rightsquigarrow {}^1\!/_6\, R_3 \quad \begin{bmatrix} 1 & -3 & 4 & \Big\| & 5 \\ 0 & 1 & -1 & \Big\| & 4 \\ 0 & 0 & 1 & \Big\| & -1 \end{bmatrix}$$

$$R_2 \rightsquigarrow R_2 + R_3 \quad \begin{bmatrix} 1 & -3 & 4 & \Big\| & 5 \\ 0 & 1 & 0 & \Big\| & 3 \\ 0 & 0 & 1 & \Big\| & -1 \end{bmatrix}$$

$$R_1 \rightsquigarrow R_1 + 3R_2 - 4R_3 \quad \begin{bmatrix} 1 & 0 & 0 & \Big\| & 18 \\ 0 & 1 & 0 & \Big\| & 3 \\ 0 & 0 & 1 & \Big\| & -1 \end{bmatrix}$$

Hence, $L = \{(18, 3, -1)\}$.

(ii)

$$\text{augmented coefficient matrix} \qquad \begin{bmatrix} 1 & 3 & 1 & 1 & \bigm\| & 3 \\ 2 & -1 & 1 & 2 & \bigm\| & 8 \\ 1 & -5 & 0 & 1 & \bigm\| & 5 \end{bmatrix}$$

$$R_2 \rightsquigarrow R_2 - 2R_1,\ R_3 \rightsquigarrow R_3 - R_1 \qquad \begin{bmatrix} 1 & 3 & 1 & 1 & \bigm\| & 3 \\ 0 & -7 & -1 & 0 & \bigm\| & 2 \\ 0 & -8 & -1 & 0 & \bigm\| & 2 \end{bmatrix}$$

$$R_3 \rightsquigarrow R_3 - {}^8\!/_7\, R_2 \qquad \begin{bmatrix} 1 & 3 & 1 & 1 & \bigm\| & 3 \\ 0 & -7 & -1 & 0 & \bigm\| & 2 \\ 0 & 0 & {}^1\!/_7 & 0 & \bigm\| & -{}^2\!/_7 \end{bmatrix}$$

$$R_2 \rightsquigarrow {}^1\!/_{(-7)}\, R_2,\ R_3 \rightsquigarrow 7R_3 \qquad \begin{bmatrix} 1 & 3 & 1 & 1 & \bigm\| & 3 \\ 0 & 1 & {}^1\!/_7 & 0 & \bigm\| & -{}^2\!/_7 \\ 0 & 0 & 1 & 0 & \bigm\| & -2 \end{bmatrix}$$

$$R_2 \rightsquigarrow R_2 - {}^1\!/_7\, R_3 \qquad \begin{bmatrix} 1 & 3 & 1 & 1 & \bigm\| & 3 \\ 0 & 1 & 0 & 0 & \bigm\| & 0 \\ 0 & 0 & 1 & 0 & \bigm\| & -2 \end{bmatrix}$$

$$R_1 \rightsquigarrow R_1 - 3R_2 - R_3 \qquad \begin{bmatrix} 1 & 0 & 0 & 1 & \bigm\| & 5 \\ 0 & 1 & 0 & 0 & \bigm\| & 0 \\ 0 & 0 & 1 & 0 & \bigm\| & -2 \end{bmatrix}$$

Hence, $L = \{(5 - t_4, 0, -2, t_4) \mid t_4 \in \mathbb{R}\}$ or in parameter representation

$$F: \mathbb{R} \to \mathbb{R}^4, \quad t_4 \mapsto \begin{pmatrix} 5 \\ 0 \\ -2 \\ 0 \end{pmatrix} + t_4 \begin{pmatrix} -1 \\ 0 \\ 0 \\ 1 \end{pmatrix}, \text{ which represents a line in } \mathbb{R}^4.$$

3. Consider the linear system given by the following augmented coefficient matrix

$$\begin{bmatrix} 1 & 1 & 3 & \bigm\| & 2 \\ 1 & 2 & 4 & \bigm\| & 3 \\ 1 & 3 & \alpha & \bigm\| & \beta \end{bmatrix}.$$

(i) For which values of α and β in $\mathbb{R}$ does the system have infinitely many solutions?

(ii) For which values of α and β in $\mathbb{R}$ does the system have no solutions?

Solutions By using Gaussian elimination, we bring the augmented coefficient matrix into *row echelon form*.

$$\text{augmented coefficient matrix} \quad \begin{bmatrix} 1 & 1 & 3 & \bigg\| & 2 \\ 1 & 2 & 4 & \bigg\| & 3 \\ 1 & 3 & \alpha & \bigg\| & \beta \end{bmatrix}$$

$$R_2 \rightsquigarrow R_2 - R_1, R_3 \rightsquigarrow R_3 - R_1 \quad \begin{bmatrix} 1 & 1 & 3 & \bigg\| & 2 \\ 0 & 1 & 1 & \bigg\| & 1 \\ 0 & 2 & \alpha-3 & \bigg\| & \beta-2 \end{bmatrix}$$

$$R_3 \rightsquigarrow R_3 - 2R_2 \quad \begin{bmatrix} 1 & 0 & 2 & \bigg\| & 1 \\ 0 & 1 & 1 & \bigg\| & 1 \\ 0 & 0 & \alpha-5 & \bigg\| & \beta-4 \end{bmatrix}$$

$\alpha \neq 5$: only for $\alpha \neq 5$ can the augmented coefficient matrix be divided by $\alpha - 5$ and hence be transformed to *reduced row echelon form*.

$$\begin{bmatrix} 1 & 0 & 0 & \bigg\| & 1 - 2\,{}^{(\beta-4)}/_{(\alpha-5)} \\ 0 & 1 & 0 & \bigg\| & 1 - {}^{(\beta-4)}/_{(\alpha-5)} \\ 0 & 0 & 1 & \bigg\| & {}^{(\beta-4)}/_{(\alpha-5)} \end{bmatrix}$$

The unique solution of the system is thus given by

$$L = \left\{ \left(1 - 2\frac{(\beta-4)}{(\alpha-5)}, 1 - \frac{(\beta-4)}{(\alpha-5)}, \frac{(\beta-4)}{(\alpha-5)} \right) \right\}.$$

(i) $\alpha = 5, \beta = 4$: for $\alpha = 5$ and $\beta = 4$, we get the following augmented coefficient matrix

$$\begin{bmatrix} 1 & 0 & 2 & \bigg\| & 1 \\ 0 & 1 & 1 & \bigg\| & 1 \\ 0 & 0 & 0 & \bigg\| & 0 \end{bmatrix}.$$

The set of solutions thus consists of infinitely many solutions given by

$$L = \left\{ (1 - 2t, 1 - t, t) \mid t \in \mathbb{R} \right\}$$

with t a free variable.

(ii) $\alpha = 5, \beta \neq 4$: for $\alpha = 5$ and $\beta \neq 4$, we get

$$\begin{bmatrix} 1 & 0 & 2 & \bigg\| & 1 \\ 0 & 1 & 1 & \bigg\| & 1 \\ 0 & 0 & 0 & \bigg\| & \beta-4 \end{bmatrix}$$

and the set of solutions is empty, $L = \emptyset$.

4. Determine the set of solutions of the following linear system of n equations and n unknowns

$$\begin{cases} x_1 + 5x_2 & = 0 \\ \quad x_2 + 5x_3 & = 0 \\ \quad \vdots \\ \quad x_{n-1} + 5x_n = 0 \\ 5x_1 \qquad\qquad + x_n = 1 \end{cases}$$

Solutions By using Gaussian elimination iteratively, we get

augmented coefficient matrix
$$\begin{bmatrix} 1 & 5 & & & & & 0 \\ & 1 & 5 & & & & 0 \\ & & \vdots & & & & \vdots \\ & & & & 1 & 5 & 0 \\ 5 & & & & & 1 & 1 \end{bmatrix}$$

$R_n \rightsquigarrow R_n - 5R_1$
$$\begin{bmatrix} 1 & 5 & & & & & 0 \\ & 1 & 5 & & & & 0 \\ & & \vdots & & & & \vdots \\ & & & & 1 & 5 & 0 \\ 0 & -5^2 & & & & 1 & 1 \end{bmatrix}$$

$R_n \rightsquigarrow R_n + 5^2 R_2$
$$\begin{bmatrix} 1 & 5 & & & & & 0 \\ & 1 & 5 & & & & 0 \\ & & \vdots & & & & \vdots \\ & & & & 1 & 5 & 0 \\ 0 & 0 & 5^3 & & & 1 & 1 \end{bmatrix}$$

$$\vdots$$

$R_n \rightsquigarrow R_n + (-1)^{n-1}5^{n-1}R_{n-1}$
$$\begin{bmatrix} 1 & 5 & & & & & 0 \\ & 1 & 5 & & & & 0 \\ & & \vdots & & & & \vdots \\ & & & 1 & & 5 & 0 \\ 0 & 0 & 0 & \dots & & 1+(-1)^{n-1}5^n & 1 \end{bmatrix}$$

Finally, we get a solution for x_n: $x_n = \dfrac{1}{1+(-1)^{n-1}5^n}$.

By inserting the solutions backwards into the equations described by the augmented coefficient matrix, one obtains

$$x_{n-1} = (-5) \cdot \frac{1}{1+(-1)^{n-1}5^n}, \dots, x_1 = (-5)^{n-1} \cdot \frac{1}{1+(-1)^{n-1}5^n}.$$

Hence the solution is unique,

$$L = \left\{ \left(\frac{(-5)^{n-1}}{1+(-1)^{n-1}5^n}, \frac{(-5)^{n-2}}{1+(-1)^{n-1}5^n}, \dots, \frac{(-5)}{1+(-1)^{n-1}5^n}, \frac{1}{1+(-1)^{n-1}5^n} \right) \right\}.$$

5. Decide whether the following assertions are true or false and justify your answers.

   (i) There exist linear systems with three equations and three unknowns, which have precisely three solutions due to special symmetry properties.

   (ii) Every linear system with two equations and three unknowns has infinitely many solutions.

Solutions

(i) False. The set of solutions is either empty, a single point, a line, a plane, or the whole space. Thus there is either no solution, a unique solution, or there are infinitely many solutions.

(ii) False. It can have no solutions as well. Consider for example the system of equations with associated augmented coefficient matrix

$$\text{augmented coefficient matrix} \quad \left[\begin{array}{ccc|c} 1 & 1 & 1 & 0 \\ 1 & 1 & 1 & 1 \end{array}\right]$$

$$R_2 \rightsquigarrow R_2 - R_1 \quad \left[\begin{array}{ccc|c} 1 & 1 & 1 & 0 \\ 0 & 0 & 0 & 1 \end{array}\right]$$

which has no solutions.

Solutions of Problems of Sect. 2.1

1. Let A, B, C be the following matrices

$$A = \begin{pmatrix} 2 & -1 & 2 \\ 4 & -2 & 4 \end{pmatrix}, \quad B = \begin{pmatrix} -1 & 0 \\ 2 & 2 \\ 2 & 1 \end{pmatrix}, \quad C = \begin{pmatrix} -1 & 2 \\ 0 & 2 \end{pmatrix}.$$

(i) Determine which product $Q \cdot P$ are defined for $Q, P \in \{A, B, C\}$ and which are not (the matrices Q and P do not have to be different).

(ii) Compute AB and BA.

(iii) Compute $3C^5 + 2C^2$.

(iv) Compute ABC.

Solutions

(i) For the product $Q \cdot P$ to be defined, the number of columns of Q has to be the same as the number of rows of P. Since A has 2 rows and 3 columns, $A \in \mathbb{R}^{2 \times 3}$, B has 3 rows and 2 columns, $B \in \mathbb{R}^{3 \times 2}$ and C has 2 rows and 2 columns, $C \in \mathbb{R}^{2 \times 2}$, AB, BA, BC, CA and CC are defined.

(ii) $AB = \begin{pmatrix} 2 & -1 & 2 \\ 4 & -2 & 4 \end{pmatrix} \begin{pmatrix} -1 & 0 \\ 2 & 2 \\ 2 & 1 \end{pmatrix} = \begin{pmatrix} 0 & 0 \\ 0 & 0 \end{pmatrix}$ and

$$BA = \begin{pmatrix} -1 & 0 \\ 2 & 2 \\ 2 & 1 \end{pmatrix} \begin{pmatrix} 2 & -1 & 2 \\ 4 & -2 & 4 \end{pmatrix} = \begin{pmatrix} -2 & 1 & -2 \\ 12 & -6 & 12 \\ 8 & -4 & 8 \end{pmatrix}.$$

(iii) $C^2 = \begin{pmatrix} -1 & 2 \\ 0 & 2 \end{pmatrix} \begin{pmatrix} -1 & 2 \\ 0 & 2 \end{pmatrix} = \begin{pmatrix} 1 & 2 \\ 0 & 4 \end{pmatrix}$,

$$C^4 = C^2 \cdot C^2 = \begin{pmatrix} 1 & 2 \\ 0 & 4 \end{pmatrix} \begin{pmatrix} 1 & 2 \\ 0 & 4 \end{pmatrix} = \begin{pmatrix} 1 & 10 \\ 0 & 16 \end{pmatrix},$$

$$C^5 = C \cdot C^4 = \begin{pmatrix} -1 & 2 \\ 0 & 2 \end{pmatrix} \begin{pmatrix} 1 & 10 \\ 0 & 16 \end{pmatrix} = \begin{pmatrix} -1 & 22 \\ 0 & 32 \end{pmatrix},$$

$$3C^5 + 2C^2 = 3 \begin{pmatrix} -1 & 22 \\ 0 & 32 \end{pmatrix} + 2 \begin{pmatrix} 1 & 2 \\ 0 & 4 \end{pmatrix} = \begin{pmatrix} -3 & 66 \\ 0 & 96 \end{pmatrix} + \begin{pmatrix} 2 & 4 \\ 0 & 8 \end{pmatrix} = \begin{pmatrix} -1 & 70 \\ 0 & 104 \end{pmatrix}.$$

(iv) From the solution of (i), it follows that $AB = \begin{pmatrix} 0 & 0 \\ 0 & 0 \end{pmatrix}$,

hence $ABC = \begin{pmatrix} 0 & 0 \\ 0 & 0 \end{pmatrix} \begin{pmatrix} -1 & 2 \\ 0 & 2 \end{pmatrix} = \begin{pmatrix} 0 & 0 \\ 0 & 0 \end{pmatrix}.$

2. Determine which of the following matrices are regular and if so, determine their inverses.

(i) $A = \begin{pmatrix} 1 & 2 & -2 \\ 0 & -1 & 1 \\ 2 & 3 & 0 \end{pmatrix}$

(ii) $B = \begin{pmatrix} 1 & 2 & 2 \\ 0 & 2 & -1 \\ -1 & 0 & -3 \end{pmatrix}$

Solutions

(i) Using Gaussian elimination, we decide whether A is regular and if so, compute its inverse.

$$\text{augmented coefficient matrix} \quad \left[\begin{array}{ccc|ccc} 1 & 2 & -2 & 1 & 0 & 0 \\ 0 & -1 & 1 & 0 & 1 & 0 \\ 2 & 3 & 0 & 0 & 0 & 1 \end{array} \right]$$

$$R_3 \rightsquigarrow R_3 - 2R_1 \quad \left[\begin{array}{ccc|ccc} 1 & 2 & -2 & 1 & 0 & 0 \\ 0 & -1 & 1 & 0 & 1 & 0 \\ 0 & -1 & 4 & -2 & 0 & 1 \end{array} \right]$$

$$R_2 \rightsquigarrow -R_2 \quad \left[\begin{array}{ccc|ccc} 1 & 2 & -2 & 1 & 0 & 0 \\ 0 & 1 & -1 & 0 & -1 & 0 \\ 0 & -1 & 4 & -2 & 0 & 1 \end{array} \right]$$

$$R_3 \rightsquigarrow R_3 + R_2 \quad \left[\begin{array}{ccc|ccc} 1 & 2 & -2 & 1 & 0 & 0 \\ 0 & 1 & -1 & 0 & -1 & 0 \\ 0 & 0 & 3 & -2 & -1 & 1 \end{array} \right]$$

$$R_3 \rightsquigarrow {}^1\!/_3 \, R_3 \quad \left[\begin{array}{ccc|ccc} 1 & 2 & -2 & 1 & 0 & 0 \\ 0 & 1 & -1 & 0 & -1 & 0 \\ 0 & 0 & 1 & -{}^2\!/_3 & -{}^1\!/_3 & {}^1\!/_3 \end{array} \right]$$

It follows that A is regular. Its inverse can be computed as follows.

$$R_1 \rightsquigarrow R_1 - 2R_2 \quad \left[\begin{array}{ccc|ccc} 1 & 0 & 0 & 1 & 2 & 0 \\ 0 & 1 & -1 & 0 & -1 & 0 \\ 0 & 0 & 1 & -2/3 & -1/3 & 1/3 \end{array}\right]$$

$$R_2 \rightsquigarrow R_2 + R_3 \quad \left[\begin{array}{ccc|ccc} 1 & 0 & 0 & 1 & 2 & 0 \\ 0 & 1 & 0 & -2/3 & -4/3 & 1/3 \\ 0 & 0 & 1 & -2/3 & -1/3 & 1/3 \end{array}\right]$$

Hence, the inverse of A is $A^{-1} = \begin{pmatrix} 1 & 2 & 0 \\ -2/3 & -4/3 & 1/3 \\ -2/3 & -1/3 & 1/3 \end{pmatrix}$.

(ii) Using Gaussian elimination, we decide whether B is regular or not.

augmented coefficient matrix $\quad \left[\begin{array}{ccc|ccc} 1 & 2 & 2 & 1 & 0 & 0 \\ 0 & 2 & -1 & 0 & 1 & 0 \\ -1 & 0 & -3 & 0 & 0 & 1 \end{array}\right]$

$$R_3 \rightsquigarrow R_3 + R_1 \quad \left[\begin{array}{ccc|ccc} 1 & 2 & 2 & 1 & 0 & 0 \\ 0 & 2 & -1 & 0 & 1 & 0 \\ 0 & 2 & -1 & 1 & 0 & 1 \end{array}\right]$$

$$R_3 \rightsquigarrow R_3 - R_2 \quad \left[\begin{array}{ccc|ccc} 1 & 2 & 2 & 1 & 0 & 0 \\ 0 & 2 & -1 & 0 & 1 & 0 \\ 0 & 0 & 0 & 1 & -1 & 1 \end{array}\right]$$

Hence B is not regular.

3. (i) Determine all real numbers α, β for which the 2×2 matrix

$$A := \begin{pmatrix} \alpha & \beta \\ \beta & \alpha \end{pmatrix}$$

is invertible and compute for those numbers the inverse of A.

(ii) Determine all real numbers a, b, c, d, e, f for which the matrix

$$B := \begin{pmatrix} a & d & e \\ 0 & b & f \\ 0 & 0 & c \end{pmatrix}$$

is invertible and compute for those numbers the inverse of B.

Solutions

(i) The matrix is invertible if and only if $\det(A) = \alpha^2 - \beta^2 \neq 0$. In the latter case, A^{-1} is given by Eq. (2.1),

$$A^{-1} = \frac{1}{\alpha^2 - \beta^2} \begin{pmatrix} \alpha & -\beta \\ -\beta & \alpha \end{pmatrix}.$$

(ii) From

$$\left[\begin{array}{ccc|ccc} a & d & e & 1 & 0 & 0 \\ 0 & b & f & 0 & 1 & 0 \\ 0 & 0 & c & 0 & 0 & 1 \end{array}\right]$$

we see that B is invertible if and only if $abc \neq 0$. If $abc \neq 0$, we proceed by Gaussian elimination.

augmented coefficient matrix

$$\left[\begin{array}{ccc|ccc} a & d & e & 1 & 0 & 0 \\ 0 & b & f & 0 & 1 & 0 \\ 0 & 0 & c & 0 & 0 & 1 \end{array}\right]$$

$R_1 \rightsquigarrow {}^1/_a R_1,\ R_2 \rightsquigarrow {}^1/_b R_2,\ R_3 \rightsquigarrow {}^1/_c R_3$

$$\left[\begin{array}{ccc|ccc} 1 & {}^d/_a & {}^e/_a & {}^1/_a & 0 & 0 \\ 0 & 1 & {}^f/_b & 0 & {}^1/_b & 0 \\ 0 & 0 & 1 & 0 & 0 & {}^1/_c \end{array}\right]$$

$R_2 \rightsquigarrow R_2 - {}^f/_b R_3$

$$\left[\begin{array}{ccc|ccc} 1 & {}^d/_a & {}^e/_a & {}^1/_a & 0 & 0 \\ 0 & 1 & 0 & 0 & {}^1/_b & -{}^f/_{bc} \\ 0 & 0 & 1 & 0 & 0 & {}^1/_c \end{array}\right]$$

$R_1 \rightsquigarrow R_1 - {}^d/_a R_2 - {}^e/_a R_3$

$$\left[\begin{array}{ccc|ccc} 1 & 0 & 0 & {}^1/_a & -{}^d/_{ab} & {}^{df}/_{abc} - {}^e/_{ac} \\ 0 & 1 & 0 & 0 & {}^1/_b & -{}^f/_{bc} \\ 0 & 0 & 1 & 0 & 0 & {}^1/_c \end{array}\right]$$

Hence, the inverse matrix ist given by

$$\begin{pmatrix} {}^1/_a & -{}^d/_{ab} & {}^{df}/_{abc} - {}^e/_{ac} \\ 0 & {}^1/_b & -{}^f/_{bc} \\ 0 & 0 & {}^1/_c \end{pmatrix}.$$

4. (i) Find symmetric 2×2 matrices A, B so that the product AB is not symmetric.

   (ii) Verify: for any 4×4 matrices of the form

$$A = \begin{pmatrix} A_1 & A_2 \\ 0 & A_3 \end{pmatrix}, \qquad B = \begin{pmatrix} B_1 & B_2 \\ 0 & B_3 \end{pmatrix}$$

   where A_1, A_2, A_3 and B_1, B_2, B_3 are 2×2 matrices, the 4×4 matrix AB is given by

$$AB = \begin{pmatrix} A_1 B_1 & A_1 B_2 + A_2 B_3 \\ 0 & A_3 B_3 \end{pmatrix}.$$

Solutions

(i) Consider the following general 2×2 matrices with $a, b, c, d, a', b', c', d' \in \mathbb{R}$,

$$\begin{pmatrix} a & b \\ c & d \end{pmatrix} \begin{pmatrix} a' & b' \\ c' & d' \end{pmatrix} = \begin{pmatrix} aa' + bb' & ab' + bc' \\ ba' + cb' & bb' + cc' \end{pmatrix}.$$

Hence, for the symmetric matrices $A = \begin{pmatrix} 1 & 1 \\ 1 & 0 \end{pmatrix}$ and $B = \begin{pmatrix} 2 & 1 \\ 1 & 0 \end{pmatrix}$, one has $AB = \begin{pmatrix} 3 & 1 \\ 2 & 1 \end{pmatrix}$.

(ii) For any $1 \le i, j \le 2$:

$$(AB)_{ij} = R_i(A)C_j(B) = R_i(A_1)C_j(B_1) = (A_1 B_1)_{ij},$$

$$(AB)_{(i+2)j} = R_{i+2}(A)C_j(B) = \begin{pmatrix} 0 & 0 \end{pmatrix} C_j(B_1) + R_i(A_3) \begin{pmatrix} 0 \\ 0 \end{pmatrix} = 0,$$

$$(AB)_{i(j+2)} = R_i(A)C_{j+2}(B) = R_i(A_1)C_j(B_2) + R_i(A_2)C_j(B_3) = (A_1 B_2)_{ij} + (A_2 B_3)_{ij},$$

$$(AB)_{(i+2)(j+2)} = R_{i+2}(A)C_{j+2}(B) = \begin{pmatrix} 0 & 0 \end{pmatrix} C_j(B_2) + R_i(A_3)C_j(B_3) = (A_3 B_3)_{ij}.$$

5. Decide whether the following assertions are true or false and justify your answers.

(i) For arbitrary matrices A, B in $\mathbb{R}^{2\times2}$,

$$(A + B)^2 = A^2 + 2AB + B^2.$$

(ii) Let A be the 2×2 matrix $A = \begin{pmatrix} 1 & 2 \\ 3 & 5 \end{pmatrix}$. Then for any $k \in \mathbb{N}$, A^k is invertible and for any $n, m \in \mathbb{Z}$,

$$A^{n+m} = A^n A^m.$$

(Recall that $A^0 = \mathrm{Id}_{2\times2}$ and for any $k \in \mathbb{N}$, A^{-k} is defined as $A^{-k} = (A^{-1})^k$.)

Solutions

(i) False. We first note that

$$(A + B)^2 = A^2 + AB + BA + B^2,$$

thus if $AB \ne BA$, the claim is false. Consider for example

$$A = \begin{pmatrix} 2 & 0 \\ 0 & 3 \end{pmatrix}, \quad B = \begin{pmatrix} 0 & 1 \\ -1 & 0 \end{pmatrix}, \quad AB = \begin{pmatrix} 0 & 2 \\ -3 & 0 \end{pmatrix}, \quad BA = \begin{pmatrix} 0 & 3 \\ -2 & 0 \end{pmatrix}.$$

Then

$$(A + B)^2 = \begin{pmatrix} 4 & 0 \\ 0 & 9 \end{pmatrix} + \begin{pmatrix} 0 & 2 \\ -3 & 0 \end{pmatrix} + \begin{pmatrix} 0 & 3 \\ -2 & 0 \end{pmatrix} + \begin{pmatrix} -1 & 0 \\ 0 & -1 \end{pmatrix} = \begin{pmatrix} 3 & 5 \\ -5 & 8 \end{pmatrix},$$

while

$$A^2 + 2AB + B^2 = \begin{pmatrix} 4 & 0 \\ 0 & 9 \end{pmatrix} + \begin{pmatrix} 0 & 4 \\ -6 & 0 \end{pmatrix} + \begin{pmatrix} -1 & 0 \\ 0 & -1 \end{pmatrix} = \begin{pmatrix} 3 & 4 \\ -6 & 8 \end{pmatrix}.$$

(ii) True. First note that A is invertible, since $\det(A) = 5 - 6 = -1 \neq 0$. By Eq. (2.1) one therefore has

$$A^{-1} = \frac{1}{\det(A)} \begin{pmatrix} 5 & -2 \\ -3 & 1 \end{pmatrix} = \begin{pmatrix} -5 & 2 \\ 3 & -1 \end{pmatrix}.$$

This implies that A^k is invertible, since $A A^{-1} = \mathrm{Id}_{2 \times 2}$ and thus $A^k A^{-k} = A^k (A^{-1})^k = \underbrace{A \cdots A \cdot A \cdot A^{-1} \cdot A^{-1} \cdots A^{-1}}_{= \mathrm{Id}_{2 \times 2}} = \mathrm{Id}_{2 \times 2}$. Since A^k and A^{-k} are defined for any $k \in \mathbb{N}$ and $A^0 = \mathrm{Id}_{2 \times 2}$, the exponentiation law $A^{n+m} = A^n A^m$ holds for any $n, m \in \mathbb{Z}$. More precisely, for any $n, m \in \mathbb{N}$ one has

$$A^n A^m = \underbrace{A \cdots A}_{n} \cdot \underbrace{A \cdots A}_{m} = A^{n+m},$$

$$A^{-n} A^{-m} = \underbrace{A^{-1} \cdots A^{-1}}_{n} \cdot \underbrace{A^{-1} \cdots A^{-1}}_{m} = A^{-n-m},$$

$$A^n A^0 = A^n = A^0 A^n,$$

$$A^{-n} A^m = \begin{cases} A^{-r} A^{-m} A^m = A^{-n+m} & \text{if } r := n - m \geq 0 \\ A^{-n} A^n A^{-r} = A^{-n+m} & \text{if } r := n - m < 0 \end{cases}.$$

Solutions of Problems of Sect. 2.2

1. (i) Decide whether $a^{(1)} = (2, 5)$, $a^{(2)} = (5, 2)$ are linearly dependent in $\mathbb{R}^2$.
    (ii) If possible, represent $b = (1, 9)$ as a linear combination of $a^{(1)} = (1, 1)$ and $a^{(2)} = (3, -1)$.
    Solutions

(i) $a^{(1)}$ and $a^{(2)}$ are linearly independent if the matrix $\begin{pmatrix} 2 & 5 \\ 5 & 2 \end{pmatrix}$ is regular. Since $\det(A) = 4 - 10 \neq 0$, A is regular.

(ii) To write b, if possible, as a linear combination of $a^{(1)}$ and $a^{(2)}$, we need to solve the linear system

$$Ax = b, \qquad A = \begin{pmatrix} 1 & 3 \\ 1 & -1 \end{pmatrix}, \qquad x = \begin{pmatrix} x_1 \\ x_2 \end{pmatrix}.$$

The corresponding augmented coefficient matrix can be transformed by Gaussian elimination into row echelon form.

augmented coefficient matrix $\qquad \begin{bmatrix} 1 & 3 & \| & 1 \\ 1 & -1 & \| & 9 \end{bmatrix}$

$R_2 \rightsquigarrow R_2 - R_1 \qquad \begin{bmatrix} 1 & 3 & \| & 1 \\ 0 & -4 & \| & 8 \end{bmatrix}$

$R_2 \rightsquigarrow -{}^1\!/_4\, R_2 \qquad \begin{bmatrix} 1 & 3 & \| & 1 \\ 0 & 1 & \| & -2 \end{bmatrix}$

$R_1 \rightsquigarrow R_1 - 3R_2 \qquad \begin{bmatrix} 1 & 0 & \| & 7 \\ 0 & 1 & \| & -2 \end{bmatrix}$

Hence, $b = 7a^{(1)} - 2a^{(2)}$.

2. Decide whether the following vectors in $\mathbb{R}^3$ are linearly dependent or not.

   (i) $a^{(1)} = (1, 2, 1)$, $a^{(2)} = (-1, 1, 3)$.
   (ii) $a^{(1)} = (1, 1, -1)$, $a^{(2)} = (0, 4, -3)$, $a^{(3)} = (1, 0, 3)$.

Solutions

(i) Let us assume that the vectors are linearly dependent and then either show that it is true or show that the assumption leads to a contradiction.

   Case 1. Let us assume there exists $\alpha_1 \in \mathbb{R}$ with $a^{(1)} = \alpha_1 a^{(2)}$. This leads to $1 = -1\alpha_1$, $2 = 1\alpha_1$ and $1 = 3\alpha_1$, clearly a contradiction.

   Case 2. Let us assume there exists $\alpha_2 \in \mathbb{R}$ with $a^{(2)} = \alpha_2 a^{(1)}$. This leads to $-1 = 1\alpha_2$, $1 = 2\alpha_2$ and $3 = 1\alpha_2$, also a contradiction.

   Hence, the vectors $a^{(1)}$ and $a^{(2)}$ are linearly independent. Alternatively, the linear independence of $a^{(1)}$ and $a^{(2)}$ can be established by showing that the linear system

   $$Ax = \begin{pmatrix} 0 \\ 0 \\ 0 \end{pmatrix} \text{ has only the trivial solution } x = \begin{pmatrix} 0 \\ 0 \end{pmatrix}. \text{ Here } A \text{ is the } 3 \times 2 \text{ matrix, whose}$$

   columns are $C_1(A) = a^{(1)}$, $C_2(A) = a^{(2)}$. By Gaussian elimination we transform the augmented coefficient matrix into row echelon form,

   $$\text{augmented coefficient matrix} \quad \left[\begin{array}{cc|c} 1 & 1 & 0 \\ 1 & 1 & 0 \\ 1 & 2 & 0 \end{array}\right]$$

   $$R_2 \rightsquigarrow R_2 - R_1, R_3 \rightsquigarrow R_3 - R_1 \quad \left[\begin{array}{cc|c} 1 & 1 & 0 \\ 0 & 0 & 0 \\ 0 & 1 & 0 \end{array}\right]$$

   $$R_{2\leftrightarrow 3} \quad \left[\begin{array}{cc|c} 1 & 1 & 0 \\ 0 & 1 & 0 \\ 0 & 0 & 0 \end{array}\right].$$

   Hence, $Ax = \begin{pmatrix} 0 \\ 0 \\ 0 \end{pmatrix}$ has only the trivial solution.

(ii) The vectors $a^{(1)}, a^{(2)}$ and $a^{(3)}$ are linearly independent if and only if for any real numbers x_1, x_3, x_3 with $x_1 a^{(1)} + x_2 a^{(2)} + x_3 a^{(3)} = 0$, it follows that $x_1 = 0$, $x_2 = 0$, and $x_3 = 0$. It means that the homogenous linear system $Ax = 0$ has only the trivial solution $x = 0$. Here A is the 3×3 matrix with columns $C_1(A) = a^{(1)}$, $C_2(A) = a^{(2)}$ and $C_3(A) = a^{(3)}$. By Gaussian elimination, we transform the augmented coefficient matrix into row echelon form,

$$\text{augmented coefficient matrix} \quad \begin{bmatrix} 1 & 0 & 1 & \Big\| & 0 \\ 1 & 4 & 0 & \Big\| & 0 \\ -1 & -3 & 3 & \Big\| & 0 \end{bmatrix}$$

$$R_2 \rightsquigarrow R_2 - R_1, \ R_3 \rightsquigarrow R_3 + R_1 \quad \begin{bmatrix} 1 & 0 & 1 & \Big\| & 0 \\ 0 & 4 & -1 & \Big\| & 0 \\ 0 & -3 & 4 & \Big\| & 0 \end{bmatrix}$$

$$R_2 \rightsquigarrow \tfrac{1}{4} R_2 \quad \begin{bmatrix} 1 & 0 & 1 & \Big\| & 0 \\ 0 & 1 & -\tfrac{1}{4} & \Big\| & 0 \\ 0 & -3 & 4 & \Big\| & 0 \end{bmatrix}$$

$$R_3 \rightsquigarrow R_3 + 3R_2 \quad \begin{bmatrix} 1 & 0 & 1 & \Big\| & 0 \\ 0 & 1 & -\tfrac{1}{4} & \Big\| & 0 \\ 0 & 0 & \tfrac{13}{4} & \Big\| & 0 \end{bmatrix}.$$

Hence, $Ax = 0$ admits only the trivial solution $x = 0$.

3. (i) Let $a^{(1)} := (1, 1, 1)$, $a^{(2)} := (1, 1, 2)$. Find a vector $a^{(3)} \in \mathbb{R}^3$ so that $a^{(1)}$, $a^{(2)}$, and $a^{(3)}$ form a basis of $\mathbb{R}^3$.

(ii) Let $[a] = [a^{(1)}, a^{(2)}]$ be the basis of $\mathbb{R}^2$, given by

$$a^{(1)} = (1, -1), \qquad a^{(2)} = (2, 1).$$

Compute $\mathrm{Id}_{[e] \to [a]}$ and $\mathrm{Id}_{[a] \to [e]}$.

Solutions

(i) First, we note that $a^{(1)}$ and $a^{(2)}$ arc lincarly independent. Indeed, transforming, the augmented coefficient matrix into row echelon form, one gets

$$\text{augmented coefficient matrix} \quad \begin{bmatrix} 1 & 1 & \Big\| & 0 \\ 1 & 1 & \Big\| & 0 \\ 1 & 2 & \Big\| & 0 \end{bmatrix}$$

$$R_2 \rightsquigarrow R_2 - R_1, \ R_3 \rightsquigarrow R_3 - R_1 \quad \begin{bmatrix} 1 & 1 & \Big\| & 0 \\ 0 & 0 & \Big\| & 0 \\ 0 & 1 & \Big\| & 0 \end{bmatrix}$$

$$R_{2 \leftrightarrow 3} \quad \begin{bmatrix} 1 & 1 & \Big\| & 0 \\ 0 & 1 & \Big\| & 0 \\ 0 & 0 & \Big\| & 0 \end{bmatrix}.$$

Hence, $a^{(1)}$ and $a^{(2)}$ are linearly independent. To find $a^{(3)} = (x_1, x_2, x_3)$ so that $a^{(1)}, a^{(2)}, a^{(3)}$ form a basis, we consider

$$\begin{bmatrix} 1 & 1 & x_1 & \Big\| & 0 \\ 1 & 1 & x_2 & \Big\| & 0 \\ 1 & 2 & x_3 & \Big\| & 0 \end{bmatrix}$$

and transform it into row echelon form. As before

$$\text{augmented coefficient matrix} \qquad \begin{bmatrix} 1 & 1 & x_1 & \Big\| & 0 \\ 1 & 1 & x_2 & \Big\| & 0 \\ 1 & 2 & x_3 & \Big\| & 0 \end{bmatrix}$$

$$R_2 \rightsquigarrow R_2 - R_1,\ R_3 \rightsquigarrow R_3 - R_1 \qquad \begin{bmatrix} 1 & 1 & x_1 & \Big\| & 0 \\ 0 & 0 & x_2 - x_1 & \Big\| & 0 \\ 0 & 1 & x_3 - x_1 & \Big\| & 0 \end{bmatrix}$$

$$R_{2 \leftrightarrow 3} \qquad \begin{bmatrix} 1 & 1 & x_1 & \Big\| & 0 \\ 0 & 1 & x_3 - x_1 & \Big\| & 0 \\ 0 & 0 & x_2 - x_1 & \Big\| & 0 \end{bmatrix}.$$

Hence, a possible solution is $(x_1, x_2, x_3) = (0, 1, 0)$. Note that there are many solutions. In geometric terms, we need to choose $a^{(3)}$ in such a way that it is not contained in the plane spanned by $a(1)$ and $a^{(2)}$, $\{sa^{(1)} + ta^{(2)} \mid s, t \in \mathbb{R}\} \subseteq \mathbb{R}^3$.

(ii) Since the coefficients of $a^{(1)}$ and $a^{(2)}$ with respect to the standard basis $[e] = [e^{(1)}, e^{(2)}]$ are the components of $a^{(1)}$ and $a^{(2)}$, respectively, we first compute $\text{Id}_{[a] \to [e]}$.

As $a^{(1)} = 1 \cdot e^{(1)} - 1 \cdot e^{(2)}$ and $a^{(2)} = 2 \cdot e^{(1)} + 1 \cdot e^{(2)}$, we get

$$S := \text{Id}_{[a] \to [e]} = \begin{pmatrix} 1 & 2 \\ -1 & 1 \end{pmatrix}.$$

Since $T := \text{Id}_{[e] \to [a]} = S^{-1}$, we need to find the inverse of S. We use Eq. (2.1) for the inverse of a 2×2 matrix,

$$S^{-1} = \frac{1}{\det(S)} \begin{pmatrix} 1 & -2 \\ 1 & 1 \end{pmatrix} = \frac{1}{3} \begin{pmatrix} 1 & -2 \\ 1 & 1 \end{pmatrix} = \begin{pmatrix} 1/3 & -2/3 \\ 1/3 & 1/3 \end{pmatrix}.$$

4. Consider the basis $[a] = [a^{(1)}, a^{(2)}, a^{(3)}]$ of $\mathbb{R}^3$, given by $a^{(1)} = (1, 1, 0)$, $a^{(2)} = (1, 0, 1)$ and $a^{(3)} = (0, 1, 1)$, and denote by $[e] = [e^{(1)}, e^{(2)}, e^{(3)}]$ the standard basis of $\mathbb{R}^3$.

(i) Compute $S := \text{Id}_{[a] \to [e]}$ and $T := \text{Id}_{[e] \to [a]}$.
(ii) Compute the coordinates $\alpha_1, \alpha_2, \alpha_3$ of the vector $b = (1, 2, 3)$ with respect to the basis $[a]$, $b = \alpha_1 a^{(1)} + \alpha_2 a^{(2)} + \alpha_3 a^{(3)}$, and determine the coefficients $\beta_1, \beta_2, \beta_3$ of the vector $a^{(1)} + 2a^{(2)} + 3a^{(3)}$ with respect to the standard basis $[e]$.

Solutions

(i) Since the coefficients of $a^{(1)}, a^{(2)}, a^{(3)}$ with respect to the standard basis $[e]$ are known, it is convenient to first compute S. One has

$$a^{(1)} = 1 \cdot e^{(1)} + 1 \cdot e^{(2)} + 0 \cdot e^{(3)}, \qquad a^{(2)} = 1 \cdot e^{(1)} + 0 \cdot e^{(2)} + 1 \cdot e^{(3)},$$

$$a^{(3)} = 0 \cdot e^{(1)} + 1 \cdot e^{(2)} + 1 \cdot e^{(3)},$$

hence

$$S = \begin{pmatrix} 1 & 1 & 0 \\ 1 & 0 & 1 \\ 0 & 1 & 1 \end{pmatrix}.$$

Since $T = S^{-1}$, we find T by transforming the augmented coefficient matrix into reduced row echelon form,

augemented coefficient matrix
$$\left[\begin{array}{ccc|ccc} 1 & 1 & 0 & 1 & 0 & 0 \\ 1 & 0 & 1 & 0 & 1 & 0 \\ 0 & 1 & 1 & 0 & 0 & 1 \end{array}\right]$$

$R_2 \rightsquigarrow R_2 - R_1$
$$\left[\begin{array}{ccc|ccc} 1 & 1 & 0 & 1 & 0 & 0 \\ 0 & -1 & 1 & -1 & 1 & 0 \\ 0 & 1 & 1 & 0 & 0 & 1 \end{array}\right]$$

$R_3 \rightsquigarrow R_3 + R_2$
$$\left[\begin{array}{ccc|ccc} 1 & 1 & 0 & 1 & 0 & 0 \\ 0 & -1 & 1 & -1 & 1 & 0 \\ 0 & 0 & 2 & -1 & 1 & 1 \end{array}\right]$$

$R_2 \rightsquigarrow -R_2, R_3 \rightsquigarrow \frac{1}{2} R_3$
$$\left[\begin{array}{ccc|ccc} 1 & 1 & 0 & 1 & 0 & 0 \\ 0 & 1 & -1 & 1 & -1 & 0 \\ 0 & 0 & 1 & -\frac{1}{2} & \frac{1}{2} & \frac{1}{2} \end{array}\right]$$

$R_2 \rightsquigarrow R_2 + R_3$
$$\left[\begin{array}{ccc|ccc} 1 & 1 & 0 & 1 & 0 & 0 \\ 0 & 1 & 0 & \frac{1}{2} & -\frac{1}{2} & \frac{1}{2} \\ 0 & 0 & 1 & -\frac{1}{2} & \frac{1}{2} & \frac{1}{2} \end{array}\right]$$

$R_1 \rightsquigarrow R_1 - R_2$
$$\left[\begin{array}{ccc|ccc} 1 & 0 & 0 & \frac{1}{2} & \frac{1}{2} & -\frac{1}{2} \\ 0 & 1 & 0 & \frac{1}{2} & -\frac{1}{2} & \frac{1}{2} \\ 0 & 0 & 1 & -\frac{1}{2} & \frac{1}{2} & \frac{1}{2} \end{array}\right]$$

Hence, $T = \frac{1}{2} \begin{pmatrix} 1 & 1 & -1 \\ 1 & -1 & 1 \\ -1 & 1 & 1 \end{pmatrix}.$

(ii) The coefficients $\alpha_1, \alpha_2, \alpha_3$ of $b = (1, 2, 3)$ with respect to basis $[a]$ can be computed as

$$\alpha = \begin{pmatrix} \alpha_1 \\ \alpha_2 \\ \alpha_3 \end{pmatrix} = \mathrm{Id}_{[e] \to [a]} \begin{pmatrix} 1 \\ 2 \\ 3 \end{pmatrix} = T \begin{pmatrix} 1 \\ 2 \\ 3 \end{pmatrix} = \frac{1}{2} \begin{pmatrix} 1 & 1 & -1 \\ 1 & -1 & 1 \\ -1 & 1 & 1 \end{pmatrix} \begin{pmatrix} 1 \\ 2 \\ 3 \end{pmatrix} = \begin{pmatrix} 0 \\ 1 \\ 2 \end{pmatrix}.$$

The coefficients $\beta_1, \beta_2, \beta_3$ of the vector $1a^{(1)} + 2^{(2)} + 3a^{(3)}$ with respect to the basis $[e]$ can be computed as

$$\beta = \begin{pmatrix} \beta_1 \\ \beta_2 \\ \beta_3 \end{pmatrix} = \mathrm{Id}_{[a] \to [e]} \begin{pmatrix} 1 \\ 2 \\ 3 \end{pmatrix} = S \begin{pmatrix} 1 \\ 2 \\ 3 \end{pmatrix} = \begin{pmatrix} 1 & 1 & 0 \\ 1 & 0 & 1 \\ 0 & 1 & 1 \end{pmatrix} \begin{pmatrix} 1 \\ 2 \\ 3 \end{pmatrix} = \begin{pmatrix} 3 \\ 4 \\ 5 \end{pmatrix}.$$

5. Decide whether the following assertions are true or false and justify your answers.

(i) Let $n, m \geq 2$. If one of the vectors $a^{(1)}, \ldots, a^{(n)} \in \mathbb{R}^m$ is the null vector, then $a^{(1)}, \ldots, a^{(n)}$ are linearly dependent.

(ii) Assume that $a^{(1)}, \ldots, a^{(n)}$ are vectors in $\mathbb{R}^2$. If $n \geq 3$, then any vector $b \in \mathbb{R}^2$ can be written as a linear combination of $a^{(1)}, \ldots, a^{(n)}$.

Solutions

(i) True. Suppose $a^{(k)} = 0$ in $\mathbb{R}^m$ with $1 \leq k \leq n$. Then

$$a^{(k)} = 0 = \sum_{j \neq k} 0 \cdot a^{(j)}.$$

(ii) False. Suppose $a^{(1)} = a^{(2)} = a^{(3)} = 0$, then the claim is false for any $b \in \mathbb{R}^2 \setminus \{0\}$.

Solutions of Problems of Sect. 2.3

1. Decide whether the following vectors in $\mathbb{R}^3$ form a basis of $\mathbb{R}^3$ and if so, represent $b = (1, 0, 1)$ as a linear combination of the basis vectors.

   (i) $a^{(1)} = (1, 0, 0)$, $a^{(2)} = (0, 4, -1)$, $a^{(3)} = (2, 2, -3)$,
   (ii) $a^{(1)} = (2, -4, 5)$, $a^{(2)} = (1, 5, 6)$, $a^{(3)} = (1, 1, 1)$.

Solutions

(i) We simultaneously check $a^{(1)}$, $a^{(2)}$, and $a^{(3)}$ for linear independence and solve $Ax = b$ where $A = \begin{pmatrix} 1 & 0 & 2 \\ 0 & 4 & 2 \\ 0 & -1 & -3 \end{pmatrix}$. To this end, we transform the augmented coefficient matrix in row echelon form.

augmented coefficient matrix $\qquad \begin{bmatrix} 1 & 0 & 2 & \Big\| & 1 \\ 0 & 4 & 2 & \Big\| & 0 \\ 0 & -1 & -3 & \Big\| & 1 \end{bmatrix}$

$R_3 \rightsquigarrow R_3 + \frac{1}{4} R_2 \qquad \begin{bmatrix} 1 & 0 & 2 & \Big\| & 1 \\ 0 & 4 & 2 & \Big\| & 0 \\ 0 & 0 & -\frac{5}{2} & \Big\| & 1 \end{bmatrix}$

$R_2 \rightsquigarrow \frac{1}{4} R_2, R_3 \rightsquigarrow -\frac{2}{5} R_3 \qquad \begin{bmatrix} 1 & 0 & 2 & \Big\| & 1 \\ 0 & 1 & \frac{1}{2} & \Big\| & 0 \\ 0 & 0 & 1 & \Big\| & -\frac{2}{5} \end{bmatrix}$

$R_2 \rightsquigarrow R_2 - \frac{1}{2} R_3, R_1 \rightsquigarrow R_1 - 2R_3 \qquad \begin{bmatrix} 1 & 0 & 0 & \Big\| & \frac{9}{5} \\ 0 & 1 & 0 & \Big\| & \frac{1}{5} \\ 0 & 0 & 1 & \Big\| & -\frac{2}{5} \end{bmatrix}$

Consequently, A is regular, hence $a^{(1)}, a^{(2)}, a^{(3)}$ is a basis of $\mathbb{R}^3$ and

$$b = \frac{9}{5} a^{(1)} + \frac{1}{5} a^{(2)} - \frac{2}{5} a^{(3)}.$$

(ii) We simultaneously check $a^{(1)}$, $a^{(2)}$, and $a^{(3)}$ for linear independence and solve $Ax = b$

where $A = \begin{pmatrix} 2 & 1 & 2 \\ -4 & 5 & 1 \\ 5 & 6 & 1 \end{pmatrix}$. Again, we transform the augmented coefficient matrix into row echelon form.

augmented coefficient matrix $\qquad \left[\begin{array}{ccc|c} 2 & 1 & 1 & 1 \\ -4 & 5 & 1 & 0 \\ 5 & 6 & 1 & 1 \end{array} \right]$

$R_2 \rightsquigarrow R_2 + 2R_1,\ R_3 - {}^5\!/_2\, R_1 \qquad \left[\begin{array}{ccc|c} 2 & 1 & 1 & 1 \\ 0 & 7 & 3 & 2 \\ 0 & {}^7\!/_2 & {}^3\!/_2 & -{}^3\!/_2 \end{array} \right]$

$R_3 \rightsquigarrow R_3 - {}^1\!/_2\, R_2 \qquad \left[\begin{array}{ccc|c} 2 & 1 & 1 & 1 \\ 0 & 7 & 3 & 2 \\ 0 & 0 & -3 & -{}^5\!/_2 \end{array} \right]$

$R_1 \rightsquigarrow {}^1\!/_2\, R_1,\ R_2 \rightsquigarrow {}^1\!/_7\, R_2,\ R_3 \rightsquigarrow -{}^1\!/_3\, R_2 \qquad \left[\begin{array}{ccc|c} 1 & {}^1\!/_2 & {}^1\!/_2 & {}^1\!/_2 \\ 0 & 1 & {}^3\!/_7 & {}^2\!/_7 \\ 0 & 0 & 1 & {}^5\!/_6 \end{array} \right]$

$R_1 \rightsquigarrow R_1 - {}^1\!/_2\, R_3,\ R_2 \rightsquigarrow R_2 - {}^3\!/_7\, R_3 \qquad \left[\begin{array}{ccc|c} 1 & {}^1\!/_2 & 0 & {}^1\!/_{12} \\ 0 & 1 & 0 & -{}^1\!/_{14} \\ 0 & 0 & 1 & {}^5\!/_6 \end{array} \right]$

$R_1 \rightsquigarrow R_1 - {}^1\!/_2\, R_2 \qquad \left[\begin{array}{ccc|c} 1 & 0 & 0 & {}^5\!/_{42} \\ 0 & 1 & 0 & -{}^1\!/_{14} \\ 0 & 0 & 1 & {}^5\!/_6 \end{array} \right]$

Consequently, A is regular, hence $a^{(1)}$, $a^{(2)}$, $a^{(3)}$ is a basis of $\mathbb{R}^3$ and

$$b = \frac{5}{42}\, a^{(1)} - \frac{1}{14}\, a^{(2)} + \frac{5}{6}\, a^{(3)}.$$

2. Compute the determinants of the following 3×3 matrices

(i) $A = \begin{pmatrix} -1 & 2 & 3 \\ 4 & 5 & 6 \\ 7 & 8 & 9 \end{pmatrix}$,

(ii) $B = \begin{pmatrix} 1 & 2 & 3 \\ 4 & 5 & 6 \\ 7 & 8 & 9 \end{pmatrix}$.

Solutions

(i) We compute the determinant using Gaussian elimination.

$$\det \begin{pmatrix} -1 & 2 & 3 \\ 4 & 5 & 6 \\ 7 & 8 & 9 \end{pmatrix} \overset{R_2 \rightsquigarrow R_2 + 4R_1}{\underset{R_3 \rightsquigarrow R_3 + 7R_1}{=}} \det \begin{pmatrix} -1 & 2 & 3 \\ 0 & 13 & 18 \\ 0 & 22 & 30 \end{pmatrix} = (-1)(13 \cdot 30 - 18 \cdot 22)$$

$$= 18 \cdot 22 - 30 \cdot 13 = 396 - 390 = 6.$$

(ii) We compute the determinant using Gaussian elimination.

$$\det \begin{pmatrix} 1 & 2 & 3 \\ 4 & 5 & 6 \\ 7 & 8 & 9 \end{pmatrix} = \det \begin{pmatrix} 1 & 2 & 3 \\ 0 & -3 & -6 \\ 0 & -6 & -12 \end{pmatrix} = \det \begin{pmatrix} 1 & 2 & 3 \\ 0 & -3 & -6 \\ 0 & 0 & 0 \end{pmatrix} = 0.$$

3. (i) Compute the determinant of the 3×3 matrix

$$A = \begin{pmatrix} 1 & 0 & 1 \\ -1 & 1 & 0 \\ 1 & 1 & 1 \end{pmatrix}^{25}.$$

(ii) Determine all numbers $a \in \mathbb{R}$ for which the determinant of the 2×2 matrix

$$B = \begin{pmatrix} 4 & 3 \\ 1 & 0 \end{pmatrix} + a \begin{pmatrix} -2 & 1 \\ -1 & -1 \end{pmatrix}$$

vanishes.

Solutions

(i) Since $\det(AB) = \det(A)\det(B)$, one has $\det(A^{25}) = (\det(A))^{25}$. Furthermore, by expanding $\det(A)$ with respect to the first row, we get

$$\det \begin{pmatrix} 1 & 0 & 1 \\ -1 & 1 & 0 \\ 1 & 1 & 1 \end{pmatrix} = \det \begin{pmatrix} 1 & 0 \\ 1 & 1 \end{pmatrix} + \det \begin{pmatrix} -1 & 1 \\ 1 & 1 \end{pmatrix} = 1 - 2 = -1$$

and hence $\det(A) = (-1)^{25} = -1$.

(ii) One has $B = \begin{pmatrix} 4 - 2a & 3 + a \\ 1 - a & -a \end{pmatrix}$ and hence

$$\det(B) = (4 - 2a)(-a) - (3 + a)(1 - a) = 3a^2 - 2a - 3.$$

Solving $3a^2 - 2a - 3 = 0$, we find that the determinant vanishes for $a = \frac{1}{3} \pm \frac{\sqrt{10}}{3}$.

4. Verify that for any basis $[a^{(1)}, a^{(2)}, a^{(3)}]$ of $\mathbb{R}^3$, $[-a^{(1)}, 2a^{(2)}, a^{(1)} + a^{(3)}]$ is also a basis.

Solutions Let $b^{(1)} = -a^{(1)}$, $b^{(2)} = 2a^{(2)}$, $b^{(3)} = a^{(1)} + a^{(3)}$. Then

$$\begin{pmatrix} b^{(1)} & b^{(2)} & b^{(3)} \end{pmatrix} = \begin{pmatrix} a^{(1)} & a^{(2)} & a^{(3)} \end{pmatrix} \begin{pmatrix} -1 & 0 & 1 \\ 0 & 2 & 0 \\ 0 & 0 & 1 \end{pmatrix}.$$

The vectors $b^{(1)}$, $b^{(2)}$, $b^{(3)}$ form a basis if the determinant of the left hand side of the latter identity is nonzero. Since $\det(AB) = \det(A)\det(B)$, we conclude

$$\det \begin{pmatrix} b^{(1)} & b^{(2)} & b^{(3)} \end{pmatrix} = \det \begin{pmatrix} a^{(1)} & a^{(2)} & a^{(3)} \end{pmatrix} \det \begin{pmatrix} -1 & 0 & 1 \\ 0 & 2 & 0 \\ 0 & 0 & 1 \end{pmatrix}$$

$$= -2 \det \begin{pmatrix} a^{(1)} & a^{(2)} & a^{(3)} \end{pmatrix}$$

and the right hand side is nonzero since $a^{(1)}, a^{(2)}, a^{(3)}$ form a basis.

5. Decide whether the following assertions are true or false and justify your answers.

   (i) $\det(\lambda A) = \lambda \det(A)$ for any $\lambda \in \mathbb{R}$, $A \in \mathbb{R}^{n \times n}$, $n \geq 1$.
   (ii) Let $A \in \mathbb{R}^{n \times n}$, $n \geq 1$, have the property that $\det(A^k) = 0$ for some $k \in \mathbb{N}$. Then $\det(A) = 0$.

Solutions

   (i) False. Since $\lambda A = \left(\lambda C_1 \; \lambda C_2 \; \ldots \; \lambda C_n\right)$ with $C_j \equiv C_j(A)$, $1 \leq j \leq n$, one has by Theorem 2.2.3

$$\det \lambda A = \lambda \det \left(C_1 \; \lambda C_2 \; \ldots \; \lambda C_n\right)$$

$$= \lambda^2 \det \left(C_1 \; C_2 \; \lambda C_3 \; \ldots \; \lambda C_n\right) = \ldots$$

$$= \lambda^n \det \left(C_1 \; C_2 \; \ldots \; C_n\right) = \lambda^n \det(A)$$

   Hence, for $n \geq 2$ and $\det(A) \neq 0$, it follows that for $\lambda = 2$, $\det(2A) = 2 \det(A)$ is *not* correct.
   (ii) True. Since $\det(AB) = \det(A)\det(B)$, $\det(A^k) = \left(\det(A)\right)^k$. Hence, if $\left(\det(A)\right)^k = 0$, then $\det(A) = 0$.

Solutions of Problems of Sect. 3.1

1. Compute the real and the imaginary part of the following complex numbers.

   (i) $\dfrac{1+i}{2+3i}$ $\qquad$ (ii) $(2+3i)^2$ $\qquad$ (iii) $\dfrac{1}{(1-i)^3}$ $\qquad$ (iv) $\dfrac{1 + \frac{1-i}{1+i}}{1 + \frac{1}{1+2i}}$

Solutions

   (i) $\dfrac{1+i}{2+3i} = \dfrac{(1+i)(2-3i)}{(2+3i)(2-3i)} = \dfrac{2+3+2i-3i}{4+9} = \dfrac{5-i}{13} = \dfrac{5}{13} - \dfrac{i}{13}$

   (ii) $(2+3i)^3 = 2^3 + 3 \cdot 2^2 \cdot (3i) + 3 \cdot 2 \cdot (3i)^2 + (3i)^3 = 8 - 54 + (36 - 27)i = -46 + 9i$

   (iii) $\dfrac{1}{(1-i)^3} = \dfrac{(1+i)^3}{2^3} = \dfrac{1+3i+3i^2+i^3}{8} = \dfrac{-2+2i}{8} = -\dfrac{1}{4} + \dfrac{i}{4}$

   (iv) $\dfrac{1 + \frac{1-i}{1+i}}{1 + \frac{1}{1+2i}} = \dfrac{\frac{2}{1+i}}{\frac{2+2i}{1+2i}} = \dfrac{2}{1+i} \cdot \dfrac{1+2i}{2+2i} = \dfrac{1+2i}{(1+i)^2} = \dfrac{1+2i}{2i} = 1 - \dfrac{1}{2}i$

2. (i) Compute the polar coordinates r, φ of the complex number $z = \sqrt{3} + i$ and find all possible values of $z^{1/3}$.
   (ii) Compute the polar coordinates r, φ of the complex number $z = 1 + i$ and find all possible values of $z^{1/5}$.
   (iii) Find all solutions of $z^4 = 16$.

Solutions

(i) Note that $r = |z| = \sqrt{3+1} = 2$ and that $\cos\varphi = \sqrt{3}/2$, $\sin\varphi = 1/2$, hence $\varphi = \pi/6$ (mod 2π) and $z = 2e^{i\pi/6}$. The third root $z^{1/3}$ takes therefore the values

$$2^{\frac{1}{3}} \cdot e^{i(\frac{\pi}{18}+\frac{2\pi j}{3})}, \quad 0 \le j \le 2.$$

(ii) One has $z = \sqrt{2}\,e^{i\pi/4}$ so that $r = \sqrt{2}$ and $\varphi = \pi/4$ (mod 2π). The fifth root $z^{1/5}$ then takes the values

$$2^{\frac{1}{10}} \cdot e^{(\frac{i\pi}{20}+\frac{i2\pi j}{5})}, \quad 0 \le j \le 4.$$

(iii) One has $z_j = 2e^{i\,2\pi j/4}$, $0 \le j \le 3$, that is $z_0 = 2$, $z_1 = 2\,i$, $z_2 = -2$ and $z_3 = -2\,i$.

3. (i) Compute $\left|\dfrac{2-3\,i}{3+4\,i}\right|$.

    (ii) Compute real and imaginary part of the complex number $\sum_{n=1}^{16} i^n$.

    (iii) Express $\sin^3\varphi$ in terms of sin and cos of multiples of the angle φ.

    Solutions

    (i) We compute

    $$\frac{2-3\,i}{3+4\,i} = \frac{(2-3\,i)(3-4\,i)}{9+16} = \frac{-6-17\,i}{25}$$

    so that

    $$\left|\frac{2-3\,i}{3+4\,i}\right| = \frac{1}{25}\sqrt{6^2+17^2} = \frac{1}{25}\sqrt{325} = \frac{1}{5}\sqrt{13}.$$

    (ii) One has $\displaystyle\sum_{n=1}^{16} i^n = \frac{i^{17}-i}{i-1} = i \cdot \frac{i^{16}-1}{i-1} = 0.$

    (iii) Let $z := e^{i\varphi} = \cos\varphi + i\sin\varphi$. Then $z - \bar{z} = 2\,i\sin\varphi$ and

    $$-8\,i\sin^3\varphi = (2\,i\sin\varphi)^3 = (z-\bar{z})^3 = z^3 - 3z^2\bar{z} + 3z\bar{z}^2 - \bar{z}^3$$

    $$= z^3 - \bar{z}^3 - 3(z-\bar{z}).$$

    Since

    $$z^3 - \bar{z}^3 = \cos(3\varphi) + i\sin(3\varphi) - \big(\cos(3\varphi) - i\sin(3\varphi)\big) = 2\,i\sin(3\varphi)$$

    it follows that

    $$-8\,i\sin^3\varphi = 2\,i\sin(3\varphi) - 6\,i\sin\varphi,$$

    yielding $\sin^3\varphi = \dfrac{3}{4}\sin\varphi - \dfrac{1}{4}\sin(3\varphi).$

4. Sketch the following subsets of the complex plane $\mathbb{C}$.

    (i) $M_1 = \big\{z \in \mathbb{C} \mid |z-1+2\,i| \ge |z+1|\big\}$
    (ii) $M_2 = \big\{z \in \mathbb{C} \mid |z+i| \ge 2;\ |z-2| \le 1\big\}$

Solutions

(i) The set of points $z \in \mathbb{C}$ satisfying $|z - 1 + 2i| = |z + 1|$ is the set of all points in $\mathbb{C}$, having the same distance from -1 and $1 - 2i$. This set is given by the straight line of slope 1 through $-i$. The set M_1 is given by the closed half plane above this line. In particular, $0 \in M_1$.

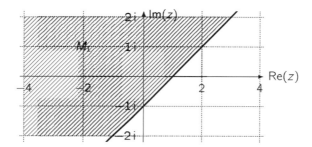

(ii) M_2 is given by the intersection of the complement of the open disk of radius 2, centered at $-i$, and the closed disk of radius 1, centered at 2.

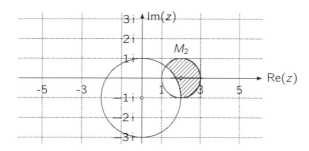

5. Decide whether the following assertions are true or false and justify your answers.

(i) There are complex numbers $z_1 \neq 0$ and $z_2 \neq 0$ so that $z_1 z_2 = 0$.
(ii) The identity $i^0 = i$ holds.
(iii) The identity $i = e^{-i\pi/2}$ holds.

Solutions

(i) False. Indeed, suppose $z_1 \neq 0$, then z_1^{-1} exists and we can multiply the equation $z_1 z_2 = 0$ by z_1^{-1} to obtain $0 = z_1^{-1} z_1 z_2 = z_2$, which contradicts $z_2 \neq 0$.
(ii) False. This would imply that $-1 = i^2 = i \cdot i = i^0 \cdot i = i^{0+1} = i$.
(iii) False. By Euler's formula $e^{-i\pi/2} = \cos(-\pi/2) + i\sin(-\pi/2) = -i$.

Solutions of Problems of Sect. 3.2

1. Find the roots of the following polynomials and write the latter as a product of linear factors.

(i) $p(z) = 2z^2 - 4z - 5$

(ii) $p(z) = z^2 + 2z + 3$

Solutions

(i) The discriminant of p is given by $\frac{b^2}{4a^2} - \frac{c}{a} = \frac{16}{16} - \frac{-5}{2} = \frac{7}{2} > 0$. Hence p has two real roots. We compute these roots by completing the square. First devide $2z^2 - 4z - 5 = 0$ by 2 to get $z^2 - 2z - \frac{5}{2} = 0$. Completing the square then yields

$$z^2 - 2z - \frac{5}{2} = z^2 - 2z + 1 - 1 - \frac{5}{2} = (z-1)^2 - \frac{7}{2}.$$

Hence $z - 1 = \pm\sqrt{7/2}$ or $z = 1 \pm \sqrt{7/2}$ and p can be factorized as

$$p(z) = 2(z - 1 + \sqrt{7/2})(z - 1 - \sqrt{7/2}).$$

(ii) Since the discriminant of p is given by $\frac{b^2}{4a^2} - \frac{c}{a} = \frac{4}{4} - \frac{3}{1} = -2 < 0$, p has two complex roots. By the formula for the roots of a polynomial of degree two, one has

$$z_{1,2} = \left(-\frac{b}{2a} \pm \frac{i}{2a}\sqrt{4ac - b^2}\right) = -1 \pm i\sqrt{2}$$

and hence p factors as $p(z) = (z + 1 - i\sqrt{2})(z + 1 + i\sqrt{2})$.

2. Let $p(z) = z^3 - 5z^2 + z - 5$.

   (i) Verify that i is a root of $p(z)$.

   (ii) Write $p(z)$ as a product of linear factors.

Solutions

(i) One computes $p(i) = i^3 - 5i^2 + i - 5 = -i + 5 + i - 5 = 0$.

(ii) Since all coefficients of $p(z)$ are real, $-i$ must be a root as well and therefore $p(z) = q(z)(z + i)(z - i) = q(z)(z^2 + 1)$ where $q(z)$ is a polynomial of the form $(z - a)$. By comparison of coefficients, one infers that $a = 5$ and hence $p(z) = (z - 5)(z + i)(z - i)$.

3. (i) Find all roots of $p(z) = z^4 + 2z^2 - 5$ and write $p(z)$ as a product of linear factors.

   (ii) Find all roots of $p(z) = z^5 + 3z^3 + z$ and write $p(z)$ as a product of linear factors.

Solutions Both equations are first reduced to polynomials of degree two by a substitution and then analyzed by the methods discussed.

(i) With the substitution $z^2 = y$, one obtains $y^2 + 2y - 5 = 0$, which has the following roots,

$$y_1 = -\frac{b}{a} + \frac{1}{2a}\sqrt{b^2 - 4ac} = -1 + \sqrt{6},$$

$$y_2 = -\frac{b}{a} - \frac{1}{2a}\sqrt{b^2 - 4ac} = -1 - \sqrt{6}.$$

Hence, the four roots of p are given by

$$z_{1,2} = \pm\sqrt{-1 + \sqrt{6}}, \qquad z_{3,4} = \pm i\sqrt{1 + \sqrt{6}}.$$

(Two roots lie on the real axis and two on the imaginary one.) Thus p factors as

$$p(z) = \left(z - \sqrt{-1 + \sqrt{6}}\,\right)\left(z + \sqrt{-1 + \sqrt{6}}\,\right)\left(z - i\sqrt{1 + \sqrt{6}}\,\right)\left(z + i\sqrt{1 + \sqrt{6}}\,\right).$$

(ii) The polynomial $p(z)$ factors as $p(z) = zq(z)$ where $q(z) := (z^4 + 3z^2 + 1)$. Hence $z_5 := 0$ is a root of p. With the substitution $z^2 = y$, $q(z)$ leads to the polynomial $y^2 + 3y + 1$, whose roots are

$$y_1 = -\frac{3}{2} + \frac{1}{2}\sqrt{5} = -\frac{1}{2}(3 - \sqrt{5}), \qquad y_2 = -\frac{3}{2} - \frac{1}{2}\sqrt{5} = -\frac{1}{2}(3 + \sqrt{5}).$$

Hence,

$$z_{1,2} = \pm i\,\frac{1}{\sqrt{2}}\sqrt{3 - \sqrt{5}}, \qquad z_{3,4} = \pm i\,\frac{1}{\sqrt{2}}\sqrt{3 + \sqrt{5}}.$$

(All roots lie on the imaginary axis.) Therefore, p factorizes as

$$p(z) = z\left(z - i\,\frac{1}{\sqrt{2}}\sqrt{3 - \sqrt{5}}\,\right)\left(z + i\,\frac{1}{\sqrt{2}}\sqrt{3 - \sqrt{5}}\,\right)$$
$$\cdot\left(z - i\,\frac{1}{\sqrt{2}}\sqrt{3 + \sqrt{5}}\,\right)\left(z + i\,\frac{1}{\sqrt{2}}\sqrt{3 + \sqrt{5}}\,\right).$$

4. Consider the function

$$f : \mathbb{R} \to \mathbb{R}, \quad x \mapsto x^3 + 5x^2 + x - 1.$$

(i) Compute the values of f at $x = -1$, $x = 0$ and $x = 1$.
(ii) Conclude that f has three real roots.

Solutions

(i) $f(-1) = 2$, $f(0) = -1$ and $f(1) = 6$.
(ii) Since $f(-1) > 0$, $f(0) < 0$ and $f(1) > 0$, f has at least one root in the open interval $(-1, 0)$ and at least one in the open interval $(0, 1)$. Since all coefficients of the polynomial f are real, it then follows that the third root of f also must be real.

5. Decide whether the following assertions are true or false and justify your answers.

(i) Any polynomial of degree 5 with real coefficients has at least three real roots.
(ii) Any polynomial $p(z)$ of degree $n \geq 1$ with real coefficients can be written as a product of polynomials with real coefficients, each of which has degree one or two.

Solutions

(i) False. The polynomial $p(z) = z(z^2 + 1)(z^2 + 4)$ is of degree 5 and all its coefficients are real. It has the four purely imaginary roots $i, -i, 2i, -2i$.

(ii) True. Without loss of generality we assume that $p(z) = z^n + a_{n-1}z^{n-1} + \ldots$. First, we consider the case where p has at least one real root. Denote the real roots of p by $r_j, 1 \leq j \leq k$ where $1 \leq j \leq k \leq n$. In the case where $k = n$, one has $p(z) = \prod_{1 \leq j \leq n}(z - r_j)$. In case where $1 \leq k < n$, $p(z)$ factors as $p(z) = q(z)\prod_{1 \leq j \leq k}(z - r_j)$ where $q(z)$ is a polynomial with real coefficients of degree $n - k$, which has no real roots. It therefore must be of even degree, $2m = n - k$. Since the roots come in pairs of complex numbers which are complex conjugate to each other, we can list the roots of q as $z_j, \overline{z_j}, 1 \leq j \leq m$ where $\text{Im}(z_j) > 0$. It then follows that

$$q(z) = \prod_{1 \leq j \leq m} (z - z_j)(z - \overline{z_j}) = \prod_{1 \leq j \leq m} (z^2 - |z_j|^2).$$

Hence the claim follows in the case where $p(z)$ has at least one real root. The case where $p(z)$ has no real roots follows by the same arguments. Actually, it is simpler, since in this case $q(z) = p(z)$.

Solutions of Problems of Sect. 3.3

1. Find the set of solutions of the following complex linear systems.

(i) $\begin{cases} z_1 - i z_2 = 2 \\ (-1 + i)z_1 + (2 + i)z_2 = 0 \end{cases}$

(ii) $\begin{cases} z_1 + i z_2 - (1 + i)z_3 = 0 \\ i z_1 + z_2 + (1 + i)z_3 = 0 \\ (1 + 2i)z_1 + (1 + i)z_2 + 2z_3 = 0 \end{cases}$

Solutions We solve these complex linear systems with Gaussian elimination.

(i)

$$\text{augmented coefficient matrix} \quad \begin{bmatrix} 1 & -i & \Big\| & 2 \\ (-1+i) & (2+i) & \Big\| & 0 \end{bmatrix}$$

$$R_2 \rightsquigarrow R_2 - (-1+i)R_1 \quad \begin{bmatrix} 1 & -i & \Big\| & 2 \\ 0 & 1 & \Big\| & 2 - 2i \end{bmatrix}$$

Hence, $z_2 = 2 - 2i$ and in turn $z_1 = 2 - (-i)(2 - 2i) = 4 + 2i$.

(ii)

$$
\text{augmented coefficient matrix} \quad
\begin{bmatrix}
1 & i & -(1+i) & \Big\| & 0 \\
i & 1 & (1+i) & \Big\| & 0 \\
(1+2i) & (1+i) & 2 & \Big\| & 0
\end{bmatrix}
$$

$$
R_2 \rightsquigarrow R_2 - i\,R_1,\; R_3 \rightsquigarrow R_3 - (1+2i)R_1 \quad
\begin{bmatrix}
1 & i & -(1+i) & \Big\| & 0 \\
0 & 2 & 2i & \Big\| & 0 \\
0 & 3 & (1+3i) & \Big\| & 0
\end{bmatrix}
$$

$$
R_3 \rightsquigarrow R_3 - {}^3\!/_2\, R_2 \quad
\begin{bmatrix}
1 & i & -(1+i) & \Big\| & 0 \\
0 & 2 & 2i & \Big\| & 0 \\
0 & 0 & 1 & \Big\| & 0
\end{bmatrix}
$$

Hence, $z_3 = z_2 = z_1 = 0$.

2. (i) Compute the determinant of the following complex 3×3 matrix,

$$
A = \begin{pmatrix}
1 & i & 1+i \\
0 & -1+i & 2 \\
i & 2 & 1+2i
\end{pmatrix}.
$$

(ii) Find all complex numbers $z \in \mathbb{C}$ with the property that $\det(B) = 0$ where

$$
B = \begin{pmatrix}
1+2i & 3+4i \\
z & 1-2i
\end{pmatrix}^{25}.
$$

Solutions

(i) We expand with respect to the first column.

$$
\det(A) = 1 \cdot \det\begin{pmatrix} (-1+i) & 2 \\ 2 & (1+2i) \end{pmatrix} + i \det\begin{pmatrix} i & (1+i) \\ (-1+i) & 2 \end{pmatrix}
$$

$$
= (-1+i)(1+2i) - 4 + i\left(2i - (-1+i)(1+i)\right) = -9 + i.
$$

(ii) Since $\det(B) = \left((1+2i)(1-2i) - (3+4i)z\right)^{25} = \left(5 - (3+4i)z\right)^{25}$.
Therefore, $\det(B) = 0$, if

$$
z = \frac{5}{3+4i} = \frac{5(3-4i)}{(3+4i)(3-4i)} = \frac{15-20i}{25} = \frac{3}{5} - \frac{4}{5}i.
$$

3. (i) Compute AA^{T} and $A^{\mathsf{T}}A$ where A^{T} is the transpose of A and $A \in \mathbb{C}^{2\times 3}$ is given by

$$
A = \begin{pmatrix} 1 & i & 2 \\ i & -2 & i \end{pmatrix}.
$$

(ii) Compute the inverse of the following complex 2×2 matrix

$$A = \begin{pmatrix} 2i & 1 \\ 1+i & 1-i \end{pmatrix}.$$

Solutions

(i) Since $A^T = \begin{pmatrix} 1 & i \\ i & -2 \\ 2 & i \end{pmatrix}$, we get

$$AA^T = \begin{pmatrix} 1 & i & 2 \\ i & -2 & i \end{pmatrix} \begin{pmatrix} 1 & i \\ i & -2 \\ 2 & i \end{pmatrix} = \begin{pmatrix} 4 & i \\ i & 2 \end{pmatrix}$$

and

$$A^T A = \begin{pmatrix} 1 & i \\ i & -2 \\ 2 & i \end{pmatrix} \begin{pmatrix} 1 & i & 2 \\ i & -2 & i \end{pmatrix} = \begin{pmatrix} 0 & -i & 1 \\ -i & 3 & 0 \\ 1 & 0 & 3 \end{pmatrix}.$$

(ii) The inverse of a regular 2×2 matrix $B = \begin{pmatrix} a & b \\ c & d \end{pmatrix} \in \mathbb{C}^{2 \times 2}$ can be computed as

$$B^{-1} = \frac{1}{\det(B)} \begin{pmatrix} d & -b \\ -c & a \end{pmatrix}.$$

In the case at hand, $\det(A) = 1 + i$. Hence $\frac{1}{\det(A)} = \frac{1-i}{2}$ and in turn

$$A^{-1} = \frac{1-i}{2} \begin{pmatrix} 1-i & -1 \\ -(1+i) & 2i \end{pmatrix} = \begin{pmatrix} -i & (-1+i)/2 \\ -1 & 1+i \end{pmatrix}.$$

4. Decide whether the following vectors in $\mathbb{C}^3$ are $\mathbb{C}$-linearly independent or $\mathbb{C}$-linearly dependent.

(i) $a^{(1)} = (1, 2+i, i)$, $a^{(2)} = (-1+3i, 1+i, 3+i)$
(ii) $b^{(1)} = (1+i, 1-i, -1+i)$, $b^{(2)} = (0, 4-2i, -3+5i)$, $b^{(3)} = (1+i, 0, 3)$

Solutions

(i) It is straightforward to verify that there is no complex number λ so that $a^{(1)} = \lambda a^{(2)}$ or $a^{(2)} = \lambda a^{(1)}$. Hence the two vectors are $\mathbb{C}$-linearly independent.

(ii) We compute the determinant of the 3×3 matrix B, whose rows are given by the transposes of the three vectors $b^{(1)}$, $b^{(2)}$, and $b^{(3)}$, by expanding $\det(B)$ with respect to the first row,

$$\det \begin{pmatrix} 1+i & 0 & 1+i \\ 1-i & 4-2i) & 0 \\ -1+i & -3+5i & 3 \end{pmatrix}$$

$$= (1+i)(4-2i)3 + (1+i)(1-i)(-3+5i) - (1+i)(-1+i)(4-2i)$$

$$= (6 + 2\,\mathrm{i})3 + 2(-3 + 5\,\mathrm{i}) + 2(4 - 2\,\mathrm{i}) = (18 - 6 + 8) + (6 + 10 - 4)\,\mathrm{i}$$
$$= 20 + 12\,\mathrm{i} \neq 0.$$

Hence, the three vectors are linearly independent.

5. Decide whether the following assertions are true or false and justify your answers.

(i) There exists vectors $a^{(1)}$, $a^{(2)}$ in $\mathbb{C}^2$ with the following two properties:
(P1) For any $\alpha_1, \alpha_2 \in \mathbb{R}$ with $\alpha_1 a^{(1)} + \alpha_2 a^{(2)} = 0$, it follows that $\alpha_1 = 0$ and $\alpha_2 = 0$ (i.e. $a^{(1)}$, $a^{(2)}$ are $\mathbb{R}$-linearly independent).
(P2) There exist $\beta_1, \beta_2 \in \mathbb{C} \setminus \{0\}$ so that $\beta_1 a^{(1)} + \beta_2 a^{(2)} = 0$ (i.e. $a^{(1)}$, $a^{(2)}$ are $\mathbb{C}$-linearly dependent).

(ii) Any basis $[a] = [a^{(1)}, \ldots, a^{(n)}]$ of $\mathbb{R}^n$ gives rise to a basis of $\mathbb{C}^n$, when $a^{(1)}, \ldots, a^{(n)}$ are considered as vectors in $\mathbb{C}^n$.

(iii) The vectors $a^{(1)} = (\mathrm{i}, 1, 0)$, $a^{(2)} = (\mathrm{i}, -1, 0)$ are $\mathbb{C}$-linearly independent in $\mathbb{C}^3$ and there exists $a^{(3)} \in \mathbb{C}^3$ so that $[a] = [a^{(1)}, a^{(2)}, a^{(3)}]$ is a basis of $\mathbb{C}^3$.

Solutions

(i) True. Let $a^{(1)} = (1, 0)$ and $a^{(2)} = (\mathrm{i}, 0)$. These vectors in $\mathbb{C}^2$ are $\mathbb{C}$-linearly dependent, since $a^{(2)} = \mathrm{i}\,a^{(1)}$ where $\beta_1 = \mathrm{i}$ and $\beta_2 = -1$. Hence, (P2) is satisfied. On the other hand, for any $\alpha_1, \alpha_2 \in \mathbb{R}$ with $\alpha_1 a^{(1)} + \alpha_2 a^{(2)} = 0$, it follows that

$$z := \alpha_1 + \alpha_2\,\mathrm{i} = 0, \qquad \alpha_1 \cdot 0 + \alpha_2 \cdot 0 = 0.$$

Since α_1 and α_2 are real, $z = \alpha_1 + \alpha_2\,\mathrm{i} = 0$ implies that $\alpha_1 = 0$ and $\alpha_2 = 0$. Hence (P1) is satisfied as well.

(ii) True. Let b be an arbitrary vector in C^n. Then $b = b^{(1)} + \mathrm{i}\,b^{(2)}$ where $b^{(1)} = (b + \bar{b})/2 \in \mathbb{R}^n$ and $b^{(2)} = (b - \bar{b})/(2\,\mathrm{i}) \in \mathbb{R}^n$. It follows that there are uniquely determined real numbers $\lambda_j^{(1)}, \lambda_j^{(2)}$, $1 \leq j \leq n$, so that

$$b^{(1)} = \sum_{j=1}^{n} \lambda_j^{(1)} a^{(j)}, \qquad b^{(2)} = \sum_{j=1}^{n} \lambda_j^{(2)} a^{(j)}.$$

Hence $b = b^{(1)} + \mathrm{i}\,b^{(2)} = \sum_{j=1}^{n}(\lambda_j^{(1)} + \mathrm{i}\lambda_j^{(2)})a^{(j)}$ and $\lambda_j^{(1)} + \mathrm{i}\lambda_j^{(2)} \in \mathbb{C}$, $1 \leq j \leq n$, are uniquely determined. This shows that $[a] = [a^{(1)}, \ldots, a^{(n)}]$ is a basis of $\mathbb{C}^n$.

(iii) True. Let $a^{(3)} = (0, 0, 1)$. The 3×3 matrix $A \in \mathbb{C}^{3 \times 3}$, whose rows are given by $a^{(1)}$, $a^{(2)}$, and $a^{(3)}$, is regular since $\det(A) = -2\,\mathrm{i} \neq 0$. It follows that $a^{(1)}$, $a^{(2)}$, and $a^{(3)}$ are linearly independent (and so are $a^{(1)}$ and $a^{(2)}$) and hence form a basis of $\mathbb{C}^3$.

Solutions of Problems of Sect. 4.1

1. Decide which of the following subsets are linear subspaces of the corresponding $\mathbb{R}$-vector spaces.

(i) $W = \{(x_1, x_2, x_3) \in \mathbb{R}^3 \mid 2x_1 + 3x_2 + x_3 = 0\}$.
(ii) $V = \{(x_1, x_2, x_3, x_4) \in \mathbb{R}^4 \mid 4x_2 + 3x_3 + 2x_4 = 7\}$.
(iii) $\mathrm{GL}_{\mathbb{R}}(3) = \{A \in \mathbb{R}^{3 \times 3} \mid A \text{ regular}\}$.
(iv) $L = \{(x_1, x_2) \in \mathbb{R}^2 \mid x_1 x_2 = 0\}$.

Solutions

(i) Note that $W = \{x \in \mathbb{R}^3 \mid Ax = 0\}$ where $A = (2\ 3\ 1) \in \mathbb{R}^{1 \times 3}$. Therefore, W equals the space of solutions of the homogenous system $Ax = 0$ and hence is a linear subspace of $\mathbb{R}^3$. Indeed, for $u, v \in W$ and $\lambda \in \mathbb{R}$, we have $A(u + v) = Au + Av = 0$ and $A(\lambda u) = \lambda Au = 0$.

(ii) Note that $V = \{x \in \mathbb{R}^4 \mid Ax = 7\}$ with $A = (0\ 4\ 3\ 2) \in \mathbb{R}^{1 \times 4}$. In particular, $0 \notin V$ and hence V is not a linear subspace of $\mathbb{R}^4$. (It is however an affine subspace of $\mathbb{R}^4$).

(iii) Clearly, $0 \notin GL_{\mathbb{R}}(3)$ and hence $GL_{\mathbb{R}}(3)$ is not a linear subspace of $\mathbb{R}^{3 \times 3}$.

(iv) Note that $e^{(1)} = (1, 0)$ and $e^{(2)} = (0, 1)$ are both in L, but $e^{(1)} + e^{(2)} = (1, 1)$ is not. Hence L is not a linear subspace of $\mathbb{R}^2$.

2. Consider the following linear system (S),

$$\begin{cases} 3x_1 + x_2 - 3x_3 & = & 4 \\ x_1 + 2x_2 + 5x_3 & = & -2 \end{cases}.$$

(i) Determine the vector space of solutions $L_{\text{hom}} \subseteq \mathbb{R}^3$ of the corresponding homogenous system

$$\begin{cases} 3x_1 + x_2 - 3x_3 & = 0 \\ x_1 + 2x_2 + 5x_3 & = 0 \end{cases}$$

and compute its dimension.

(ii) Determine the affine space L of solutions of (S) by finding a particular solution of (S).

Solutions

(i) We consider the augmented coefficient matrix and use Gaussian elimination.

$$\text{augmented coefficient matrix} \quad \begin{bmatrix} 3 & 1 & -3 & \big\| & 0 \\ 1 & 2 & 5 & \big\| & 0 \end{bmatrix}$$

$$R_2 \rightsquigarrow \tfrac{1}{3} R_2 - R_1 \quad \begin{bmatrix} 3 & 1 & -3 & \big\| & 0 \\ 0 & \tfrac{5}{3} & 6 & \big\| & 0 \end{bmatrix}$$

$$R_1 \rightsquigarrow \tfrac{1}{3} R_1, R_2 \rightsquigarrow \tfrac{3}{5} \cdot R_2 \quad \begin{bmatrix} 1 & \tfrac{1}{3} & -1 & \big\| & 0 \\ 0 & 1 & \tfrac{18}{5} & \big\| & 0 \end{bmatrix}$$

$$R_1 \rightsquigarrow R_1 - \tfrac{1}{3} R_2 \quad \begin{bmatrix} 1 & 0 & -\tfrac{11}{5} & \big\| & 0 \\ 0 & 1 & \tfrac{18}{5} & \big\| & 0 \end{bmatrix}.$$

Consequently, $L_{\text{hom}} = \left\{ \left(\frac{11}{5} t, -\frac{18}{5} t, t \right) \mid t \in \mathbb{R} \right\}$ and $\dim(L_{\text{hom}} = 1$.

(ii) Setting $x_3 = 0$, one finds by Gaussian elimination the particular solution $y_{\text{part}} = (2, -2, 0)$. The space L of solutions of the inhomogenous system (S) is thus given by

$$L = y_{\text{part}} + L_{\text{hom}} = \left\{ (2, -2, 0) + \left(\frac{11}{5} t, -\frac{18}{5} t, t \right) \mid t \in \mathbb{R} \right\}.$$

3. Let $\mathcal{P}_3(\mathbb{C})$ denote the $\mathbb{C}$-vector space of polynomials of degree at most three in one complex variable,

$$\mathcal{P}_3(\mathbb{C}) = \left\{ p(z) = a_3 z^3 + a_2 z^2 + a_1 z + a_0 \mid a_0, a_1, a_2, a_3 \in \mathbb{C} \right\},$$

and by $E_3(\mathbb{C})$ the subset

$$E_3(\mathbb{C}) = \left\{ p \in \mathcal{P}_3(\mathbb{C}) \mid p(-z) = p(z), z \in \mathbb{C} \right\}.$$

(i) Find a basis of $\mathcal{P}_3(\mathbb{C})$ and compute $\dim \mathcal{P}_3(\mathbb{C})$.
(ii) Verify that $E_3(\mathbb{C})$ is a $\mathbb{C}$-subspace of $\mathcal{P}_3(\mathbb{C})$ and compute its dimension.

Solutions

(i) The standard basis of $\mathcal{P}_3(\mathbb{C})$ is given by $[p_3, p_2, p_1, p_0]$ where $p_k(z) = z^k, 0 \le k \le 3$, and hence $\dim \mathcal{P}_3(\mathbb{C}) = 4$.
(ii) By comparison of coefficients, for any polynomial $p \in \mathcal{P}_3(\mathbb{C})$, the condition

$$p(-z) - p(z) = 0, \quad z \in \mathbb{C},$$

implies that $-a_3 - a_3 = 0$ and $-a_1 - a_1 = 0$, implying that $a_3 = 0$ and $a_1 = 0$. Hence, $E_3(\mathbb{C}) = \left\{ a_2 z^2 + a_0 \mid a_0, a_2 \in \mathbb{C} \right\}$ which is a subspace of $\mathcal{P}_3(\mathbb{C})$ with basis $[p_2, p_0]$ and thus $\dim E_3(\mathbb{C}) = 2$.

4. Consider the subset W of $\mathbb{C}^{3 \times 3}$,

$$W = \left\{ (a_{ij})_{1 \le i, j \le 3} \in \mathbb{C}^{3 \times 3} \mid a_{ij} = 0, 1 \le i < j \le 3 \right\}.$$

(i) Find a basis of $\mathbb{C}^{3 \times 3}$ and compute its dimension.
(ii) Verify that W is a linear subspace of $\mathbb{C}^{3 \times 3}$ and compute its dimension.

Solutions

(i) The standard basis of $\mathbb{C}^{3 \times 3}$ is given by the nine 3×3-matrices $E_{i,j}, 1 \le i, j \le 3$ where $E_{i,j} = (a_{mn}^{(i,j)})_{1 \le m, n \le 3}$ is the matrix with $a_{mn}^{(i,j)} = 0$ if $(m, n) \ne (i, j)$ and $a_{ij}^{(i,j)} = 1$. Hence $\dim(\mathbb{C}^{3 \times 3}) = 9$.
(ii) In view of Item (i),

$$W = \left\{ \sum_{1 \le i < j \le 3} b_{i,j} E_{i,j} \mid b_{i,j} \in \mathbb{C} \right\}.$$

Hence W is a linear subspace of $\mathbb{C}^{3\times3}$ and $\dim(W) = 6$.

5. Let $A = \begin{pmatrix} a_1 \\ a_2 \end{pmatrix} \in \mathbb{R}^{2\times1}$.

(i) Determine the rank of $A^T A \in \mathbb{R}^{1\times1}$ and of $AA^T \in \mathbb{R}^{2\times2}$ in terms of A.
(ii) Decide for which A the matrix $A^T A$ and for which A the matrix AA^T is regular.

Solutions

(i) We compute

$$A^T A = \begin{pmatrix} a_1 & a_2 \end{pmatrix} \begin{pmatrix} a_1 \\ a_2 \end{pmatrix} = a_1^2 + a_2^2, \quad AA^T = \begin{pmatrix} a_1 \\ a_2 \end{pmatrix} \begin{pmatrix} a_1 & a_2 \end{pmatrix} = \begin{pmatrix} a_1^2 & a_1 a_2 \\ a_2 a_1 & a_2^2 \end{pmatrix}.$$

If $(a_1, a_2) \neq (0, 0)$, then $a_1^2 + a_2^2 \neq 0$, hence $\mathrm{rank}(A^T A) = 1$. Conversely, if $a_1 = 0$ and $a_2 = 0$, then $A^T A$ is the zero matrix and $\mathrm{rank}(A^T A) = 0$.

Solutions of Problems of Sect. 4.2

1. Verify that the following maps $f \colon \mathbb{R}^n \to \mathbb{R}^m$ are linear and determine their matrix representations $f_{[e]\to[e]}$ with respect to the standard bases.

(i) $f \colon \mathbb{R}^4 \to \mathbb{R}^2$, $(x_1, x_2, x_3, x_4) \mapsto (x_1 + 3x_2, -x_1 + 4x_2)$.
(ii) Let $f \colon \mathbb{R}^2 \to \mathbb{R}^2$ be the map acting on a vector $x = (x_1, x_2) \in \mathbb{R}^2$ as follows: first x is scaled by the factor 5 and then it is reflected at the x_2-axis.

Solutions

(i) Let $x, y \in \mathbb{R}^4$ and $\lambda \in \mathbb{R}$. Then

$$f(x + y) = \begin{pmatrix} (x+y)_1 + 3(x+y)_2 \\ -(x+y)_1 + 4(x+y)_2 \end{pmatrix} = \begin{pmatrix} x_1 + 3x_2 \\ -x_1 + 4x_2 \end{pmatrix} + \begin{pmatrix} y_1 + 3y_2 \\ -y_1 + 4y_2 \end{pmatrix}$$
$$= f(x) + f(y),$$

$$f(\lambda x) = \begin{pmatrix} (\lambda x)_1 + 3(\lambda x)_2 \\ -(\lambda x)_1 + 4(\lambda x)_2 \end{pmatrix} = \lambda \begin{pmatrix} x_1 + 3x_2 \\ -x_1 + 4x_2 \end{pmatrix} = \lambda f(x).$$

So, f is $\mathbb{R}$-linear. Moreover,

$$f(e^{(1)}) = \begin{pmatrix} 1 \\ -1 \end{pmatrix} = e^{(1)} - e^{(2)}, \qquad f(e^{(2)}) = \begin{pmatrix} 3 \\ 4 \end{pmatrix} = 3e^{(1)} + 4e^{(2)},$$

$$f(e^{(3)}) = \begin{pmatrix} 0 \\ 0 \end{pmatrix} = 0 \cdot e^{(1)} + 0 \cdot e^{(2)}, \qquad f(e^{(4)}) = \begin{pmatrix} 0 \\ 0 \end{pmatrix} = 0 \cdot e^{(1)} + 0 \cdot e^{(2)}.$$

Hence,

$$f_{[e] \to [e]} = \begin{pmatrix} 1 & 3 & 0 & 0 \\ -1 & 4 & 0 & 0 \end{pmatrix}.$$

(ii) $f : \mathbb{R}^2 \to \mathbb{R}^2$ is the composition of the two operations

$$\begin{pmatrix} x_1 \\ x_2 \end{pmatrix} \mapsto \begin{pmatrix} 5x_1 \\ 5x_2 \end{pmatrix}, \qquad \text{and} \qquad \begin{pmatrix} x_1 \\ x_2 \end{pmatrix} \mapsto \begin{pmatrix} -x_1 \\ x_2 \end{pmatrix}.$$

Consequently,

$$f(x) = \begin{pmatrix} -5x_1 \\ 5x_2 \end{pmatrix} = A \begin{pmatrix} x_1 \\ x_2 \end{pmatrix}, \qquad A = \begin{pmatrix} -5 & 0 \\ 0 & 5 \end{pmatrix}.$$

Hence, f is $\mathbb{R}$-linear and $f_{[e] \to [e]} = A$.

2. Let $\mathcal{P}_7$ be the $\mathbb{R}$-vector space of polynomials of degree at most 7 with real coefficients. Denote by p' the first and by p'' the second derivative of $p \in \mathcal{P}_7$ with respect to x. Show that the following maps map $\mathcal{P}_7$ into itself and decide whether they are linear.

(i) $T(p)(x) := 2p''(x) + 2p'(x) + 5p(x)$.
(ii) $S(p)(x) := x^2 p''(x) + p(x)$.

Solutions First note that for any polynomial $p(x) = \sum_{k=0}^{7} a_k x^k$ in $\mathcal{P}_7$, the derivative p' of p is given by

$$p'(x) = \sum_{k=1}^{7} a_k k x^{k-1} = a_1 + 2a_2 x + \cdots + 7a_7 x^6 \tag{A.1}$$

and hence a polynomial of degree at most 6, implying that p' is an element in $\mathcal{P}_7$. Similarly, the second derivative p'' of p is given by

$$p''(x) = \sum_{k=2}^{7} a_k k(k-1) x^{k-2} = 2a_2 + 3 \cdot 2a_3 x + \cdots + 7 \cdot 6a_7 x^5, \tag{A.2}$$

which is a polynomial of degree at most 5 and thus also in $\mathcal{P}_7$. Furthermore, it is easy to see that for any polynomials p, q in $\mathcal{P}_7$ and any $\lambda \in \mathbb{R}$, one has

$$(p+q)'(x) = p'(x) + q'(x), \qquad (p+q)''(x) = p''(x) + q''(x), \tag{A.3}$$

$$(\lambda p)'(x) = \lambda p'(x), \qquad (\lambda p)''(x) = \lambda p''(x). \tag{A.4}$$

(i) By (A.1)–(A.2) it follows that for any $p \in \mathcal{P}_7$, $T(p)$ is in $\mathcal{P}_7$ and by (A.3)–(A.4) one has for any $p, q \in \mathcal{P}_7$ and $\lambda \in \mathbb{R}$,

$$T(p+q)(x) = 2(p+q)''(x) + 2(p+q)'(x) + 5(p+q)(x)$$
$$= 2p''(x) + 2p'(x) + 5p(x) + 2q''(x) + 2q'(x) + 5q(x)$$
$$= T(p)(x) + T(q)(x),$$

and

$$T(\lambda p)(x) = 2\lambda p''(x) + 2\lambda p'(x) + 5\lambda p(x) = \lambda(2p''(x) + 2p'(x) + 5p(x))$$
$$= \lambda T(p)(x).$$

Altogether, this shows that T is linear map from $\mathcal{P}_7$ into itself.

(ii) By (A.2) it follows that for any $p \in \mathcal{P}_7$, $x^2 p''(x)$ is a polynomial of degree at most 7 and hence in $\mathcal{P}_7$. One then concludes that for any $p \in \mathcal{P}_7$, $S(p)$ is in $\mathcal{P}_7$. Furthermore, by (A.3)–(A.4) one has for any $p, q \in \mathcal{P}_7$ and $\lambda \in \mathbb{R}$,

$$S(p + q)(x) = x^2(p + q)''(x) + (p + q)(x)$$
$$= x^2 p''(x) + p(x) + x^2 q''(x) + q(x)$$
$$= S(p)(x) + S(q)(x)$$

and

$$S(\lambda p)(x) = x^2(\lambda p)''(x) + (\lambda p)(x) = \lambda(x^2 p''(x) + p(x)) = \lambda S(p)(x).$$

This establishes that S is a linear map from $\mathcal{P}_7$ into itself.

3. Consider the bases in $\mathbb{R}^2$, $[e] = [e^{(1)}, e^{(2)}]$, $[v] = [e^{(2)}, e^{(1)}]$, and

$$[w] = [w^{(1)}, w^{(2)}], \qquad w^{(1)} = (1, 1), \quad w^{(2)} = (1, -1),$$

and let $f : \mathbb{R}^2 \to \mathbb{R}^2$ be the linear map with matrix representation

$$f_{[e] \to [e]} = \begin{pmatrix} 1 & 2 \\ 3 & 4 \end{pmatrix}.$$

(i) Determine $\mathrm{Id}_{[v] \to [e]}$ and $\mathrm{Id}_{[e] \to [v]}$.
(ii) Determine $f_{[v] \to [e]}$ and $f_{[v] \to [v]}$.
(iii) Determine $f_{[w] \to [w]}$.

Solutions

(i) Since $v^{(1)} = e^{(2)}$ and $v^{(2)} = e^{(1)}$,

$$\mathrm{Id}_{[v] \to [e]} = \begin{pmatrix} 0 & 1 \\ 1 & 0 \end{pmatrix}.$$

Conversely, $e^{(1)} = v^{(2)}$ and $e^{(2)} = v^{(1)}$, hence

$$\mathrm{Id}_{[e] \to [v]} = \begin{pmatrix} 0 & 1 \\ 1 & 0 \end{pmatrix}.$$

(Note that $\mathrm{Id}_{[e] \to [v]} \, \mathrm{Id}_{[v] \to [e]} = \mathrm{Id}_{[v] \to [v]}$ and $\mathrm{Id}_{[v] \to [e]} \, \mathrm{Id}_{[e] \to [v]} = \mathrm{Id}_{[e] \to [e]}$. In particular, one has $\mathrm{Id}_{[e] \to [v]} = \mathrm{Id}_{[v] \to [e]}^{-1}$.)

(ii) A straightforward computation gives

$$f_{[v]\to[e]} = f_{[e]\to[e]} \, \mathrm{Id}_{[v]\to[e]} = \begin{pmatrix} 1 & 2 \\ 3 & 4 \end{pmatrix} \begin{pmatrix} 0 & 1 \\ 1 & 0 \end{pmatrix} = \begin{pmatrix} 2 & 1 \\ 4 & 3 \end{pmatrix},$$

$$f_{[v]\to[v]} = \mathrm{Id}_{[e]\to[v]} \, f_{[e]\to[e]} \, \mathrm{Id}_{[v]\to[e]} = \begin{pmatrix} 0 & 1 \\ 1 & 0 \end{pmatrix} \begin{pmatrix} 2 & 1 \\ 4 & 3 \end{pmatrix} = \begin{pmatrix} 4 & 3 \\ 2 & 1 \end{pmatrix}.$$

(iii) We first note that

$$\mathrm{Id}_{[w]\to[e]} = \begin{pmatrix} 1 & 1 \\ 1 & -1 \end{pmatrix}.$$

To compute $\mathrm{Id}_{[e]\to[w]}$, we invert $\mathrm{Id}_{[w]\to[e]}$,

$$\mathrm{Id}_{[e]\to[w]} = \begin{pmatrix} 1 & 1 \\ 1 & -1 \end{pmatrix}^{-1} = -\frac{1}{2}\begin{pmatrix} -1 & -1 \\ -1 & 1 \end{pmatrix} = \frac{1}{2}\begin{pmatrix} 1 & 1 \\ 1 & -1 \end{pmatrix}.$$

Consequently,

$$f_{[w]\to[w]} = \mathrm{Id}_{[e]\to[w]} \, f_{[e]\to[e]} \, \mathrm{Id}_{[w]\to[e]}$$

$$= \frac{1}{2}\begin{pmatrix} 1 & 1 \\ 1 & -1 \end{pmatrix}\begin{pmatrix} 1 & 2 \\ 3 & 4 \end{pmatrix}\begin{pmatrix} 1 & 1 \\ 1 & -1 \end{pmatrix} = \frac{1}{2}\begin{pmatrix} 1 & 1 \\ 1 & -1 \end{pmatrix}\begin{pmatrix} 3 & -1 \\ 7 & -1 \end{pmatrix}$$

$$= \frac{1}{2}\begin{pmatrix} 10 & -2 \\ -4 & 0 \end{pmatrix} = \begin{pmatrix} 5 & -1 \\ -2 & 0 \end{pmatrix}.$$

4. Consider the basis $[v] = [v^{(1)}, v^{(2)}, v^{(3)}]$ of $\mathbb{R}^3$, defined as

$$v^{(1)} = (1, 0, -1), \qquad v^{(2)} = (1, 2, 1), \qquad v^{(3)} = (-1, 1, 1),$$

and the basis $[w] = [w^{(1)}, w^{(2)}]$ of $\mathbb{R}^2$, given by

$$w^{(1)} = (1, -1), \qquad w^{(2)} = (2, -1).$$

Determine the matrix representations $T_{[v]\to[w]}$ of the following linear maps.

(i) $T: \mathbb{R}^3 \to \mathbb{R}^2$, $(x_1, x_2, x_3) \mapsto (2x_3, x_1)$.
(ii) $T: \mathbb{R}^3 \to \mathbb{R}^2$, $(x_1, x_2, x_3) \mapsto (x_1 - x_2, x_1 + x_3)$.

Solutions

(i) We first note that

$$T_{[e]\to[e]} = \begin{pmatrix} 0 & 0 & 2 \\ 1 & 0 & 0 \end{pmatrix}, \quad \mathrm{Id}_{[v]\to[e]} = \begin{pmatrix} 1 & 1 & -1 \\ 0 & 2 & 1 \\ -1 & 1 & 1 \end{pmatrix}, \quad \mathrm{Id}_{[w]\to[e]} = \begin{pmatrix} 1 & 2 \\ -1 & -1 \end{pmatrix}.$$

We further compute

$$\mathrm{Id}_{[e]\to[w]} = \mathrm{Id}_{[w]\to[e]}^{-1} = \begin{pmatrix} -1 & -2 \\ 1 & 1 \end{pmatrix},$$

hence

$$T_{[v]\to[w]} = \mathrm{Id}_{[e]\to[w]}\, T_{[e]\to[e]}\, \mathrm{Id}_{[v]\to[e]}$$

$$= \begin{pmatrix} -1 & -2 \\ 1 & 1 \end{pmatrix} \begin{pmatrix} 0 & 0 & 2 \\ 1 & 0 & 0 \end{pmatrix} \begin{pmatrix} 1 & 1 & -1 \\ 0 & 2 & 1 \\ -1 & 1 & 1 \end{pmatrix}$$

$$= \begin{pmatrix} -1 & -2 \\ 1 & 1 \end{pmatrix} \begin{pmatrix} -2 & 2 & 2 \\ 1 & 1 & -1 \end{pmatrix}$$

$$= \begin{pmatrix} 0 & -4 & 0 \\ -1 & 3 & 1 \end{pmatrix}.$$

(ii) With

$$T_{[e]\to[e]} = \begin{pmatrix} 1 & -1 & 0 \\ 1 & 0 & 1 \end{pmatrix}$$

and the computations of $\mathrm{Id}_{[v]\to[e]}$ and $\mathrm{Id}_{[e]\to[w]}$ of (i), we obtain

$$T_{[v]\to[w]} = \mathrm{Id}_{[e]\to[w]}\, T_{[e]\to[e]}\, \mathrm{Id}_{[e]\to[w]}$$

$$= \begin{pmatrix} -1 & -2 \\ 1 & 1 \end{pmatrix} \begin{pmatrix} 1 & -1 & 0 \\ 1 & 0 & 1 \end{pmatrix} \begin{pmatrix} 1 & 1 & -1 \\ 0 & 2 & 1 \\ -1 & 1 & 1 \end{pmatrix}$$

$$= \begin{pmatrix} -1 & -2 \\ 1 & 1 \end{pmatrix} \begin{pmatrix} 1 & -1 & -2 \\ 0 & 2 & 0 \end{pmatrix}$$

$$= \begin{pmatrix} -1 & -3 & 2 \\ 1 & 1 & -2 \end{pmatrix}.$$

5. Decide whether the following assertions are true or false and justify your answers.

   (i) There exists a linear map $T: \mathbb{R}^3 \to \mathbb{R}^7$ so that $\{T(x) \mid x \in \mathbb{R}^3\} = \mathbb{R}^7$.
   (ii) For any linear map $f: \mathbb{R}^n \to \mathbb{R}^n$, f is bijective if and only if $\det(f_{[e]\to[e]}) \neq 0$.

Solutions

   (i) False. Suppose the claim was true. Then, there exist $x^{(i)} \in \mathbb{R}^3$, $1 \leq i \leq 7$, with $f(x^{(i)}) = e^{(i)}$. Since the dimension of $\mathbb{R}^3$ is three, the vectors $x^{(i)}$ are linearly dependent. Hence, there exist $(\alpha_1, \ldots, \alpha_7) \in \mathbb{R}^7 \setminus \{(0, \cdots, 0)\}$ so that $0 = \sum_{i=1}^{7} \alpha_i x^{(i)}$, implying that

   $$0 = f\left(\sum_{i=1}^{7} \alpha_i x^{(i)}\right) = \sum_{i=1}^{7} \alpha_i f(x^{(i)}) = \sum_{i=1}^{7} \alpha_i e^{(i)},$$

   which contradicts the fact that $[e^{(1)}, \ldots, e^{(7)}]$ is a basis of $\mathbb{R}^7$.

(ii) True. By Theorem 2.2.5, the matrix $f_{[e]\to[e]} \in \mathbb{R}^{n\times n}$ is invertible if and only if $\det(f_{[e]\to[e]}) \neq 0$. Furthermore, since $f(x) = f_{[e]\to[e]}x$ for any $x \in \mathbb{R}^n$ (here we identify x with the corresponding $n \times 1$ matrix), f is bijective if and only if $f_{[e]\to[e]}$ is invertible. It then follows that f is bijective if and only if $\det(f_{[e]\to[e]}) \neq 0$.

Solutions of Problems of Sect. 4.3

1. Let $B\colon \mathbb{R}^2 \times \mathbb{R}^2 \to \mathbb{R}$ be given by

$$B(x, y) = 3x_1y_1 + x_1y_2 + x_2y_1 + 2x_2y_2$$

where $x = (x_1, x_2)$, $y = (y_1, y_2) \in \mathbb{R}^2$.

(i) Verify that B is an inner product on $\mathbb{R}^2$.
(ii) Verify that the vectors $a = (\frac{1}{\sqrt{3}}, 0)$ and $b = (0, \frac{1}{\sqrt{2}})$ have length 1 with respect to the inner product B.
(iii) Compute the cosine of the (unoriented) angle between the vectors a and b of Item (ii).

Solutions

(i) It is to check that B satisfies $(IP1)$, $(IP2)$, and $(IP3)$, i.e. that B is symmetric, linear (in the first component), and positive definite.

$(IP1)$: For any x, y in $\mathbb{R}^2$,

$$B(y, x) = 3y_1x_1 + y_1x_2 + y_2x_1 + 2y_2x_2 = 3x_1y_1 + x_2y_1 + x_1y_2 + 2x_2y_2 = B(x, y).$$

$(IP2)$: For any x, y, z in $\mathbb{R}^2$, and $\alpha \in \mathbb{R}$

$$\begin{aligned} B(x + y, z) &= 3(x_1 + y_1)z_1 + (x_1 + y_1)z_2 + (x_2 + y_2)z_1 + 2(x_2 + y_2)z_2 \\ &= B(x, z) + B(y, z), \end{aligned}$$

$$B(\alpha x, y) = 3\alpha x_1y_1 + \alpha x_1y_2 + \alpha x_2y_1 + 2\alpha x_2y_2 = \alpha B(x, y).$$

$(IP3)$: For any $x \in \mathbb{R}^2$, one has $B(x, x) = 3x_1^2 + 2x_1x_2 + 2x_2^2 = (x_1+x_2)^2 + 2x_1^2 + x_2^2 \geq 0$.

If $x = 0$, then $B(x, x) = 0$. Conversely, if $B(x, x) = 0$, then $(x_1 + x_2)^2 = 0$, $x_1^2 = 0$ and $x_2^2 = 0$, hence $x = 0$.

(ii) We compute

$$\left\|(\frac{1}{\sqrt{3}}, 0)\right\| := B\left((\frac{1}{\sqrt{3}}, 0), (\frac{1}{\sqrt{3}}, 0)\right)^{\frac{1}{2}} = \sqrt{3 \cdot \frac{1}{3}} = 1,$$

and similarly

$$\left\|(0, \frac{1}{\sqrt{2}})\right\| := B\left((0, \frac{1}{\sqrt{2}}), (0, \frac{1}{\sqrt{2}})\right)^{\frac{1}{2}} = \sqrt{2 \cdot \frac{1}{2}} = 1.$$

(iii) We find

$$\cos \varphi := \frac{B\big((\tfrac{1}{\sqrt{3}}, 0), (0, \tfrac{1}{\sqrt{2}})\big)}{\|(\tfrac{1}{\sqrt{3}}, 0)\| \|(0, \tfrac{1}{\sqrt{2}})\|} = \frac{1}{\sqrt{3}} \frac{1}{\sqrt{2}} = \frac{1}{\sqrt{6}}.$$

2. Let $\langle \cdot, \cdot \rangle$ be the Euclidean inner product of $\mathbb{R}^3$ and $v^{(1)}, v^{(2)}, v^{(3)}$ be the following vectors in $\mathbb{R}^3$, $v^{(1)} = \big(-\tfrac{1}{\sqrt{6}}, -\tfrac{2}{\sqrt{6}}, \tfrac{1}{\sqrt{6}}\big)$, $v^{(2)} = \big(\tfrac{1}{\sqrt{2}}, 0, \tfrac{1}{\sqrt{2}}\big)$, $v^{(3)} = \big(-\tfrac{1}{\sqrt{3}}, \tfrac{1}{\sqrt{3}}, \tfrac{1}{\sqrt{3}}\big)$.

   (i) Verify that $[v] = [v^{(1)}, v^{(2)}, v^{(3)}]$ is an orthonormal basis of $\mathbb{R}^3$.
   (ii) Compute $\mathrm{Id}_{[v] \to [e]}$, verify that it is an orthonormal 3×3 matrix, and then compute $\mathrm{Id}_{[e] \to [v]}$. Here $[e] = [e^{(1)}, e^{(2)}, e^{(3)}]$ denotes the standard basis of $\mathbb{R}^3$.
   (iii) Represent the vectors $a = (1, 2, 1)$ and $b = (1, 0, 1)$ as linear combinations of $v^{(1)}, v^{(2)},$ and $v^{(3)}$.

Solutions

   (i) We need to check that $\langle v^{(i)}, v^{(j)} \rangle = \delta_{ij}$. By a straightforward computation we get

$$\langle v^{(1)}, v^{(1)} \rangle = \tfrac{1}{6} + \tfrac{4}{6} + \tfrac{1}{6} = 1,$$
$$\langle v^{(2)}, v^{(2)} \rangle = \tfrac{1}{2} + \tfrac{1}{2} = 1,$$
$$\langle v^{(3)}, v^{(3)} \rangle = \tfrac{1}{3} + \tfrac{1}{3} + \tfrac{1}{3} = 1,$$

   and by symmetry of the inner product, it suffices to further check that

$$\langle v^{(1)}, v^{(2)} \rangle = -\tfrac{1}{\sqrt{12}} + \tfrac{1}{\sqrt{12}} = 0,$$
$$\langle v^{(1)}, v^{(3)} \rangle = \tfrac{1}{\sqrt{18}} - \tfrac{2}{\sqrt{18}} + \tfrac{1}{\sqrt{18}} = 0,$$
$$\langle v^{(2)}, v^{(3)} \rangle = -\tfrac{1}{\sqrt{6}} + \tfrac{1}{\sqrt{6}} = 0.$$

   (ii) One has

$$\mathrm{Id}_{[v] \to [e]} = \begin{pmatrix} -1/\sqrt{6} & 1/\sqrt{2} & -1/\sqrt{3} \\ -2/\sqrt{6} & 0 & 1/\sqrt{3} \\ 1/\sqrt{6} & 1/\sqrt{2} & 1/\sqrt{3} \end{pmatrix}.$$

   Since $[v]$ is an orthonormal basis of $\mathbb{R}^3$, $\mathrm{Id}_{[v] \to [e]}$ is an orthogonal matrix. Indeed, one computes

$$\mathrm{Id}^{\mathsf{T}}_{[v] \to [e]} \, \mathrm{Id}_{[v] \to [e]} = \begin{pmatrix} -1/\sqrt{6} & -2/\sqrt{6} & 1/\sqrt{6} \\ 1/\sqrt{2} & 0 & 1/\sqrt{2} \\ -1/\sqrt{3} & 1/\sqrt{3} & 1/\sqrt{3} \end{pmatrix} \begin{pmatrix} -1/\sqrt{6} & 1/\sqrt{2} & -1/\sqrt{3} \\ -2/\sqrt{6} & 0 & 1/\sqrt{3} \\ 1/\sqrt{6} & 1/\sqrt{2} & 1/\sqrt{3} \end{pmatrix}$$

$$= \begin{pmatrix} 1 & 0 & 0 \\ 0 & 1 & 0 \\ 0 & 0 & 1 \end{pmatrix}.$$

   So, we verified $\mathrm{Id}_{[e] \to [v]} x = \mathrm{Id}^{-1}_{[v] \to [e]} = \mathrm{Id}^{\mathsf{T}}_{[v] \to [e]}$.

(iii) Since $[v]$ is an orthonormal basis of $\mathbb{R}^3$, one has for any vector $x \in \mathbb{R}^3$, $x = \sum_{j=1}^{3} d_j v^{(j)}$ where $d_j = \langle x, v^{(j)} \rangle$. Alternatively, $d = (d_1, d_2, d_3)$ can be computed as $d = \mathrm{Id}_{[e] \to [v]} x$ with $x = (x_1, x_2, x_3)$. In this way one obtains for a and b as given,

$$a = -\frac{4}{\sqrt{6}} v^{(1)} + \sqrt{2} v^{(2)} + \frac{2}{\sqrt{3}} v^{(3)}, \qquad b = \sqrt{2} v^{(2)}.$$

3. Let $\langle \cdot, \cdot \rangle$ be the Euclidean inner product on $\mathbb{R}^2$ and $v^{(1)} = \left(\frac{1}{\sqrt{2}}, -\frac{1}{\sqrt{2}} \right)$.

  (i) Determine all possible vectors $v^{(2)} \in \mathbb{R}^2$ so that $[v^{(1)}, v^{(2)}]$ is an orthonormal basis of $\mathbb{R}^2$.
  (ii) Determine the matrix representation $R(\varphi)_{[v] \to [e]}$ of the linear map

$$R(\varphi) : \mathbb{R}^2 \to \mathbb{R}^2, \quad (x_1, x_2) \mapsto (\cos\varphi \cdot x_1 - \sin\varphi \cdot x_2, \ \sin\varphi \cdot x_1 + \cos\varphi \cdot x_2)$$

  with respect to an orthonormal basis $[v] = [v^{(1)}, v^{(2)}]$ of $\mathbb{R}^2$, found in Item (i).

Solutions

  (i) An orthonormal basis has to satisfy $\langle v^{(i)}, v^{(j)} \rangle = \delta_{ij}$. Denote $v^{(2)} = (x_1, x_2)$, then

$$x_1^2 + x_2^2 = 1, \quad \text{and} \quad \frac{1}{\sqrt{2}}(x_1 - x_2) = 0.$$

  This gives $x_1 = x_2$ and hence $2x_1^2 = 1$. So $v^{(2)}$ equals either $\left(\frac{1}{\sqrt{2}}, \frac{1}{\sqrt{2}} \right)$ or $\left(-\frac{1}{\sqrt{2}}, -\frac{1}{\sqrt{2}} \right)$.
  (ii) Let $[v] = [v^{(1)}, v^{(2)}]$ denote the orthonormal basis of $\mathbb{R}^2$ with $v^{(2)}$ given by $\left(\frac{1}{\sqrt{2}}, \frac{1}{\sqrt{2}} \right)$. Note that

$$R(\varphi)_{[e] \to [e]} = \begin{pmatrix} \cos\varphi & -\sin\varphi \\ \sin\varphi & \cos\varphi \end{pmatrix}, \qquad \mathrm{Id}_{[v^+] \to [e]} = \begin{pmatrix} 1/\sqrt{2} & 1/\sqrt{2} \\ -1/\sqrt{2} & 1/\sqrt{2} \end{pmatrix}.$$

  Hence $R(\varphi)_{[v] \to [e]} = R(\varphi)_{[e] \to [e]} \, \mathrm{Id}_{[v] \to [e]}$ can be computed as

$$R(\varphi)_{[v] \to [e]} = \frac{1}{\sqrt{2}} \begin{pmatrix} \cos\varphi + \sin\varphi & \cos\varphi - \sin\varphi \\ \sin\varphi - \cos\varphi & \sin\varphi + \cos\varphi \end{pmatrix}.$$

4. (i) Let $T : \mathbb{R}^3 \to \mathbb{R}^3$ be the rotation by the angle $\frac{\pi}{3}$ in clockwise direction in the $x_1 x_2$-plane with rotation axis $\{0\} \times \{0\} \times \mathbb{R}$. Determine $T_{[e] \to [e]}$ where $[e] = [e^{(1)}, e^{(2)}, e^{(3)}]$ is the standard basis of $\mathbb{R}^3$.
  (ii) Let $S : \mathbb{R}^3 \to \mathbb{R}^3$ be the rotation by the angle $\frac{\pi}{3}$ in counterclockwise direction in the $x_2 x_3$-plane with rotation axis $\mathbb{R} \times \{0\} \times \{0\}$. Determine $S_{[e] \to [e]}$ and verify that it is an orthogonal 3×3 matrix.
  (iii) Compute $(S \circ T)_{[e] \to [e]}$ and $(T \circ S)_{[e] \to [e]}$.

Solutions

(i) Clearly, $T e^{(3)} = e^{(3)}$. Further, the vectors $e^{(1)}$ and $e^{(2)}$ are rotated in the $x_1 x_2$-plane by $-\pi/3$. In particular, their third component stays constant. As in the case of $\mathbb{R}^2$, we compute

$$T e^{(1)} = \big(\cos(\pi/3), -\sin(\pi/3), 0 \big), \qquad T e^{(2)} = \big(\sin(\pi/3), \cos(\pi/3), 0 \big).$$

Altogether we thus obtain

$$T_{[e] \to [e]} = \begin{pmatrix} \cos(\pi/3) & \sin(\pi/3) & 0 \\ -\sin(\pi/3) & \cos(\pi/3) & 0 \\ 0 & 0 & 1 \end{pmatrix}.$$

(ii) We have $S e^{(1)} = e^{(1)}$, $S e^{(2)} = \big(0, \cos(\pi/3), \sin(\pi/3) \big)$, and $S e^{(3)} = \big(0, -\sin(\pi/3), \cos(\pi/3) \big)$, so that

$$S_{[e] \to [e]} = \begin{pmatrix} 1 & 0 & 0 \\ 0 & \cos(\pi/3) & -\sin(\pi/3) \\ 0 & \sin(\pi/3) & \cos(\pi/3) \end{pmatrix}.$$

S is orthogonal since for any vectors $e^{(i)}, e^{(j)}$ of the standard basis $[e] = [e^{(1)}, \dots, e^{(n)}]$, one sees by straightforward computations that

$$\langle S e^{(i)}, S e^{(j)} \rangle = 0, \quad i \neq j \qquad \text{and} \qquad \langle S e^{(i)}, S e^{(i)} \rangle = 1.$$

(iii) Recall that the matrix representation of a composition of two linear maps is given by the matrix multiplication of the appropriate matrix representations of the two linear maps. Hence

$$(S \circ T)_{[e] \to [e]} = S_{[e] \to [e]} \cdot T_{[e] \to [e]}, \qquad (T \circ S)_{[e] \to [e]} = T_{[e] \to [e]} \cdot S_{[e] \to [e]}.$$

It thus follows that

$$(S \circ T)_{[e] \to [e]} = \begin{pmatrix} 1 & 0 & 0 \\ 0 & \cos(\pi/3) & -\sin(\pi/3) \\ 0 & \sin(\pi/3) & \cos(\pi/3) \end{pmatrix} \begin{pmatrix} \cos(\pi/3) & \sin(\pi/3) & 0 \\ -\sin(\pi/3) & \cos(\pi/3) & 0 \\ 0 & 0 & 1 \end{pmatrix}$$

$$= \begin{pmatrix} \cos(\pi/3) & \sin(\pi/3) & 0 \\ -\cos(\pi/3)\sin(\pi/3) & \cos^2(\pi/3) & -\sin(\pi/3) \\ -\sin(\pi/3) & \cos(\pi/3)\sin(\pi/3) & \cos(\pi/3) \end{pmatrix}$$

and

$$(T \circ S)_{[e] \to [e]} = \begin{pmatrix} \cos(\pi/3) & \sin(\pi/3) & 0 \\ -\sin(\pi/3) & \cos(\pi/3) & 0 \\ 0 & 0 & 1 \end{pmatrix} \begin{pmatrix} 1 & 0 & 0 \\ 0 & \cos(\pi/3) & -\sin(\pi/3) \\ 0 & \sin(\pi/3) & \cos(\pi/3) \end{pmatrix}$$

$$= \begin{pmatrix} \cos(\pi/3) & \cos(\pi/3)\sin(\pi/3) & -\sin^2(\pi/3) \\ -\sin(\pi/3) & \cos^2(\pi/3) & -\cos(\pi/3)\sin(\pi/3) \\ 0 & \sin(\pi/3) & \cos(\pi/3) \end{pmatrix}$$

5. Decide whether the following assertions are true or false and justify your answers.

(i) There exists an orthogonal 2×2 matrix A with $\det(A) = -1$.
(ii) Let $v^{(1)}, v^{(2)}, v^{(3)}, v^{(4)}$ be vectors in $\mathbb{R}^4$ so that

$$\langle v^{(i)}, v^{(j)} \rangle = \delta_{ij}$$

for any $1 \leq i, j \leq 4$ where $\langle \cdot, \cdot \rangle$ denotes the Euclidean inner product in $\mathbb{R}^4$ and δ_{ij} is the *Kronecker delta*, defined as

$$\delta_{ij} = \begin{cases} 0 & \text{if } i \neq j, \\ 1 & \text{if } i = j. \end{cases}$$

Then $[v^{(1)}, v^{(2)}, v^{(3)}, v^{(4)}]$ is a basis of $\mathbb{R}^4$.

Solutions

(i) True. Consider $A = \begin{pmatrix} 1 & 0 \\ 0 & -1 \end{pmatrix}$. Then

$$A^T A = \begin{pmatrix} 1 & 0 \\ 0 & -1 \end{pmatrix} \begin{pmatrix} 1 & 0 \\ 0 & -1 \end{pmatrix} = \begin{pmatrix} 1 & 0 \\ 0 & 1 \end{pmatrix}.$$

(ii) True. Since the dimension of $\mathbb{R}^4$ equals 4, it remains to verify that $v^{(1)}, v^{(2)}, v^{(3)}, v^{(4)}$ are linearly independent. Suppose this is not the case. Then one of the vectors, let us say $v^{(4)}$, can be written as a linear combination of the others,

$$v^{(4)} = \alpha_1 v^{(1)} + \alpha_2 v^{(2)} + \alpha_3 v^{(3)}, \quad \alpha_1, \alpha_2, \alpha_3 \in \mathbb{R},$$

implying that $\langle v^{(4)}, v^{(4)} \rangle = \sum_{j=1}^3 \alpha_j \langle v^{(j)}, v^{(4)} \rangle$. Since by assumption $\langle v^{(4)}, v^{(4)} \rangle = 1$ and $\langle v^{(4)}, v^{(j)} \rangle = 0$ for any $1 \leq j \leq 3$, one then obtains $1 = 0$. Hence $v^{(1)}, v^{(2)}, v^{(3)}, v^{(4)}$ are indeed linearly independent.

Solutions of Problems of Sect. 5.1

1. Let $A = \begin{pmatrix} -1+2i & 1+i \\ 2+2i & 2-i \end{pmatrix} \in \mathbb{C}^{2 \times 2}$.

(i) Compute the eigenvalues of A.
(ii) Compute the eigenspaces of the eigenvalues of A.
(iii) Find a regular 2×2 matrix $S \in \mathbb{C}^{2 \times 2}$ so that $S^{-1} A S$ is diagonal.

Solutions

(i) The characteristic polynomial $\chi_A(z) = \det(A - z \operatorname{Id}_{2 \times 2})$ of A can be computed as

$$\chi_A(z) = (-1+2i-z)(2-i-z) - (1+i)(2+2i)$$

$$= z^2 - z - iz + i = (z-1)(z-i).$$

Hence the eigenvalues of A, given by the roots of the characteristic polynomial, are $\lambda_1 = 1$ and $\lambda_2 = i$. (Note that both eigenvalues have arithmetic multiplicity 1, hence the spectrum of A is simple.)

(ii) The eigenspace $E_{\lambda_1}(A)$ of A for the eigenvalue λ_1 is defined as the nullspace of $A - \lambda_1 \operatorname{Id}_{2\times2}$. One obtains by Gaussian elimination

$$A - \lambda_1 \operatorname{Id}_{2\times2} = \begin{pmatrix} -2+2i & 1+i \\ 2+2i & 1-i \end{pmatrix} \rightsquigarrow \begin{pmatrix} 1 & -i/2 \\ 1 & -i/2 \end{pmatrix} \rightsquigarrow \begin{pmatrix} 1 & -i/2 \\ 0 & 0 \end{pmatrix}.$$

Hence $v^{(1)} = (i, 2)$ is an eigenvector of A for λ_1. Since λ_1 is simple, the eigenspace $E_{\lambda_1}(A)$ is given by $E_{\lambda_1}(A) = \{\alpha(i, 2) \mid \alpha \in \mathbb{C}\}$. Similarly, one computes the eigenspace $E_{\lambda_2}(A)$ of A for λ_2.

$$A - \lambda_2 \operatorname{Id}_{2\times2} = \begin{pmatrix} -1+i & 1+i \\ 2+2i & 2-2i \end{pmatrix} \rightsquigarrow \begin{pmatrix} 1 & -i \\ 1 & -i \end{pmatrix} \rightsquigarrow \begin{pmatrix} 1 & -i \\ 0 & 0 \end{pmatrix}.$$

Hence $v^{(2)} = (i, 1)$ is an eigenvector of A for λ_2. Since λ_2 is simple, the eigenspace $E_{\lambda_2}(A)$ is given by $E_{\lambda_2}(A) = \{\alpha(i, 1) \mid \alpha \in \mathbb{C}\}$.

(iii) Since $v^{(1)}$ and $v^{(2)}$ are eigenvectors of A for distinct eigenvalues, they are linearly independent and hence $[v] = [v^{(1)}, v^{(2)}]$ is a basis of $\mathbb{C}^2$. It follows that

$$(\operatorname{Id}_{[v]\to[e]})^{-1} A \operatorname{Id}_{[v]\to[e]} = \begin{pmatrix} 1 & 0 \\ 0 & i \end{pmatrix}$$

and

$$S := \operatorname{Id}_{[v]\to[e]} = \begin{pmatrix} i & i \\ 2 & 1 \end{pmatrix}.$$

2. Let $A = \begin{pmatrix} 1 & -1 \\ 2 & -1 \end{pmatrix}$, viewed as an element in $\mathbb{C}^{2\times2}$.

(i) Compute the eigenvalues of A.
(ii) Compute the eigenspaces (in $\mathbb{C}^2$) of the eigenvalues of A.
(iii) Find a regular 2×2 matrix $S \in \mathbb{C}^{2\times2}$ so that $S^{-1}AS$ is diagonal.

Solutions

(i) The characteristic polynomial $\chi_A(z) = \det(A - z\operatorname{Id}_{2\times2})$ of A can be computed as

$$\chi_A(z) = (1 - z)(-1 - z) + 2 = z^2 + 1.$$

Hence the eigenvalues of A are $\lambda_1 = i$ and $\lambda_2 = -i$. (Note that the spectrum of A is simple.)

(ii) The eigenspace $E_{\lambda_1}(A)$ of A for the eigenvalue λ_1 is defined as the nullspace of $A - \lambda_1 \operatorname{Id}_{2\times2}$. One obtains by Gaussian elimination

$$A - \lambda_1 \operatorname{Id}_{2\times2} = \begin{pmatrix} 1-i & -1 \\ 2 & -1-i \end{pmatrix} \rightsquigarrow \begin{pmatrix} 1 & -(1+i)/2 \\ 1 & -(1+i)/2 \end{pmatrix} \rightsquigarrow \begin{pmatrix} 1 & -(1+i)/2 \\ 0 & 0 \end{pmatrix}.$$

Hence $v^{(1)} = (1 + i, 2)$ is an eigenvector of A for λ_1. Since λ_1 is simple, the eigenspace $E_{\lambda_1}(A)$ is given by $E_{\lambda_1}(A) = \{\alpha(1 + i, 2) \mid \alpha \in \mathbb{C}\}$. Similarly, one computes the eigenspace $E_{\lambda_2}(A)$ of A for λ_2.

$$A - \lambda_2 \,\mathrm{Id}_{2 \times 2} = \begin{pmatrix} 1 + i & -1 \\ 2 & -1 + i \end{pmatrix} \rightsquigarrow \begin{pmatrix} 1 & {(-1+i)}/{2} \\ 1 & {(-1+i)}/{2} \end{pmatrix} \rightsquigarrow \begin{pmatrix} 1 & {(-1+i)}/{2} \\ 0 & 0 \end{pmatrix}.$$

Hence $v^{(2)} = (1 - i, 2)$ is an eigenvector of A for λ_2. Since λ_2 is simple, the eigenspace $E_{\lambda_2}(A)$ is given by $E_{\lambda_2}(A) = \{\alpha(1 - i, 2) \mid \alpha \in \mathbb{C}\}$.

(iii) Since $v^{(1)}$ and $v^{(2)}$ are eigenvectors of A for distinct eigenvalues, they are linearly independent and hence $[v] = [v^{(1)}, v^{(2)}]$ is a basis of $\mathbb{C}^2$. It follows that

$$(\mathrm{Id}_{[v] \to [e]})^{-1} A \,\mathrm{Id}_{[v] \to [e]} = \begin{pmatrix} i & 0 \\ 0 & -i \end{pmatrix}$$

and

$$S := \mathrm{Id}_{[v] \to [e]} = \begin{pmatrix} 1 + i & 1 - i \\ 2 & 2 \end{pmatrix}$$

3. Let $A = \begin{pmatrix} {1}/{2} + {i}/{2} & {1}/{2} + {i}/{2} \\ -{1}/{2} - {i}/{2} & {1}/{2} + {i}/{2} \end{pmatrix} \in \mathbb{C}^{2 \times 2}$.

(i) Verify that A is unitary.
(ii) Compute the spectrum of A.
(iii) Find an orthonormal basis of $\mathbb{C}^2$, consisting of eigenvectors of A.

Solutions

(i) The matrix A is unitary if and only if $\overline{A}^{\mathrm{T}} A = \mathrm{Id}_{2 \times 2}$. We find

$$\overline{A} = \begin{pmatrix} {1}/{2} - {i}/{2} & {1}/{2} - {i}/{2} \\ -{1}/{2} + {i}/{2} & {1}/{2} - {i}/{2} \end{pmatrix}$$

and hence

$$\overline{A}^{\mathrm{T}} = \begin{pmatrix} {1}/{2} - {i}/{2} & -{1}/{2} + {i}/{2} \\ {1}/{2} - {i}/{2} & {1}/{2} - {i}/{2} \end{pmatrix}.$$

By a straightforward computation one finds that $\overline{A}^{\mathrm{T}} A = \mathrm{Id}_{2 \times 2}$ and hence A is indeed unitary.

(ii) The characteristic polynomial $\chi_A(z) = \det(A - z \,\mathrm{Id}_{2 \times 2})$ of A can be computed as

$$\chi_A(z) = (a - z)(a - z) - (-a)a = (a - z)^2 + a^2, \qquad a := (1 + i)/2.$$

Hence the roots of $\chi_A(z)$ satisfy $a - z = \sqrt{-a^2} = \pm i a$ and hence the eigenvalues of A are given by

$$\lambda_1 = a - ia = (1 - i)a = (1 - i)(1 + i)/2 = 1,$$
$$\lambda_2 = a + ia = (1 + i)a = (1 + i)(1 + i)/2 = i.$$

(iii) We first compute an eigenvector of A for λ_1. By Gaussian elimination,

$$A - \lambda_1 \, \mathrm{Id}_{2\times2} = \begin{pmatrix} -{}^1\!/_2 + {}^i\!/_2 & {}^1\!/_2 + {}^i\!/_2 \\ -{}^1\!/_2 - {}^i\!/_2 & -{}^1\!/_2 + {}^i\!/_2 \end{pmatrix} \rightsquigarrow \begin{pmatrix} 1 & -i \\ 1 & -i \end{pmatrix} \rightsquigarrow \begin{pmatrix} 1 & -i \\ 0 & 0 \end{pmatrix}.$$

Thus $v^{(1)} = \frac{1}{\sqrt{2}} (i, 1)$ is an eigenvector of A for λ_1 with $\|v^{(1)}\| = 1$. Similarly, we compute an eigenvector of A for λ_2,

$$A - \lambda_2 \, \mathrm{Id}_{2\times2} = \begin{pmatrix} {}^1\!/_2 - {}^i\!/_2 & {}^1\!/_2 + {}^i\!/_2 \\ -{}^1\!/_2 - {}^i\!/_2 & {}^1\!/_2 - {}^i\!/_2 \end{pmatrix} \rightsquigarrow \begin{pmatrix} 1 & i \\ 1 & i \end{pmatrix} \rightsquigarrow \begin{pmatrix} 1 & i \\ 0 & 0 \end{pmatrix}.$$

Thus $v^{(2)} = \frac{1}{\sqrt{2}} (-i, 1)$ is an eigenvector of A for λ_2 with $\|v^{(2)}\| = 1$.

Since A is unitary and λ_1 and λ_2 are distinct eigenvalues, $v^{(1)}$ and $v^{(2)}$ are orthogonal. Indeed,

$$\langle v^{(1)}, v^{(2)} \rangle = v_1^{(1)} \cdot \overline{v_1^{(2)}} + v_2^{(1)} \cdot \overline{v_2^{(2)}} = 0$$

Taking into account that both $v^{(1)}$ and $v^{(2)}$ are of norm one, it follows that $[v^{(1)}, v^{(2)}]$ is an orthonormal basis of $\mathbb{C}^2$.

4. Let $A = \begin{pmatrix} 2 & i & 1 \\ -i & 2 & -i \\ 1 & i & 2 \end{pmatrix} \in \mathbb{C}^{3\times3}$.

   (i) Verify that A is Hermitian.
   (ii) Compute the spectrum of A.
   (iii) Find a unitary 3×3 matrix $S \in \mathbb{C}^{3\times3}$ so that $S^{-1}AS$ is diagonal.

Solutions

(i) $\overline{A^{\mathsf{T}}} = \begin{pmatrix} 2 & i & 1 \\ -i & 2 & -i \\ 1 & i & 2 \end{pmatrix} = A.$

(ii) The characteristic polynomial $\chi_A(z) = \det(A - z \, \mathrm{Id}_{3\times3})$ of A can be computed as

$$\chi_A(z) = (2 - z)\big((2 - z)^2 - 1\big) + i\big(i(2 - z) - i\big) + \big(1 - (2 - z)\big)$$
$$= (2 - z)^3 - 3(2 - z) + 2 = -z^3 + 6z^2 - 9z + 4.$$

By an educated guess, we find that $z = 1$ is a root of $\chi_A(z)$ and hence $(z - 1)$ (and therefore $-(z - 1)$) is a factor of $\chi_A(z)$. One computes

$$-z^3 + 6z^2 - 9z + 4 = -(z - 1)(z^2 - 5z + 4) = -(z - 1)(z - 1)(z - 4).$$

Therefore, the eigenvalues of A are given by $\lambda_1 = 1$, $\lambda_2 = 1$ and $\lambda_3 = 4$. Note that λ_1 has algebraic multilpicity two, whereas λ_3 is simple.

(iii) To find a basis of $\mathbb{C}^3$, consisting of eigenvectors of A, we first compute the eigenspace $E_{\lambda_1}(A) (= E_{\lambda_2}(A))$ of A. By Gaussian elimination

$$A - \lambda_1 \, \mathrm{Id}_{3\times3} = \begin{pmatrix} 1 & i & 1 \\ -i & 1 & -i \\ 1 & i & 1 \end{pmatrix} \rightsquigarrow \begin{pmatrix} 1 & i & 1 \\ 0 & 0 & 0 \\ 0 & 0 & 0 \end{pmatrix}.$$

Hence $w^{(1)} = (-i, 1, 0)$ and $w^{(2)} = (-1, 0, 1)$ are linearly independent eigenvectors of A for λ_1 and hence

$$E_{\lambda_1}(A) = \{\alpha_1 w^{(1)} + \alpha_2 w^{(2)} \mid \alpha_1, \alpha_2 \in \mathbb{C}\}.$$

We need to find an orthonormal basis of $E_{\lambda_1}(A)$. Note that $w^{(1)}$ and $w^{(2)}$ are not orthogonal. Indeed, $\langle w^{(2)}, w^{(1)} \rangle = -i \neq 0$. Define $v^{(1)} := \frac{1}{\sqrt{2}}(-i, 1, 0)$. Then $v^{(1)}$ is an eigenvector of A for λ_1 of norm one. Further set

$$\tilde{v}^{(2)} = w^{(2)} - \langle w^{(2)}, v^{(1)} \rangle v^{(1)} = (-1, 0, 1) - \frac{-i}{2}(-i, 1, 0) = (-\frac{1}{2}, \frac{i}{2}, 1).$$

Then $v^{(1)}$ and $\tilde{v}^{(2)}$ are orthogonal. Finally, to obtain an orthonormal basis of $E_{\lambda_1}(A)$ we need to normalize $\tilde{v}^{(2)}$,

$$v^{(2)} := \frac{1}{\|\tilde{v}^{(2)}\|} \tilde{v}^{(2)} = \frac{1}{\sqrt{6}}(-1, i, 2).$$

Now let us turn to the eigenvalue λ_3, which is simple. To find an eigenvector of A for λ_3, we get by Gaussian elimination

$$A - \lambda_3 \, \mathrm{Id}_{3 \times 3} = \begin{pmatrix} -2 & i & 1 \\ -i & -2 & -i \\ 1 & i & -2 \end{pmatrix} \rightsquigarrow \begin{pmatrix} 2 & -i & -1 \\ -2i & -4 & -2i \\ 2 & 2i & -4 \end{pmatrix} \rightsquigarrow \begin{pmatrix} 2 & -i & -1 \\ 0 & -3 & -3i \\ 0 & 3i & -3 \end{pmatrix}$$

$$\rightsquigarrow \begin{pmatrix} 2 & -i & -1 \\ 0 & 1 & i \\ 0 & 0 & 0 \end{pmatrix}.$$

Hence $v^{(3)} = \frac{1}{\sqrt{3}}(1, -i, 1)$ is an eigenvector of A for λ_3 of norm one.

Since A is Hermitian, and λ_1 and λ_3 are distinct eigenvalues, $v^{(3)}$ is orthogonal to $E_{\lambda_1}(A)$ and hence $[v] = [v^{(1)}, v^{(2)}, v^{(3)}]$ is an orthonormal basis of $\mathbb{C}^3$. Altogether we conclude that

$$(\mathrm{Id}_{[v] \to [e]})^{-1} A \, \mathrm{Id}_{[v] \to [e]} = \begin{pmatrix} 1 & 0 & 0 \\ 0 & 1 & 0 \\ 0 & 0 & 4 \end{pmatrix}$$

is a diagonal 3×3 matrix and

$$S = \mathrm{Id}_{[v] \to [e]} = \begin{pmatrix} -i/\sqrt{2} & -1/\sqrt{6} & 1/\sqrt{3} \\ 1/\sqrt{2} & i/\sqrt{6} & -i/\sqrt{3} \\ 0 & 2/\sqrt{6} & 1/\sqrt{3} \end{pmatrix}$$

a unitary 3×3 matrix.

5. Decide whether the following assertions are true or false and justify your answers.

   (i) Assume that $A, B \in \mathbb{C}^{2 \times 2}$ and that $\lambda \in \mathbb{C}$ is an eigenvalue of A and $\mu \in \mathbb{C}$ is an eigenvalue of B. Then $\lambda + \mu$ is an eigenvalue of $A + B$.
   (ii) For any $A \in \mathbb{C}^{2 \times 2}$, A and A^{T} have the same eigenspaces.

Solutions

(i) False. Let $A := \begin{pmatrix} 1 & 0 \\ 0 & 0 \end{pmatrix}$ and $B := \begin{pmatrix} 0 & 0 \\ 0 & 1 \end{pmatrix}$. Then $A + B = \mathrm{Id}_{2\times 2}$. Note that 1 is an eigenvalue of A and of B, but $1 + 1 = 2$ is not an eigenvalue of $\mathrm{Id}_{2\times 2}$.

(ii) False. Let $A := \begin{pmatrix} 0 & 1 \\ 0 & 0 \end{pmatrix}$. Then $A^{\mathrm{T}} = \begin{pmatrix} 0 & 0 \\ 1 & 0 \end{pmatrix}$. Note that $\lambda_1 = 0$ is the only eigenvalue of A. It has algebraic multiplicity 2 and geometric multiplicity 1. Furthermore $v^{(1)} = (1, 0)$ is an eigenvector of A and $E_{\lambda_1}(A) = \{\alpha(1, 0) \mid \alpha \in \mathbb{C}\}$ the eigenspace of λ_1. The spectrum of A^{T} coincides with the one of A. But the eigenspace $E_{\lambda_1}(A^{\mathrm{T}})$ is given by $E_{\lambda_1}(A^{\mathrm{T}}) = \{\alpha(0, 1) \mid \alpha \in \mathbb{C}\}$.

Solutions of Problems of Sect. 5.2

1. Let $A = \begin{pmatrix} 1 & 4 \\ 2 & 3 \end{pmatrix} \in \mathbb{R}^{2\times 2}$.

   (i) Compute the spectrum of A.
   (ii) Find eigenvectors $v^{(1)}, v^{(2)} \in \mathbb{R}^2$ of A, which form a basis of $\mathbb{R}^2$.
   (iii) Find a regular 2×2 matrix $S \in \mathbb{R}^{2\times 2}$ so that $S^{-1}AS$ is diagonal.

   Solutions

   (i) The characteristic polynomial $\chi_A(z) = \det(A - z\,\mathrm{Id}_{2\times 2})$ of A can be computed as

   $$\chi_A(z) = (1 - z)(3 - z) - 8 = z^2 - 4z - 5 = (z + 1)(z - 5).$$

   Hence the eigenvalues of A are $\lambda_1 = -1$ and $\lambda_2 = 5$. Note that both eigenvalues are simple and real.

   (ii) Let us first find an eigenvector of A for λ_1. By Gaussian elimination

   $$A - \lambda_1\,\mathrm{Id}_{2\times 2} = \begin{pmatrix} 2 & 4 \\ 2 & 4 \end{pmatrix} \rightsquigarrow \begin{pmatrix} 1 & 2 \\ 0 & 0 \end{pmatrix},$$

   Hence $v^{(1)} = (-2, 1) \in \mathbb{R}^2$ is an eigenvector of A for λ_1. To find an eigenvector of A for λ_2, one argues similarly. By Gaussian elimination

   $$A - \lambda_2\,\mathrm{Id}_{2\times 2} = \begin{pmatrix} -4 & 4 \\ 2 & -2 \end{pmatrix} \rightsquigarrow \begin{pmatrix} 1 & -1 \\ 0 & 0 \end{pmatrix},$$

   Hence $v^{(2)} = (1, 1) \in \mathbb{R}^2$ is an eigenvector of A for λ_2.

   (iii) Since $v^{(1)}$ and $v^{(2)}$ are eigenvectors of A for distinct eigenvalues and in addition vectors in $\mathbb{R}^2$, they are linearly independent in $\mathbb{R}^2$. Hence $[v] = [v^{(1)}, v^{(2)}]$ is a basis of $\mathbb{R}^2$. It follows that

   $$(\mathrm{Id}_{[v]\to[e]})^{-1}A\,\mathrm{Id}_{[v]\to[e]} = \begin{pmatrix} -1 & 0 \\ 0 & 5 \end{pmatrix}$$

and

$$S := \mathrm{Id}_{[v] \to [e]} = \begin{pmatrix} -2 & 1 \\ 1 & 1 \end{pmatrix}.$$

2. Let $A = \begin{pmatrix} 2 & 1 & 0 \\ 0 & 1 & -1 \\ 0 & 2 & -1 \end{pmatrix}$, viewed as an element in $\mathbb{C}^{3 \times 3}$.

   (i) Compute the spectrum of A.
   (ii) For each eigenvalue of A, compute the eigenspace.
   (iii) Find a regular matrix $S \in \mathbb{C}^{3 \times 3}$ so that $S^{-1} A S$ is diagonal.

Solutions

   (i) The characteristic polynomial $\chi_A(z) = \det(A - z \, \mathrm{Id}_{3 \times 3})$ of A can be computed as

$$\chi_A(z) = (2 - z)(1 - z)(-1 - z) + 2(2 - z) = -(z - 2)(z^2 + 1).$$

   Thus the eigenvalues of A are $\lambda_1 = 2$, $\lambda_2 = i$, and $\lambda_3 = -i$. Note that all the eigenvalues are simple.

   (ii) The eigenspace $E_{\lambda_1}(A)$ of A for the eigenvalue λ_1 is defined as the nullspace of $A - \lambda_1 \, \mathrm{Id}_{3 \times 3}$. One obtains by Gaussian elimination

$$A - \lambda_1 \, \mathrm{Id}_{3 \times 3} = \begin{pmatrix} 0 & 1 & 0 \\ 0 & -1 & -1 \\ 0 & 2 & -3 \end{pmatrix} \rightsquigarrow \begin{pmatrix} 0 & 1 & 0 \\ 0 & 0 & 1 \\ 0 & 0 & 1 \end{pmatrix},$$

   Hence $v^{(1)} = (1, 0, 0)$ is an eigenvector of A for λ_1. Since λ_1 is a simple eigenvalue, $E_{\lambda_1}(A) = \{\alpha(1, 0, 0) \mid \alpha \in \mathbb{C}\}$.
   The eigenspace $E_{\lambda_2}(A)$ of A for the eigenvalue λ_2 is computed in a similar fashion. By Gaussian elimination

$$A - \lambda_2 \, \mathrm{Id}_{3 \times 3} = \begin{pmatrix} 2 - i & 1 & 0 \\ 0 & 1 - i & -1 \\ 0 & 2 & -1 - i \end{pmatrix} \rightsquigarrow \begin{pmatrix} 2 - i & 1 & 0 \\ 0 & 2 & -1 - i \\ 0 & 0 & 0 \end{pmatrix},$$

   Hence $v^{(2)} = (-1 - 3i, 5 + 5i, 10)$ is an eigenvector of A for λ_2. Since λ_2 is a simple eigenvalue, $E_{\lambda_2}(A) = \{\alpha(-1 - 3i, 5 + 5i, 10) \mid \alpha \in \mathbb{C}\}$.
   Finally, we compute the eigenspace $E_{\lambda_3}(A)$ of A for the eigenvalue λ_3. By Gaussian elimination

$$A - \lambda_3 \, \mathrm{Id}_{3 \times 3} = \begin{pmatrix} 2 + i & 1 & 0 \\ 0 & 1 + i & -1 \\ 0 & 2 & -1 + i \end{pmatrix} \rightsquigarrow \begin{pmatrix} 1 & 0 & (1 - 3i)/10 \\ 0 & 2 & -1 + i \\ 0 & 0 & 0 \end{pmatrix}.$$

   Hence $v^{(3)} = (-1 + 3i, 5 - 5i, 10)$ is an eigenvector of A for λ_3. Since λ_3 is a simple eigenvalue, $E_{\lambda_3}(A) = \{\alpha(-1 + 3i, 5 - 5i, 10) \mid \alpha \in \mathbb{C}\}$.

(iii) Since $v^{(1)}$, $v^{(2)}$, and $v^{(3)}$ are eigenvectors of A in $\mathbb{C}^3$ for distinct eigenvalues, they are linearly independent in $\mathbb{C}^3$. Hence they are linearly independent in $\mathbb{C}^3$ and therefore $[v] = [v^{(1)}, v^{(2)}, v^{(3)}]$ is a basis of $\mathbb{C}^3$. It follows that

$$(\mathrm{Id}_{[v] \to [e]})^{-1} A \,\mathrm{Id}_{[v] \to [e]} = \begin{pmatrix} 2 & 0 & 0 \\ 0 & i & 0 \\ 0 & 0 & -i \end{pmatrix}$$

and

$$S := \mathrm{Id}_{[v] \to [e]} = \begin{pmatrix} 1 & -1 - 3i & -1 + 3i \\ 0 & \dfrac{5 + 5i}{10} & \dfrac{5 - 5i}{10} \\ 0 & \dfrac{10}{} & \dfrac{10}{} \end{pmatrix}.$$

3. Determine for each of the following symmetric matrices

   (i) $A = \begin{pmatrix} 2 & 1 \\ 1 & 2 \end{pmatrix}$

   (ii) $B = \begin{pmatrix} 2 & 1 & 1 \\ 1 & 3 & -2 \\ 1 & -2 & 3 \end{pmatrix}$

its spectrum and find an orthogonal matrix, which diagonalizes it.

Solutions

(i) The characteristic polynomial $\chi_A(z) = \det(A - z\,\mathrm{Id}_{2\times2})$ of A can be computed as

$$\chi_A(z) = (2 - z)^2 - 1 = z^2 - 4z + 3 = (z - 1)(z - 3).$$

Hence the eigenvalues of A are $\lambda_1 = 1$ and $\lambda_2 = 3$. Note that both eigenvalues are simple. First we determine an eigenvector of A for λ_1. By Gaussian elimination one has

$$A - \lambda_1\,\mathrm{Id}_{2\times2} = \begin{pmatrix} 1 & 1 \\ 1 & 1 \end{pmatrix} \rightsquigarrow \begin{pmatrix} 1 & 1 \\ 0 & 0 \end{pmatrix}.$$

Hence $v^{(1)} = \frac{1}{\sqrt{2}}(-1, 1)$ is an eigenvector of A for the eigenvalue λ_1 with $\|v^{(1)}\| = 1$. To determine an eigenvector of A for λ_2, we argue similarly,

$$A - \lambda_2\,\mathrm{Id}_{2\times2} = \begin{pmatrix} -1 & 1 \\ 1 & -1 \end{pmatrix} \rightsquigarrow \begin{pmatrix} 1 & -1 \\ 0 & 0 \end{pmatrix}.$$

Hence $v^{(2)} = \frac{1}{\sqrt{2}}(1, 1)$ is an eigenvector of A for the eigenvalue λ_2 with $\|v^{(2)}\| = 1$. Since $v^{(1)}$ and $v^{(2)}$ are eigenvectors of A in $\mathbb{R}^2$ for distinct eigenvalues and A is a symmetric 2×2 matrix, $v^{(1)}$ and $v^{(2)}$ are orthogonal in $\mathbb{R}^2$. Therefore $[v] = [v^{(1)}, v^{(2)}]$ is an orthonormal basis of $\mathbb{R}^2$. It follows that

$$(\mathrm{Id}_{[v] \to [e]})^{-1} A \,\mathrm{Id}_{[v] \to [e]} = \begin{pmatrix} 1 & 0 \\ 0 & 3 \end{pmatrix}$$

and that

$$S := \mathrm{Id}_{[v] \to [e]} = \begin{pmatrix} -1/\sqrt{2} & 1/\sqrt{2} \\ 1/\sqrt{2} & 1/\sqrt{2} \end{pmatrix}$$

is an orthogonal 2×2 matrix.

(ii) The characteristic polynomial $\chi_A(z) = \det(A - z\,\mathrm{Id}_{3\times3})$ of A can be computed as

$$\chi_A(z) = (2-z)\big((3-z)^2 - 4\big) - \big((3-z) + 2\big) + \big(-2 - (3-z)\big)$$

$$= (3-z)(2-z)(3-z) - 2(3-z) - 4(3-z)$$

$$= (3-z)\big((2-z)(3-z) - 6)\big) = (3-z)(z^2 - 5z) = z(3-z)(z-5).$$

Hence the eigenvalues of A are $\lambda_1 = 0$, $\lambda_2 = 3$, and $\lambda_3 = 5$. Note that each of the three eigenvalues is simple.

First we determine an eigenvector of A for λ_1. By Gaussian elimination one has

$$A - \lambda_1\,\mathrm{Id}_{3\times3} = \begin{pmatrix} 2 & 1 & 1 \\ 1 & 3 & -2 \\ 1 & -2 & 3 \end{pmatrix} \rightsquigarrow \begin{pmatrix} 2 & 1 & 1 \\ 2 & 6 & -4 \\ 2 & -4 & 6 \end{pmatrix}$$

$$\rightsquigarrow \begin{pmatrix} 2 & 1 & 1 \\ 0 & 5 & -5 \\ 0 & -5 & 5 \end{pmatrix} \rightsquigarrow \begin{pmatrix} 2 & 1 & 1 \\ 0 & 1 & -1 \\ 0 & 0 & 0 \end{pmatrix}.$$

Hence $v^{(1)} = \frac{1}{\sqrt{3}}(-1, 1, 1)$ is an eigenvector of A for λ_1 with $\|v^{(1)}\| = 1$.

Similarly, we compute an eigenvector of A for λ_2. By Gaussian elimination

$$A - \lambda_2\,\mathrm{Id}_{3\times3} = \begin{pmatrix} -1 & 1 & 1 \\ 1 & 0 & -2 \\ 1 & -2 & 0 \end{pmatrix} \rightsquigarrow \begin{pmatrix} -1 & 1 & 1 \\ 0 & 1 & -1 \\ 0 & 0 & 0 \end{pmatrix},$$

Hence $v^{(2)} = \frac{1}{\sqrt{6}}(2, 1, 1)$ is an eigenvector of A for λ_2 with $\|v^{(2)}\| = 1$.

Finally, we compute an eigenvector of A for λ_3,

$$A - \lambda_3\,\mathrm{Id}_{3\times3} = \begin{pmatrix} -3 & 1 & 1 \\ 1 & -2 & -2 \\ 1 & -2 & -2 \end{pmatrix} \rightsquigarrow \begin{pmatrix} -3 & 1 & 1 \\ 1 & -2 & -2 \\ 0 & 0 & 0 \end{pmatrix} \rightsquigarrow \begin{pmatrix} -3 & 1 & 1 \\ 3 & -6 & -6 \\ 0 & 0 & 0 \end{pmatrix}$$

$$\rightsquigarrow \begin{pmatrix} -3 & 1 & 1 \\ 0 & -5 & -5 \\ 0 & 0 & 0 \end{pmatrix} \rightsquigarrow \begin{pmatrix} -3 & 0 & 0 \\ 0 & 1 & 1 \\ 0 & 0 & 0 \end{pmatrix}.$$

Hence $v^{(3)} = \frac{1}{\sqrt{2}}(0, -1, 1)$ is an eigenvector of A for λ_3 with $\|v^{(3)}\| = 1$. Since all three eigenvalues are simple and $v^{(1)}$, $v^{(2)}$, and $v^{(3)}$ are normalized eigenvectors in $\mathbb{R}^3$, it follows that $[v] = [v^{(1)}, v^{(2)}, v^{(3)}]$ is an orthonormal basis of $\mathbb{R}^3$. Altogether we conclude that

$$(\mathrm{Id}_{[v]\to[e]})^{-1} A\,\mathrm{Id}_{[v]\to[e]} = \begin{pmatrix} 0 & 0 & 0 \\ 0 & 3 & 0 \\ 0 & 0 & 5 \end{pmatrix}$$

and that

$$S := \mathrm{Id}_{[v]\to[e]} = \begin{pmatrix} -1/\sqrt{3} & 2/\sqrt{6} & 0 \\ 1/\sqrt{3} & 1/\sqrt{6} & -1/\sqrt{2} \\ 1/\sqrt{3} & 1/\sqrt{6} & 1/\sqrt{2} \end{pmatrix}$$

is an orthogonal 3×3 matrix.

4. (i) Assume that A is a $n \times n$ matrix with real coefficients, $A \in \mathbb{R}^{n \times n}$, satisfying $A^2 = A$. Verify that each eigenvalue of A is either 0 or 1.

   (ii) Let $A \in \mathbb{R}^{n \times n}$. For any $a \in \mathbb{R}$, compute the eigenvalues and the eigenspaces of $A + a \operatorname{Id}_{n \times n}$ in terms of the eigenvalues and the eigenspaces of A.

Solutions

   (i) Suppose λ is an eigenvalue of A and v an eigenvector of A for λ, $Av = \lambda v$. Then

$$\lambda v = Av = AAv = A(\lambda v) = \lambda^2 v.$$

   Since $v \neq 0$, it follows that $\lambda^2 = \lambda$. So either $\lambda = 0$ or $\lambda = 1$.

   (ii) Let $\lambda_1, \ldots, \lambda_n$ be the eigenvalues of A, listed with their algebraic multiplicities. The characteristic polynomial $\chi_A(z) = \det(A - z \operatorname{Id}_{n \times n})$ of A the reads

$$\chi_A(z) = (-1)^n \prod_{1 \leq j \leq n} (\lambda - \lambda_j).$$

   Hence

$$\chi_{A + a \operatorname{Id}_{n \times n}}(z) = \det(A - (z - a) \operatorname{Id}_{n \times n}) = \chi_A(z - a).$$

   Therefore, the roots of $\chi_{A + a \operatorname{Id}_{n \times n}}$ are the roots of χ_A, translated by a. It means that the eigenvalues of $A + a \operatorname{Id}_{n \times n}$ are $\lambda_1 + a, \ldots, \lambda_n + a$.
   Any eigenvector of A for an eigenvalue λ of A is also an eigenvector of $A + a \operatorname{Id}_{n \times n}$ for the eigenvalue $\lambda + a$ of $A + a \operatorname{Id}_{n \times n}$. It follows that for any $1 \leq j \leq n$, $E_{\lambda_j}(A) = E_{\lambda_j + a}(A + a \operatorname{Id}_{n \times n})$.

5. Decide whether the following assertions are true or false and justify your answers.

   (i) Any matrix $A \in \mathbb{R}^{3 \times 3}$ has at least one real eigenvalue.

   (ii) There exists a symmetric matrix $A \in \mathbb{R}^{5 \times 5}$, which admits an eigenvalue, whose geometric multiplicity is 1, but its algebraic multiplicity is 2.

Solutions

   (i) True. Assume that A has not only real eigenvalues. Let $\lambda \in \mathbb{C}$ be an eigenvalue of A with $\operatorname{Im}(\lambda) \neq 0$. Then also the complex conjugate $\bar{\lambda}$ of λ is an eigenvalue of A with $\operatorname{Im} \bar{\lambda} \neq 0$. Consequently, the characteristic polynomial of A is of the form

$$\chi_A(z) = (\lambda - z)(\bar{\lambda} - z)(\mu - z) = (|\lambda|^2 - 2\operatorname{Re}(\lambda)z + z^2)(\mu - z)$$

   with μ being the third eigenvalue of A. Since A is a real matrix, $\chi_A(z) \in \mathbb{R}$ for any $z \in \mathbb{R}$. This implies that $\mu \in \mathbb{R}$.

   (ii) False. By Theorem 5.2.5, the geometric multiplicity of any eigenvalue λ of a real symmetric matrix A in $\mathbb{R}^{5 \times 5}$ is equal to the algebraic multiplicity of λ.

Solutions of Problems of Sect. 5.3

1. (i) Decide, which of the following matrices $A \in \mathbb{R}^{n \times n}$ are symmetric and which are not.

$$\text{(a)} \begin{pmatrix} 3 & 2 & 1 \\ 2 & 1 & 3 \\ 1 & 3 & 2 \end{pmatrix} \qquad \text{(b)} \begin{pmatrix} 1 & 2 \\ -2 & 1 \end{pmatrix} \qquad \text{(c)} \begin{pmatrix} 1 & 2 & 3 \\ 2 & -1 & 3 \\ 3 & 4 & 1 \end{pmatrix}$$

(ii) Determine for the following quadratic forms Q the symmetric matrices $A \in \mathbb{R}^{3 \times 3}$ so that $Q(x) = \langle x, Ax \rangle$ for any $x = (x_1, x_2, x_3) \in \mathbb{R}^3$.

(a) $Q(x_1, x_2, x_3) = 2x_1^2 + 3x_2^2 + x_3^2 + x_1 x_2 - 2x_1 x_3 + 3x_2 x_3$.

(b) $Q(x_1, x_2, x_3) = 8x_1 x_2 + 10x_1 x_3 + x_1^2 - x_3^2 + 5x_2^2 + 7x_2 x_3$.

Solutions

(i) (a) is symmetric, (b) and (c) are not.

(ii) (a) Writing $Q(x_1, x_2, x_3)$ in the form

$$Q(x_1, x_2, x_3) = x_1\left(2x_1 + \frac{1}{2}x_2 - x_3\right) + x_2\left(3x_2 + \frac{1}{2}x_1 + \frac{3}{2}x_3\right) + x_3\left(x_3 - x_1 + \frac{3}{2}x_2\right),$$

one sees that

$$A = \begin{pmatrix} 2 & 1/2 & -1 \\ 1/2 & 3 & 3/2 \\ -1 & 3/2 & 1 \end{pmatrix}.$$

(b) Similarly, write $Q(x, y, z)$ in the form

$$Q(x_1, x_2, x_3) = x_1\left(x_1 + 4x_2 + 5x_3\right) + x_2\left(4x_1 + 5x_2 + \frac{7}{2}x_3\right) + x_3\left(5x_1 + \frac{7}{2}x_2 - x_3\right),$$

yielding

$$A = \begin{pmatrix} 1 & 4 & 5 \\ 4 & 5 & 7/2 \\ 5 & 7/2 & -1 \end{pmatrix}.$$

2. Find a coordinate transformation of $\mathbb{R}^2$ (translation and/or rotation) so that the conic section $K_f = \{f(x_1, x_2) = 0\}$ is in canonical form where

$$f(x_1, x_2) = 3x_1^2 + 8x_1 x_2 - 3x_2^2 + 28.$$

Solutions Clearly

$$f(x_1, x_2) = \langle A \begin{pmatrix} x_1 \\ x_2 \end{pmatrix}, \begin{pmatrix} x_1 \\ x_2 \end{pmatrix} \rangle + 28, \qquad A = \begin{pmatrix} 3 & 4 \\ 4 & -3 \end{pmatrix}.$$

To determine the eigenvalues of A, consider the characteristic polynomial $\chi_A(z) = \det(A - z\,\mathrm{Id}_{2\times2})$ of A,

$$\chi_A(z) = (3-z)(-3-z) - 16 = z^2 - 25 = (z-5)(z+5).$$

Hence, $\lambda_1 = 5$ and $\lambda_2 = -5$. To obtain an eigenvector of A for λ_1, we compute

$$A - \lambda_1\,\mathrm{Id}_{2\times2} = \begin{pmatrix} -2 & 4 \\ 4 & -8 \end{pmatrix} \rightsquigarrow \begin{pmatrix} 1 & -2 \\ 0 & 0 \end{pmatrix}$$

Hence $v^{(1)} = \frac{1}{\sqrt5}(2,1)$ is an eigenvector of A for λ_1 with $\|v^{(1)}\| = 1$. Similarly, to find an eigenvector of A for λ_2, one computes

$$A - \lambda_2\,\mathrm{Id}_{2\times2} = \begin{pmatrix} 8 & 4 \\ 4 & 2 \end{pmatrix} \rightsquigarrow \begin{pmatrix} 1 & {}^1\!/_2 \\ 0 & 0 \end{pmatrix}$$

Hence $v^{(2)} = \frac{1}{\sqrt5}(-1,2)$ is an eigenvector of A for λ_2 with $\|v^{(2)}\| = 1$. Since $v^{(1)}$ and $v^{(2)}$ are normalized eigenvectors of A for the distinct eigenvalues λ_1 and respectively, λ_2, $[v] = [v^{(1)}, v^{(2)}]$ is an orthonormal basis of $\mathbb{R}^2$. Hence

$$S := \mathrm{Id}_{[v]\to[e]} = \frac{1}{\sqrt5}\begin{pmatrix} 2 & -1 \\ 1 & 2 \end{pmatrix}$$

is an orthogonal 2×2 matrix and $S^\top A S = \mathrm{diag}(5,-5)$ or $A = S\,\mathrm{diag}(5,-5)S^\top$. It implies that

$$f(x) = \langle Ax, x\rangle = \big(\mathrm{diag}(5,-5)S^\top x, S^\top x\big).$$

Let $y := S^\top x$. Then $x = Sy$ and

$$f(Sy) = 5y_1^2 - 5y_2^2 + 28.$$

Hence

$$K_f = \Big\{x \mid \frac{5}{28}(y_2^2 - y_1^2) = 1,\ y = S^\top x\Big\}.$$

3. Verify that the conic section $K_f = \{f(x) = 0\}$ is a parabola where f is given by

$$f(x) = x_1^2 + 2x_1x_2 + x_2^2 + 3x_1 + x_2 - 1, \quad x = (x_1, x_2) \in \mathbb{R}^2.$$

Solutions Writing $3x_1 + x_2$ as $2(x_1 + x_2) + x_1 - x_2$, one obtains by completing the square of $(x_1 + x_2)^2 + 2(x_1 + x_2)$,

$$f(x) = (x_1 + x_2)^2 + 2(x_1 + x_2) + x_1 - x_2 - 1 = (x_1 + x_2 + 1)^2 + x_1 - x_2 - 2.$$

We want to find a translation

$$T: \mathbb{R}^2 \to \mathbb{R}^2, x \mapsto y := x + a, \quad a = (a_1, a_2) \in \mathbb{R}^2,$$

so that $f(x) = (y_1 + y_2)^2 + y_1 - y_2$ or

$$(x_1 + x_2 + 1)^2 + x_1 - x_2 - 2 = ((x_1 + a_1) + (x_2 + a_2))^2 + (x_1 + a_1) - (x_2 + a_2).$$

It means that a_1, a_2 have to satisfy $a_1 + a_2 = 1$ and $a_1 - a_2 = -2$. One obtains $a_1 = -1/2$ and $a_2 = 3/2$. Note that

$$f(x) = (y_1 + y_2)^2 + y_1 - y_2, \qquad y = x + a,$$

can be written as

$$f(x) = \langle Ay, y \rangle + \langle b, y \rangle, \qquad A := \begin{pmatrix} 1 & 1 \\ 1 & 1 \end{pmatrix}, \qquad b := \begin{pmatrix} 1 \\ -1 \end{pmatrix}.$$

To bring K_f into normal form, we have to analyze the spectrum of A. The characteristic polynomial $\chi_A(z) = \det(A - z \, \mathrm{Id}_{2\times 2})$ of A is given by

$$\chi_A(z) = (1 - z)^2 - 1 = z^2 - 2z = z(z - 2),$$

hence $\lambda_1 = 2$ and $\lambda_2 = 0$ are the eigenvalues of A. Note that both eigenvalues are simple. To obtain an eigenvector of A for the eigenvalue λ_1, we use Gaussian elimination,

$$A - \lambda_1 \, \mathrm{Id}_{2\times 2} = \begin{pmatrix} -1 & 1 \\ 1 & -1 \end{pmatrix} \rightsquigarrow \begin{pmatrix} 1 & -1 \\ 0 & 0 \end{pmatrix}.$$

Hence $v^{(1)} = \frac{1}{\sqrt{2}} (1, 1) \in \mathbb{R}^2$ is a normalized eigenvector of A for the eigenvalue λ_1. Similarly, to obtain an eigenvector of A for λ_2, one computes

$$A - \lambda_2 \, \mathrm{Id}_{2\times 2} = \begin{pmatrix} 1 & 1 \\ 1 & 1 \end{pmatrix} \rightsquigarrow \begin{pmatrix} 1 & 1 \\ 0 & 0 \end{pmatrix}.$$

Hence $v^{(2)} = \frac{1}{\sqrt{2}} (-1, 1) \in \mathbb{R}^2$ is a normalized eigenvector of A for λ_2. Since $v^{(1)}$ and $v^{(2)}$ are normalized eigenvectors of A for the eigenvalues λ_1 and λ_2, the eigenvalues are simple, and $A \in \mathbb{R}^{2\times 2}$ is symmetric, $[v] = [v^{(1)}, v^{(2)}]$ is an orthonormal basis of $\mathbb{R}^2$. It implies that

$$S = \mathrm{Id}_{[v] \rightarrow [e]} = \frac{1}{\sqrt{2}} \begin{pmatrix} 1 & -1 \\ 1 & 1 \end{pmatrix}$$

is an orthogonal 2×2 matrix and $S^T A S = \mathrm{diag}(2, 0)$ or

$$\langle Ay, y \rangle + \langle b, y \rangle = \langle \mathrm{diag}(2, 0) S^T y, S^T y \rangle + \langle S^T b, S^T y \rangle, \qquad S^T b = \begin{pmatrix} 0 \\ -\sqrt{2} \end{pmatrix}.$$

Consequently,

$$K_f = \{ x \mid w_1^2 = \frac{1}{\sqrt{2}} w_2, \ w := S^T y = S^T (x + a) \}.$$

4. (i) Determine symmetric matrices $A, B \in \mathbb{R}^{2\times 2}$ so that A and B have the same eigenvalues, but not the same eigenspaces.

(ii) Assume that $A, B \in \mathbb{C}^{2 \times 2}$ have the same eigenvalues and the same eigenspaces. Decide whether in such a case $A = B$.

Solutions

(i) Consider

$$A = \begin{pmatrix} 1 & 0 \\ 0 & 0 \end{pmatrix}, \qquad B = \begin{pmatrix} 0 & 0 \\ 0 & 1 \end{pmatrix}.$$

The eigenvalues of A are $\lambda_1 = 1$ and $\lambda_2 = 0$. They coincide with the eigenvalues of B. The eigenspace $E_{\lambda_1}(A)$ of A for λ_1 is given by $E_{\lambda_1}(A) = \{\alpha(1, 0) \mid \alpha \in \mathbb{C}\}$, whereas $E_{\lambda_1}(B) = \{\alpha(0, 1) \mid \alpha \in \mathbb{C}\}$. Hence $E_{\lambda_1}(A) \neq E_{\lambda_1}(B)$.

(ii) In general, it does not follow that $A = B$. Consider

$$A = \begin{pmatrix} 0 & 1 \\ 0 & 0 \end{pmatrix}, \qquad B = \begin{pmatrix} 0 & 2 \\ 0 & 0 \end{pmatrix}.$$

The eigenvalues of A are $\lambda_1 = \lambda_2 = 0$ and coincide with the eigenvalues of B. Note that for both matrices, λ_1 has algebraic multiplicity two and geometric multiplicity one. Furthermore, $v^{(1)} = (1, 0)$ is an eigenvector of A and of B for λ_1. Hence $E_{\lambda_1}(A) = E_{\lambda_1}(B)$. However, $A \neq B$.

5. Decide whether the following assertions are true or false and justify your answers.

(i) For any eigenvalue of a symmetric matrix $A \in \mathbb{R}^{n \times n}$, its algebraic multiplicity equals its geometric multiplicity.

(ii) The linear map $T: \mathbb{R}^2 \to \mathbb{R}^2$, $(x_1, x_2) \mapsto (x_2, x_1)$ is a rotation.

(iii) The linear map $R: \mathbb{R}^2 \to \mathbb{R}^2$, $(x_1, x_2) \mapsto (-x_2, x_1)$ is orthogonal.

Solutions

(i) True. See Theorem 5.2.5. Since A is symmetric, it is diagonalizable. That is there exists an orthogonal transformation S, such that

$$S^T A S = \begin{pmatrix} \lambda_1 & & \\ & \ddots & \\ & & \lambda_n \end{pmatrix}$$

with $\lambda_1 \ldots \lambda_n$ being the possibly complex eigenvalues of A. Suppose $\lambda_i = \ldots \lambda_{i+m}$. Then, $e^{(i)}, \ldots, e^{(i+m)}$ are all eigenvectors of λ_i and the dimension of $E_{\lambda_i}(A)$ corresponds to the algebraic multiplicity of λ_i.

(ii) False. Clearly, the matrix representation $T_{[e] \to [e]}$ of T is given by

$$T_{[e] \to [e]} = \begin{pmatrix} 0 & 1 \\ 1 & 0 \end{pmatrix}.$$

Since $\det(T_{[e] \to [e]}) = -1$, T is not a rotation.

(iii) True. Indeed, for arbitrary vectors $x = (x_1, x_2) \in \mathbb{R}^2$, $y = (y_1, y_2) \in \mathbb{R}^2$ one has

$$\langle Rx, Ry \rangle = (-x_2)(-y_2) + x_1 y_1 = x_1 y_1 + x_2 y_2 = \langle x, y \rangle.$$

Hence by definition, R is orthogonal.

Solutions of Problems of Sect. 6.2

1. Find the general solution of the following linear ODEs.

   (i) $\begin{cases} y_1'(t) = \quad y_1(t) + y_2(t) \\ y_2'(t) = -2y_1(t) + 4y_2(t) \end{cases}$,

   (ii) $\begin{cases} y_1'(t) = \quad 2y_1(t) + 4y_2(t) \\ y_2'(t) = \quad -y_1(t) - 3y_2(t) \end{cases}$.

   Solutions

   (i) We write $y' = Ay$ with

   $$A = \begin{pmatrix} 1 & 1 \\ -2 & 4 \end{pmatrix}.$$

   The general solution is given by $y(t) = e^{At}y_0$. To determine $e^{At}y_0$, we compute the eigenvalues if A with

   $$\lambda_\pm = \frac{5 \pm \sqrt{25 - 24}}{2} = \frac{5 \pm 1}{2},$$

   and hence $v^{(1)} = (1, 1)$ and $v^{(2)} = (1, 2)$. The general solution is thus given by

   $$y(t) = ae^{2t}v^{(1)} + be^{3t}v^{(2)}, \quad a, b \in \mathbb{R}.$$

   (ii) Again, we write $y' = Ay$ with

   $$A = \begin{pmatrix} 2 & 4 \\ -1 & -3 \end{pmatrix}.$$

   The general solution is given by $y(t)0e^{At}y_0$. To determine $e^{At}y_0$, we compute the eigenvalues of A with

   $$\lambda_\pm = \frac{-1 \pm \sqrt{1 + 8}}{2} = \frac{-1 \pm 3}{2},$$

   and hence $v^{(1)} = (-1, 1)$ and $v^{(2)} = (-4, 1)$. The general solution is this given by

   $$y(t) = are^{-2t}v^{(1)} + be^t v^{(2)}, \quad a, b \in \mathbb{R}.$$

2. Solve the following initial value problems.

   (i) $\begin{cases} y_1'(t) = -y_1(t) + 2y_2(t) \\ y_2'(t) = \quad 2y_1(t) - y_2(t) \end{cases}$, $\quad y^{(0)} = \begin{pmatrix} 2 \\ -1 \end{pmatrix}$,

   (ii) $\begin{cases} y_1'(t) = 2y_1(t) - 6y_3(t) \\ y_2'(t) = y_1(t) - 3y_3(t) \\ y_3'(t) = y_2(t) - 2y_3(t) \end{cases}$, $\quad y^{(0)} = \begin{pmatrix} 1 \\ 0 \\ -1 \end{pmatrix}$.

Solutions

(i) We write the system in the form $y' = Ay$ with $A = \begin{pmatrix} -1 & 2 \\ 2 & -1 \end{pmatrix}$. The eigenvalues A are then given by

$$\lambda_\pm = \frac{-2 \pm \sqrt{4 + 12}}{2} = -1 \pm 2,$$

and for the eigenvectors, we obtain

$$\lambda_- : \quad \begin{pmatrix} 2 & 2 \\ 2 & 2 \end{pmatrix} \rightsquigarrow \begin{pmatrix} 1 & 1 \\ 0 & 0 \end{pmatrix},$$

$$\lambda_+ : \quad \begin{pmatrix} -2 & 2 \\ 2 & -2 \end{pmatrix} \rightsquigarrow \begin{pmatrix} 1 & -1 \\ 0 & 0 \end{pmatrix},$$

Finally, $v^{(1)} = (-1, 1)$ and $v^{(2)} = (1, 1)$. The general solution has the form

$$y(t) = ae^{-3t}v^{(1)} + be^{t}v^{(2)}, \quad a, b \in \mathbb{R}.$$

To determine (a, b), we use the initial conditions and thus solve the system

$$\text{augmented coefficient matrix} \quad \begin{bmatrix} -1 & 1 & \big| & 2 \\ 1 & 1 & \big| & -1 \end{bmatrix}$$

$$R_1 \rightsquigarrow -R_1, R_2 \rightsquigarrow R_2 + R_1 \quad \begin{bmatrix} 1 & -1 & \big| & -2 \\ 0 & 2 & \big| & 1 \end{bmatrix}$$

$$R_1 \rightsquigarrow R_1 + {}^1\!/_2 R_2, R_2 \rightsquigarrow {}^1\!/_2 R_2 \quad \begin{bmatrix} 1 & 0 & \big| & -{}^3\!/_2 \\ 0 & 1 & \big| & {}^1\!/_2 \end{bmatrix}.$$

Hence, the solution for the initial values $y^{(0)} = (2, -1)$ is

$$y(t) = -\frac{3}{2} e^{-3t} \begin{pmatrix} -1 \\ 1 \end{pmatrix} + \frac{1}{2} e^{t} \begin{pmatrix} 1 \\ 1 \end{pmatrix}.$$

(ii) We write the initial value problem in the form $y' = Ay$ with $A = \begin{pmatrix} 2 & 0 & -6 \\ 1 & 0 & -3 \\ 0 & 1 & -2 \end{pmatrix}$. The characteristic polynomial of A is given by

$$\chi_A(z) = (2 - z)(-z)(-2 - z) - 6 - (-3(2 - z))$$
$$= (2 - z)(-z)(-2 - z) - 3z = -z(z^2 - 1).$$

The eigenvalues of A thus are $\lambda_1 = 1$, $\lambda_2 = -1$, $\lambda_3 = 0$. To obtain the eigenvectors, we find for

$$\lambda_1 : \begin{pmatrix} 1 & 0 & -6 \\ 1 & -1 & -3 \\ 0 & 1 & -3 \end{pmatrix} \rightsquigarrow \begin{pmatrix} 1 & 0 & -6 \\ 0 & -1 & -3 \\ 0 & 0 & 0 \end{pmatrix},$$

$$\lambda_2 : \begin{pmatrix} 3 & 0 & -6 \\ 1 & 1 & -3 \\ 0 & 1 & -1 \end{pmatrix} \rightsquigarrow \begin{pmatrix} 1 & 0 & -2 \\ 0 & 1 & -1 \\ 0 & 0 & -0 \end{pmatrix},$$

$$\lambda_3 : \begin{pmatrix} 2 & 0 & -6 \\ 1 & 0 & -3 \\ 0 & 1 & -2 \end{pmatrix} \rightsquigarrow \begin{pmatrix} 1 & 0 & -3 \\ 0 & 0 & 0 \\ 0 & 1 & -2 \end{pmatrix},$$

which gives by choosing the third component equal to 1, $v^{(1)} = (6, 3, 1)$, $v^{(2)} = (2, 1, 1)$ and $v^{(3)} = (3, 2, 1)$.

We therefore have as a general solution

$$e^{At} y_0 = a e^t \begin{pmatrix} 6 \\ 3 \\ 1 \end{pmatrix} + b e^{-t} \begin{pmatrix} 2 \\ 1 \\ 1 \end{pmatrix} + c \begin{pmatrix} 3 \\ 2 \\ 1 \end{pmatrix}.$$

Evaluating at y_0 yields the following system.

augmented coefficient matrix $\qquad \left[\begin{array}{ccc|c} 6 & 2 & 3 & 1 \\ 3 & 1 & 2 & 0 \\ 1 & 1 & 1 & -1 \end{array} \right]$

$R_2 \rightsquigarrow 2(R_2 - \tfrac{1}{2} R_1), R_3 \rightsquigarrow 6(R_3 - \tfrac{1}{6} R_1) \qquad \left[\begin{array}{ccc|c} 6 & 2 & 3 & 1 \\ 0 & 0 & 1 & -1 \\ 0 & 4 & 3 & -7 \end{array} \right]$

$R_1 \rightsquigarrow R_1 - 3R_2, R_3 \rightsquigarrow R_3 - 4R_2, R_{2 \leftrightarrow 3} \qquad \left[\begin{array}{ccc|c} 6 & 2 & 0 & 4 \\ 0 & 4 & 0 & -4 \\ 0 & 0 & 1 & -1 \end{array} \right]$

$R_1 \rightsquigarrow \tfrac{1}{6} (R_1 - \tfrac{1}{2} R_2), R_2 \rightsquigarrow \tfrac{1}{4} R_2 \qquad \left[\begin{array}{ccc|c} 1 & 0 & 0 & 1 \\ 0 & 1 & 0 & -1 \\ 0 & 0 & 1 & -1 \end{array} \right]$

and hence

$$y(t) = e^{At} y_0 = e^t \begin{pmatrix} 6 \\ 3 \\ 1 \end{pmatrix} - e^{-t} \begin{pmatrix} 2 \\ 1 \\ 1 \end{pmatrix} - \begin{pmatrix} 3 \\ 2 \\ 1 \end{pmatrix}.$$

3. Consider the linear ODE

$$y'(t) = Ay(t), \qquad A = \begin{pmatrix} 0 & 1 & 0 \\ 0 & 0 & 1 \\ 0 & 0 & 0 \end{pmatrix} \in \mathbb{R}^{3 \times 3}.$$

(i) Compute A^2 and A^3.

(ii) Determine the general solution of $y'(t) = Ay(t)$.

Solutions

(i) $A^2 = \begin{pmatrix} 0 & 0 & 1 \\ 0 & 0 & 0 \\ 0 & 0 & 0 \end{pmatrix}$, $A^3 = \begin{pmatrix} 0 & 0 & 0 \\ 0 & 0 & 0 \\ 0 & 0 & 0 \end{pmatrix}$.

(ii) Since $A^3 = 0$, it follows that $A^n = 0$ for any $n \geq 3$. Therefore,

$$e^{tA} = \sum_{n \geq 0} \frac{t^n}{n!} A^n = \mathrm{Id} + tA + \frac{t^2}{A^2} = \begin{pmatrix} 1 & t & t^2/2 \\ 0 & 1 & t \\ 0 & 0 & 1 \end{pmatrix}.$$

The general solution is then given by $y(t) = e^{At} y_0$.

4. Let $A = \begin{pmatrix} 0 & 2 \\ -1 & 0 \end{pmatrix} \in \mathbb{R}^{2 \times 2}$. Find the solutions of the following initial value problems.

(i) $y'(t) = Ay(t) + \begin{pmatrix} e^t \\ 3 \end{pmatrix}$, $\qquad y^{(0)} = \begin{pmatrix} 1 \\ 0 \end{pmatrix}$,

(ii) $y'(t) = Ay(t) + \begin{pmatrix} 0 \\ \cos(2t) \end{pmatrix}$, $\qquad y^{(0)} = \begin{pmatrix} 0 \\ 1 \end{pmatrix}$.

Solutions

(i) Since we have an added term $(e^t, 3)$, we choose the ansatz

$$y(t) = \begin{pmatrix} a_0 + a_1 e^t \\ b_0 + b_1 e^t \end{pmatrix}.$$

Inserting this into the given equation yields

$$\begin{pmatrix} a_1 e^t \\ b_1 e^t \end{pmatrix} = \begin{pmatrix} 2b_0 + 2b_1 e^t \\ -a_0 - a_1 e^t \end{pmatrix} + \begin{pmatrix} e^t \\ 3 \end{pmatrix} = \begin{pmatrix} 2b_0 + (2b_1 + 1)e^t \\ -a_0 + 3 - a_1 e^t \end{pmatrix}.$$

This leads to the following four equations.

$$0 = 2b_0, \qquad a_1 = 2b_1 + 1, \qquad 0 = a_0 - 3, \qquad b_1 = -a_1,$$

which admit the solution

$$a_0 = 3, \qquad b_0 = 0, \qquad a_1 = \frac{1}{3}, \qquad b_1 = -\frac{1}{3}.$$

Hence,

$$y_p(t) = \frac{1}{3} \begin{pmatrix} 0 + e^t \\ -e^t \end{pmatrix}$$

is a particular solution.

To determine the general solution, we first compute the eigenvalues with

$$\lambda_\pm = \frac{0 \pm \sqrt{-8}}{2} = \pm i\sqrt{2}.$$

To obtain the eigenvectors, we get

$$\lambda_1 : \quad \begin{pmatrix} i\sqrt{2} & 2 \\ -1 & i\sqrt{2} \end{pmatrix} \rightsquigarrow \begin{pmatrix} 1 & -i\sqrt{2} \\ 0 & 0 \end{pmatrix},$$

$$\lambda_2 : \quad \begin{pmatrix} -i\sqrt{2} & 2 \\ -1 & -i\sqrt{2} \end{pmatrix} \rightsquigarrow \begin{pmatrix} 1 & i\sqrt{2} \\ 0 & 0 \end{pmatrix},$$

and thus obtain $v^{(1)} = (-i\sqrt{2}, 1)$ and $v^{(2)} = (i\sqrt{2}, 1)$. Any solution of the homogenous equation thus has the form

$$y_{\text{hom}}(t) = a \begin{pmatrix} \sqrt{2}\sin(\omega t) \\ \cos(\omega t) \end{pmatrix} + b \begin{pmatrix} -\sqrt{2}\cos(\omega t) \\ \sin(\omega t) \end{pmatrix}, \quad a, b \in \mathbb{R}.$$

The general solution of the original problem is given by

$$y(t) = y_p(t) + a \begin{pmatrix} \sqrt{2}\sin(\omega t) \\ \cos(\omega t) \end{pmatrix} + b \begin{pmatrix} -\sqrt{2}\cos(\omega t) \\ \sin(\omega t) \end{pmatrix}.$$

To find (a, b) for the initial value $(1, 0)$, we solve

$$\begin{pmatrix} 1 \\ 0 \end{pmatrix} = \frac{1}{3}\begin{pmatrix} 10 \\ -1 \end{pmatrix} + \begin{pmatrix} -\sqrt{2}b \\ a \end{pmatrix}$$

and hence obtain $a = 1/3$ and $b = 7/(3\sqrt{2})$. Finally,

$$y(t) = \frac{1}{3}\begin{pmatrix} 9 + e^t \\ -e^t \end{pmatrix} + \frac{1}{3}\begin{pmatrix} -7\cos(\sqrt{2}t) + \sqrt{2}\sin(\sqrt{2}t) \\ {}^{7}/_{\sqrt{2}}\sin(\sqrt{2}t) + \cos(\sqrt{2}t) \end{pmatrix}.$$

(ii) The added term $(0, \cos(2t))$ lets us choose the ansatz

$$y_p(t) = \begin{pmatrix} a_0\cos(2t) + a_1\sin(2t) \\ b_0\cos(2t) + b_1\sin(2t) \end{pmatrix},$$

which gives, inserted into the equation,

$$\begin{pmatrix} -2a_0\sin(2t) + 2a_1\cos(2t) \\ -2b_0\sin(2t) + 2b_1\cos(2t) \end{pmatrix} = \begin{pmatrix} 2b_0\cos(2t) + 2b_1\sin(2t) \\ -a_0\cos(2t) - a_1\sin(2t) \end{pmatrix} + \begin{pmatrix} 0 \\ \cos(2t) \end{pmatrix}.$$

This yields the following four equation.

$$-2a_0 = 2b_1, \qquad 2a_1 = 2b_0, \qquad -2b_0 = -a_1, \qquad 2b_1 = 1 - a_0$$

with the solution

$$a_0 = -1, \qquad b_1 = 1, \qquad a_1 = b_0 = 0.$$

Thus, we have the particular solution

$$y_p(t) = \begin{pmatrix} -\cos(2t) \\ \sin(2t) \end{pmatrix}.$$

The general solution is given by $y(t) = y_p(t) + y_{\text{hom}}(t)$ where y_{hom} is the solution to the homogenous equation. We thus solve with the given initial values for

$$\begin{pmatrix} 0 \\ 1 \end{pmatrix} = \begin{pmatrix} -1 \\ 0 \end{pmatrix} + \begin{pmatrix} -\sqrt{2}b \\ a \end{pmatrix}.$$

Thus, $a = 1, b = -\frac{1}{\sqrt{2}}$ and

$$y(t) = \begin{pmatrix} -\cos(2t) \\ \sin(2t) \end{pmatrix} + \begin{pmatrix} \cos(\sqrt{2}t) + \sqrt{2}\sin(\sqrt{2}t) \\ -1/\sqrt{2}\sin(\sqrt{2}t) + \cos(\sqrt{2}t) \end{pmatrix}.$$

5. Decide whether the following assertions are true or false and justify your answers.

(i) For any $A, B \in \mathbb{R}^{2\times2}$, one has $e^{A+B} = e^A e^B$.

(ii) Let $A = \begin{pmatrix} 0 & -1 \\ 1 & 0 \end{pmatrix} \in \mathbb{R}^{2\times2}$. Then the linear ODE $y'(t) = Ay(t) + \begin{pmatrix} t^2 \\ t \end{pmatrix}$

admits a particular solution of the form $\begin{pmatrix} a + bt + ct^2 \\ d + et \end{pmatrix}$.

Solutions

(i) False. Consider

$$A = \begin{pmatrix} 1 & 0 \\ 0 & 0 \end{pmatrix}, \qquad B = \begin{pmatrix} 0 & 1 \\ 0 & 0 \end{pmatrix}.$$

For the addition, we get

$$(A + B)^2 = A + B = \begin{pmatrix} 1 & 1 \\ 0 & 0 \end{pmatrix} =: C.$$

Hence,

$$e^{(A+B)t} = \text{Id} + \left(\sum_{n\geq1} \frac{t^n}{n!}\right)C = \text{Id} + (e^t - 1)C = \begin{pmatrix} e^t & e^t - 1 \\ 0 & 1 \end{pmatrix}.$$

Since $A^2 = A$ and $B^2 = 0$, we further conclude

$$e^{tA} = \begin{pmatrix} e^t & 0 \\ 0 & 1 \end{pmatrix}, \qquad e^{tB} = \text{Id} + tB = \begin{pmatrix} 1 & t \\ 0 & 1 \end{pmatrix}.$$

Consequently,

$$e^{tA}e^{tB} = \begin{pmatrix} e^t & te^t \\ 0 & 1 \end{pmatrix}.$$

(ii) False, inserting the ansatz

$$y_p(t) = \begin{pmatrix} a + bt + ct^2 \\ d + et \end{pmatrix}$$

into the equation yields

$$\begin{pmatrix} b + 2ct \\ e \end{pmatrix} = \begin{pmatrix} -d - et \\ a + bt + ct^2 \end{pmatrix} + \begin{pmatrix} t^2 \\ t \end{pmatrix},$$

which does not admit a solution due to the t^2 term in the first line.

Solutions of Problems of Sect. 6.3

1. Find the general solution of the following linear ODEs of second order.

(i) $y''(t) + 2y'(t) + 4y(t) = 0$,
(ii) $y''(t) + 2y'(t) - 4y(t) = t^2$.

Solutions

(i) We look for a solution of the form

$$y(t) = e^{\lambda t}.$$

Inserting the ansatz into the equation yields

$$0 = (\lambda^2 + 2\lambda + 4)e^{\lambda t}.$$

Hence

$$\lambda_{\pm} = \frac{-2 \pm \sqrt{4 - 16}}{2} = -1 \pm i\sqrt{3}$$

satisfies the equation. The general solution is thus given

$$y(t) = ae^{-t}\cos(\sqrt{3}\,t) + be^{-t}\sin(\sqrt{3}\,t), \quad a, b \in \mathbb{R}.$$

(ii) To obtain a particular solution, we make the ansatz

$$y(t) = at^2 + bt + c.$$

Inserting the ansatz into the equation yields

$$t^2 = -4at^2 + (-4b + 4a)t + (2a + 2b - 4c).$$

By comparison of coefficients, $a = -1/4$, $b = -1/4$, and $c = -1/4$. To obtain the general solution of the homogenous equation, we make the ansatz

$$y(t) = e^{\lambda t}.$$

Inserting this ansatz into the equation yields

$$0 = (\lambda^2 + 2\lambda - 4)e^{\lambda t}.$$

Hence

$$\lambda_\pm = \frac{-2 \pm \sqrt{4 + 16}}{2} = -1 \pm \sqrt{5}$$

and the general real solution of the homogenous equation reads

$$y_{\text{hom}}(t) = ae^{(\sqrt{5}-1)t} + be^{-(\sqrt{5}+1)t}, \quad a, b \in \mathbb{R}.$$

The general solution of the given equation is thus

$$y(t) = ae^{(\sqrt{5}-1)t} + be^{-(\sqrt{5}+1)t} - \frac{1}{4}t^2 - \frac{1}{4}t - \frac{1}{4}, \quad a, b \in \mathbb{R}.$$

2. Find the solutions of the following initial value problems.

(i) $y''(t) - y'(t) - 2y(t) = e^{-\pi t}$, $y(0) = 0$, $y'(0) = 1$.
(ii) $y''(t) + y(t) = \sin t$, $y(0) = 1$, $y'(0) = 0$.

Solutions

(i) To obtain a particular solution of $y''(t) - y'(t) - 2y(t) = e^{-\pi t}$, we look for a solution of the form

$$y_p(t) = ce^{-\pi t}.$$

Inserting the ansatz into the equation gives

$$e^{-\pi t} = c(\pi^2 + \pi - 2)e^{-\pi t},$$

and hence the choice $c = (\pi^2 + \pi - 2)^{-1}$ yields a particular solution.
To obtain the general solution of the homogenous equation $y''(t) - y'(t) - 2y(t) = 0$, we look for solutions of the form

$$y(t) = e^{\lambda t}.$$

Substituting it into $y''(t) - y'(t) - 2y(t) = 0$ one gets

$$0 = (\lambda^2 - \lambda - 2)e^{\lambda t}.$$

Hence

$$\lambda_1 = \frac{1 - 3}{2} = -1, \qquad \lambda_2 = \frac{1 + 3}{2} = 2.$$

The general solution of the homogenous equation is then given by

$$y_{\text{hom}}(t) = ae^{-t} + be^{2t}, \quad a, b \in \mathbb{R}.$$

The general solution of the given ODE is therefore

$$y(t) = ae^{-t} + be^{2t} + \frac{1}{\pi^2 + \pi - 2} e^{-\pi t}, \quad a, b \in \mathbb{R}.$$

To obtain a and b, corresponding to the initial values $y(0) = 0$, $y'(0) = 1$, we solve the linear system of equations

$$0 = y(0) = a + b + \frac{1}{\pi^2 + \pi - 2}, \qquad 1 = y'(0) = -a + 2b - \frac{\pi}{\pi^2 + \pi - 2}.$$

Adding the latter two equations, one has

$$b = \frac{1}{3}\left(1 + \frac{\pi - 1}{\pi^2 + \pi - 2}\right) = \frac{1}{3}\frac{\pi^2 + 2\pi - 3}{\pi^2 + \pi - 2} = \frac{1}{3}\frac{(\pi - 1)(\pi + 3)}{(\pi - 1)(\pi + 2)} = \frac{1}{3}\frac{\pi + 3}{\pi + 2}$$

and in turn,

$$a = -\frac{1}{3}\left(\frac{3}{\pi^2 + \pi - 2} + \frac{\pi^2 + 2\pi - 3}{\pi^2 + \pi - 2}\right) = -\frac{1}{3}\frac{\pi(\pi + 2)}{(\pi - 1)(\pi + 2)} = -\frac{1}{3}\frac{\pi}{\pi - 1}.$$

Therefore,

$$y(t) = -\frac{1}{3}\frac{\pi}{\pi - 1}e^{-t} + \frac{1}{3}\frac{\pi + 3}{\pi + 2}e^{2t} + \frac{1}{\pi^2 + \pi - 2}e^{-\pi t}.$$

(ii) To obtain a particular solution of $y''(t) + y(t) = \sin t$, we look for a solution of the form

$$y_p(t) = (a + bt)\sin t + (c + dt)\cos t.$$

Inserting the ansatz into the equation and using that $(fg)'' = f'' + 2f'g' + g''$, we get

$$\sin t = 2b \cos t - 2d \sin t.$$

By comparison of coefficients $b = 0$ and $d = -1/2$. Furthermore, we may choose $a = 0$ and $c = 0$. Hence

$$y_p(t) = -\frac{1}{2}t \cos t$$

is a particular solution. To obtain the general solution of the homogenous equation $y''(t) + y(t) = 0$, we make the ansatz

$$y(t) = e^{\lambda t},$$

from which we get the equation

$$0 = (\lambda^2 + 1)e^{\lambda t}.$$

Hence $\lambda_1 = -i$ and $\lambda_2 = i$ yield two solutions. The general real solution of the homogenous equation is thus given by

$$y_{\text{hom}}(t) = a \cos t + b \sin t, \quad a, b \in \mathbb{R}$$

and the one of the given equation is

$$a \cos t + b \sin t - \frac{1}{2} t \cos t, \quad a, b \in \mathbb{R}.$$

To find a and b, corresponding to the initial values $y(0) = 1$, $y'(0) = 0$ we solve the linear system

$$1 = y(0) = a, \qquad 0 = y'(0) = b - \frac{1}{2}.$$

It follows that $a = 1$ and $b = 1/2$. Therefore, the unique solution of the given initial value problem is

$$y(t) = \cos t + \frac{1}{2} \sin t - \frac{t}{2} \cos t.$$

3. Consider the following ODE:

$$\begin{cases} y_1'(t) = & y_1(t) + 2y_2(t) \\ y_2'(t) = & 3y_1(t) + 2y_2(t) \end{cases}.$$

(i) Find all solutions $y(t) = \big(y_1(t), y_2(t)\big)$ with the property that

$$\lim_{t \to \infty} \|y(t)\| = 0.$$

(ii) Do there exist solutions $y(t)$ so that

$$\lim_{t \to \infty} \|y(t)\| = 0 \qquad \text{and} \qquad \lim_{t \to -\infty} \|y(t)\| = 0?$$

Here $\|y(t)\| = \big(y_1(t)^2 + y_2(t)^2\big)^{1/2}$.

Solutions

(i) We write $y' = Ay$ with $A = \begin{pmatrix} 1 & 2 \\ 3 & 2 \end{pmatrix}$. Since $\det(A) = -4$ and $\text{tr}(A) = 3$, the eigenvalues of A are given by

$$\lambda_1 = \frac{3-5}{2} = -1, \qquad \lambda_2 = \frac{3+5}{2} = 4.$$

Hence the general solution is of the form

$$y(t) = ae^{-t}v^{(1)} + be^{4t}v^{(2)}, \quad a, b \in \mathbb{R}$$

where $v^{(1)}$ is an eigenvector corresponding to λ_1 and $v^{(2)}$ one corresponding to λ_2. The solutions of the form $ae^{-t}v^{(1)}$ (i.e. the solutions with $b = 0$) are thus precisely those with the property that $\lim_{t \to \infty} y(t) = 0$. To obtain an eigenvector $v^{(1)}$ of A, corresponding to λ_1, we compute

$$\lambda_1: \quad A - \lambda_1 \, \mathrm{Id}_{2 \times 2} = \begin{pmatrix} 2 & 2 \\ 3 & 3 \end{pmatrix} \rightsquigarrow \begin{pmatrix} 1 & 1 \\ 0 & 0 \end{pmatrix}.$$

Hence, $v^{(1)} = (-1, 1)$. The solutions are thus given by

$$y(t) = ae^{-t} \begin{pmatrix} -1 \\ 1 \end{pmatrix}, \quad a \in \mathbb{R}.$$

(ii) Since the general solution is given by

$$y(t) = ae^{-t} v^{(1)} + be^{4t} v^{(2)},$$

we conclude from the assumption that $a = 0$ and $b = 0$. So $y(t) \equiv 0$ is the unique solution with the property that $\lim_{t \to \infty} \|y(t)\| = 0$ and $\lim_{t \to -\infty} \|y(t)\| = 0$.

4. (i) Define for $A \in \mathbb{R}^{2 \times 2}$

$$\cos A = \sum_{k=0}^{\infty} \frac{(-1)^k}{(2k)!} A^{2k} \quad \text{and} \quad \sin A = \sum_{k=0}^{\infty} \frac{(-1)^k}{(2k+1)!} A^{2k+1}.$$

Verify that $e^{iA} = \cos A + i \sin A$.

(ii) Compute e^{tA} for $A = \begin{pmatrix} 5 & -2 \\ 2 & 5 \end{pmatrix}$.

Solutions

(i) Recall that e^{iA} is defined as a power series. We separate terms with even and odd indices,

$$e^{iA} = \sum_{k \geq 0} \frac{1}{k!} i^k A^k = \sum_{k \geq 0} \frac{1}{(2k)!} i^{2k} A^{2k} + \sum_{k \geq 0} \frac{1}{(2k+1)!} i^{2k+1} A^{2k+1}$$

$$= \sum_{k \geq 0} \frac{(-1)^k}{(2k)!} A^{2k} + i \sum_{k \geq 0} \frac{(-1)^k}{(2k+1)!} A^{2k+1} = \cos A + i \sin A.$$

(ii) We write $A = \alpha \, \mathrm{Id}_{2 \times 2} + \omega J$ with $\alpha = 5$, $\omega = 2$ and $J = \begin{pmatrix} 0 & -1 \\ 1 & 0 \end{pmatrix}$. First note that $\mathrm{Id}_{2 \times 2} J = J \, \mathrm{Id}_{2 \times 2}$ and hence

$$e^{tA} = e^{t(\alpha \, \mathrm{Id}_{2 \times 2} + \omega J)} = e^{t\alpha \, \mathrm{Id}_{2 \times 2}} e^{t\omega J}$$

and

$$e^{t\alpha \, \mathrm{Id}_{2 \times 2}} = e^{t\alpha} \, \mathrm{Id}_{2 \times 2}.$$

Moreover, $J^2 = -\operatorname{Id}_{2\times 2}$ and hence $J^{2k} = (-1)^k \operatorname{Id}_{2\times 2}$ as well as $J^{2k+1} = (-1)^k J$ for any $k \geq 0$. Therefore

$$
e^{t\omega J} = \sum_{k\geq 0} \frac{1}{k!} (t\omega)^k J^k = \sum_{k\geq 0} \frac{(-1)^k}{(2k)!} (t\omega)^{2k} \operatorname{Id}_{2\times 2} + \sum_{k\geq 0} \frac{(-1)^k}{(2k+1)!} (t\omega)^{2k+1} J
$$

$$
= \cos(t\omega) \operatorname{Id}_{2\times 2} + \sin(t\omega) J = \begin{pmatrix} \cos(t\omega) & -\sin(t\omega) \\ \sin(t\omega) & \cos(t\omega) \end{pmatrix}.
$$

Altogether, we conclude that

$$
e^{tA} = e^{\alpha t} \begin{pmatrix} \cos(t\omega) & -\sin(t\omega) \\ \sin(t\omega) & \cos(t\omega) \end{pmatrix} = e^{5t} \begin{pmatrix} \cos(2t) & -\sin(2t) \\ \sin(2t) & \cos(2t) \end{pmatrix}.
$$

5. Decide whether the following assertions are true or false and justify your answers.

   (i) The superposition principle holds for every ODE of the form $y''(t) + a(t)y(t) = b(t)$ where $a, b\colon \mathbb{R} \to \mathbb{R}$ are arbitrary continuous functions.
   (ii) Every solution $y(t) = \big(y_1(t), y_2(t)\big) \in \mathbb{R}^2$ of

$$
\begin{cases} y_1'(t) = 2y_1(t) + y_2(t) \\ y_2'(t) = 7y_1(t) - 3y_2(t) \end{cases}
$$

   is bounded, meaning that there exists a constant $C > 0$ so that

$$
\|y(t)\|^2 = y_1(t)^2 + y_2(t)^2 \leq C, \quad t \in \mathbb{R}.
$$

Solutions

   (i) False. Consider the constant functions $a = 1$ and $b = 1$. Then

$$
y'' + y = 1.
$$

   Let y_1 and y_2 denote two solutions. Then $y_3 = y_1 + y_2$ satisfies

$$
y_3'' + y_3 = y_1'' + y_1 + y_2'' + y_2 = 2 \neq 1.
$$

   (ii) False. We write $y' = Ay$ with $A = \begin{pmatrix} 3 & 1 \\ 7 & -3 \end{pmatrix}$. Since $\det(A) = -16$ and $\operatorname{tr}(A) = 0$, the eigenvalues are $\lambda_1 = 4$ and $\lambda_2 = -4$. Let $v^{(1)}$ and $v^{(2)}$ be corresponding eigenvectors. The general solution is then of the form

$$
y(t) = a e^{4t} v^{(1)} + b e^{-4t} v^{(2)}, \quad a, b \in \mathbb{R}.
$$

   Clearly, $\lim_{t\to\infty} b e^{-4t} v^{(2)} = 0$ for any choice of b. However, if $a \neq 0$, then $a e^{4t} v^{(1)}$ is unbounded as $t \to \infty$. Consequently, for any initial value of the form $a v^{(1)} + b v^{(2)}$ with $a \neq 0$, the solution is unbounded.

Bibliography

The following remarks are made by the first author after the tragic passing of the second one. This book is based on Thomas Kappeler's teaching experience of courses on linear algebra. He taught these courses many times at different universities over the last 30 years. Consequently, this material has been reworked repeatedly and contains no direct references; however, it was undoubtedly influenced by other works. The unfortunate passing of Thomas Kappeler has made it nearly impossible to identify possible sources. However, based on the books owned by Thomas Kappeler, the following alphabetically ordered list may hint at some of the influences. My apologies for not being able to make the references more explicit. In addition to these influences, the feedback of numerous colleagues, doctorates, and friends of Thomas improved the material. A collective thank you for all those contributions!

1. Bamberg, P., Sternberg, S.: A Course in Mathematics for Students of Physics. Cambridge University Press, Cambridge (1988)
2. Barnett, A.H., Gordon, C.S., Perry, P.A., Uribe, A.: Spectral Geometry. American Mathematical Society, Providence (2012)
3. Davis, B., Porta, H., Uhl, J.J.: Vector Calculus & Mathematica. CD-Rom. Addison-Wesley, Boston (1999)
4. Helffer, B., Nier, F.: Hypoelliptic Estimates and Spectral Theory for Fokker-Planck Operators and Witten Laplacians. Springer, Berlin/Heidelberg (2005)
5. Khruslov, E.Y., Pastur, L., Shepelsky, D.: Spectral Theory and Differential Equations: V.A. Marchenko's 90th Anniversary Collection. American Mathematical Society, Providence (2014)
6. Leon, S., de Pillis, L.: Linear Algebra with Applications. Pearson, London (2020)
7. S. Lipschutz, M. Lipson, Schaum's Outline of Theory and Problems of Linear Algebra. Schaum's Outline Series. McGraw-Hill, New York (2001)
8. Nipp, K., Stoffer, D.: Lineare Algebra. Eine Einführung für Ingenieure unter besonderer Berücksichtigung numerischer Aspekte. vdf Hochschulverlag ETH Zürich (2001)
9. Günter Scheja, H.-J., Vetter, U.: Algebra. Bibliographisches Institut, Gotha (1969)
10. Schaefer, H.H.: Topological Vector Spaces. Springer, New York (1971)
11. Scherfner, M., Volland, T.: Mathematik für das erste Semester. Spektrum Akademischer Verlag (2012)
12. Schroder, H.: K-Theory for Real C*-Algebras and Applications. Chapman & Hall/CRC Research Notes in Mathematics Series. Taylor & Francis, Routledge (1993)

© The Author(s), under exclusive license to Springer Nature Switzerland AG 2023   255
M. Benz, T. Kappeler, *Linear Algebra for the Sciences*, La Matematica
per il 3+2 151, https://doi.org/10.1007/978-3-031-27220-2

13. Suetin, P.K., Kostrikin, A.I., Manin, Y.I.: Linear Algebra and Geometry. CRC Press, Boca Raton (1989)
14. Wüstholz, G.: Algebra. Vieweg und Teubner Verlag, Berlin (2004)

Index

© The Author(s), under exclusive license to Springer Nature Switzerland AG 2023
M. Benz, T. Kappeler, *Linear Algebra for the Sciences*, La Matematica
per il 3+2 151, https://doi.org/10.1007/978-3-031-27220-2

Printed in the United States
by Baker & Taylor Publisher Services